Distributed Source Coding

Distributed Source Coding

Theory, Algorithms, and Applications

Pier Luigi Dragotti

Department of Electrical and Electronic Engineering
Imperial College London
London, UK

Michael Gastpar

Department of Electrical Engineering and Computer Sciences
University of California, Berkeley
Berkeley, CA

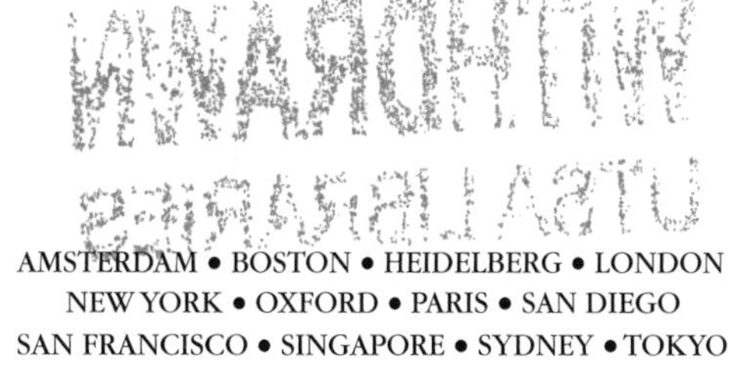

ELSEVIER

AMSTERDAM • BOSTON • HEIDELBERG • LONDON
NEW YORK • OXFORD • PARIS • SAN DIEGO
SAN FRANCISCO • SINGAPORE • SYDNEY • TOKYO

Academic Press is an imprint of Elsevier

Academic Press is an imprint of Elsevier
30 Corporate Drive, Suite 400, Burlington, MA 01803, USA
525 B Street, Suite 1900, San Diego, California 92101-4495, USA
84 Theobald's Road, London WC1X 8RR, UK

This book is printed on acid-free paper. ∞

Library of Congress Cataloging-in-Publication Data
Dragotti, Pier Luigi.
Distributed source coding : theory, algorithms, and applications / Pier Luigi Dragotti, Michael Gastpar.
p. cm.
Includes index.
ISBN 978-0-12-374485-2 (hardcover : alk. paper)
1. Data compression (Telecommunication) 2. Multisensor data fusion. 3. Coding theory.
4. Electronic data processing–Distributed processing. I. Gastpar, Michael. II. Title.
TK5102.92.D57 2009
621.382′16–dc22

2008044569

British Library Cataloguing in Publication Data
A catalogue record for this book is available from the British Library

ISBN 13: 978-0-12-374485-2

Printed in the United States of America
09 10 9 8 7 6 5 4 3 2 1

Typeset by: diacriTech, India.

Contents

List of Contributors

Chapter 1. Foundations of Distributed Source Coding

Krishnan Eswaran
Department of Electrical Engineering and Computer Sciences
University of California, Berkeley
Berkeley, CA 94720

Michael Gastpar
Department of Electrical Engineering and Computer Sciences
University of California, Berkeley
Berkeley, CA 94720

Chapter 2. Distributed Transform Coding

Varit Chaisinthop
Department of Electrical and Electronic Engineering
Imperial College London
SW7 2AZ London, UK

Pier Luigi Dragotti
Department of Electrical and Electronic Engineering
Imperial College London
SW7 2AZ London, UK

Chapter 3. Quantization for Distributed Source Coding

David Rebollo-Monedero
Department of Telematics Engineering
Universitat Politècnica de Catalunya
08034 Barcelona, Spain

Bernd Girod
Department of Electrical Engineering
Stanford University
Palo Alto, CA 94305-9515

Chapter 4. Zero-error Distributed Source Coding

Ertem Tuncel
Department of Electrical Engineering
University of California, Riverside
Riverside, CA 92521

Jayanth Nayak
Mayachitra, Inc.
Santa Barbara, CA 93111

Prashant Koulgi
Department of Electrical and Computer Engineering
University of California, Santa Barbara
Santa Barbara, CA 93106

Kenneth Rose
Department of Electrical and Computer Engineering
University of California, Santa Barbara
Santa Barbara, CA 93106

Chapter 5. Distributed Coding of Sparse Signals

Vivek K Goyal
Department of Electrical Engineering and Computer Science
Massachusetts Institute of Technology
Cambridge, MA 02139

Alyson K. Fletcher
Department of Electrical Engineering and Computer Sciences
University of California, Berkeley
Berkeley, CA 94720

Sundeep Rangan
Qualcomm Flarion Technologies
Bridgewater, NJ 08807-2856

Chapter 6. Toward Constructive Slepian–Wolf Coding Schemes

Christine Guillemot
INRIA Rennes-Bretagne Atlantique
Campus Universitaire de Beaulieu
35042 Rennes Cédex, France

Aline Roumy
INRIA Rennes-Bretagne Atlantique
Campus Universitaire de Beaulieu
35042 Rennes Cédex, France

Chapter 7. Distributed Compression in Microphone Arrays

Olivier Roy
Audiovisual Communications Laboratory
School of Computer and Communication Sciences
Ecole Polytechnique Fédérale de Lausanne
CH-1015 Lausanne, Switzerland

Thibaut Ajdler
Audiovisual Communications Laboratory
School of Computer and Communication Sciences
Ecole Polytechnique Fédérale de Lausanne
CH-1015 Lausanne, Switzerland

Robert L. Konsbruck
Audiovisual Communications Laboratory
School of Computer and Communication Sciences
Ecole Polytechnique Fédérale de Lausanne
CH-1015 Lausanne, Switzerland

Martin Vetterli
Audiovisual Communications Laboratory
School of Computer and Communication Sciences
Ecole Polytechnique Fédérale de Lausanne
CH-1015 Lausanne, Switzerland
and
Department of Electrical Engineering and Computer Sciences
University of California, Berkeley
Berkeley, CA 94720

Chapter 8. Distributed Video Coding: Basics, Codecs, and Performance

Fernando Pereira
Instituto Superior Técnico—Instituto de Telecomunicações
1049-001 Lisbon, Portugal

Catarina Brites
Instituto Superior Técnico—Instituto de Telecomunicações
1049-001 Lisbon, Portugal

João Ascenso
Instituto Superior Técnico—Instituto de Telecomunicações
1049-001 Lisbon, Portugal

Chapter 9. Model-based Multiview Video Compression Using Distributed Source Coding Principles

Jayanth Nayak
Mayachitra, Inc.
Santa Barbara, CA 93111

Bi Song
Department of Electrical Engineering
University of California, Riverside
Riverside, CA 92521

Ertem Tuncel
Department of Electrical Engineering
University of California, Riverside
Riverside, CA 92521

Amit K. Roy-Chowdhury
Department of Electrical Engineering
University of California, Riverside
Riverside, CA 92521

Chapter 10. Distributed Compression of Hyperspectral Imagery

Ngai-Man Cheung
Signal and Image Processing Institute
Department of Electrical Engineering
University of Southern California
Los Angeles, CA 90089-2564

Antonio Ortega
Signal and Image Processing Institute
Department of Electrical Engineering
University of Southern California
Los Angeles, CA 90089-2564

Chapter 11. Securing Biometric Data

Anthony Vetro
Mitsubishi Electric Research Laboratories
Cambridge, MA 02139

Shantanu Rane
Mitsubishi Electric Research Laboratories
Cambridge, MA 02139

Jonathan S. Yedidia
Mitsubishi Electric Research Laboratories
Cambridge, MA 02139

Stark C. Draper
Department of Electrical and Computer Engineering
University of Wisconsin, Madison
Madison, WI 53706

Introduction

In conventional source coding, a single encoder exploits the redundancy of the source in order to perform compression. Applications such as wireless sensor and camera networks, however, involve multiple sources often separated in space that need to be compressed independently. In such applications, it is not usually feasible to first transport all the data to a central location and compress (or further process) it there. The resulting source coding problem is often referred to as distributed source coding (DSC). Its foundations were laid in the 1970s, but it is only in the current decade that practical techniques have been developed, along with advances in the theoretical underpinnings. The practical advances were, in part, due to the rediscovery of the close connection between distributed source codes and (standard) error-correction codes for noisy channels. The latter area underwent a dramatic shift in the 1990s, following the discovery of turbo and low-density parity-check (LDPC) codes. Both constructions have been used to obtain good distributed source codes.

In a related effort, ideas from distributed coding have also had considerable impact on video compression, which is basically a centralized compression problem. In this scenario, one can consider a compression technique under which each video frame must be compressed separately, thus mimicking a distributed coding problem. The resulting algorithms are among the best-performing and have many additional features, including, for example, a shift of complexity from the encoder to the decoder.

This book summarizes the main contributions of the current decade. The chapters are subdivided into two parts. The first part is devoted to the theoretical foundations, and the second part to algorithms and applications.

Chapter 1, by Eswaran and Gastpar, summarizes the state of the art of the theory of distributed source coding, starting with classical results. It emphasizes an important distinction between direct source coding and indirect (or noisy) source coding: In the distributed setting, these two are fundamentally different. This difference is best appreciated by considering the scaling laws in the number of encoders: In the indirect case, those scaling laws are dramatically different. Historically, compression is tightly linked to transforms and thus to transform coding. It is therefore natural to investigate extensions of the traditional centralized transform coding paradigm to the distributed case. This is done by Chaisinthop and Dragotti in Chapter 2, which presents an overview of existing distributed transform coders. Rebollo-Monedero and Girod, in Chapter 3, address the important question of quantization in a distributed setting. A new set of tools is necessary to optimize quantizers, and the chapter gives a partial account of the results available to date. In the standard perspective, efficient distributed source coding always involves an error probability, even though it vanishes as the coding block length is increased. In Chapter 4, Tuncel, Nayak, Koulgi, and Rose take a more restrictive view: The error probability must be exactly zero. This is shown to lead to a strict rate penalty for many instances. Chapter 5, by Goyal, Fletcher, and Rangan, connects ideas from distributed source coding with the sparse signal models that have recently received considerable attention under the heading of compressed (or compressive) sensing.

The second part of the book focuses on algorithms and applications, where the developments of the past decades have been even more pronounced than in the theoretical foundations. The first chapter, by Guillemot and Roumy, presents an overview of practical DSC techniques based on turbo and LDPC codes, along with ample experimental illustration. Chapter 7, by Roy, Ajdler, Konsbruck, and Vetterli, specializes and applies DSC techniques to a system of multiple microphones, using an explicit spatial model to derive sampling conditions and source correlation structures. Chapter 8, by Pereira, Brites, and Ascenso, overviews the application of ideas from DSC to video coding: A single video stream is encoded, frame by frame, and the encoder treats past and future frames as side information when encoding the current frame. The chapter starts with an overview of the original distributed video coders from Berkeley (PRISM) and Stanford, and provides a detailed description of an enhanced video coder developed by the authors (and referred to as DISCOVER). The case of the multiple multiview video stream is considered by Nayak, Song, Tuncel, and Roy-Chowdhury in Chapter 9, where they show how DSC techniques can be applied to the problem of multiview video compression. Chapter 10, by Cheung and Ortega, applies DSC techniques to the problem of distributed compression of hyperspectral imagery. Finally, Chapter 11, by Vetro, Draper, Rane, and Yedidia, is an innovative application of DSC techniques to securing biometric data. The problem is that if a fingerprint, iris scan, or genetic code is used as a user password, then the password cannot be changed since users are stuck with their fingers (or irises, or genes). Therefore, biometric information should not be stored in the clear anywhere. This chapter discusses one approach to this problematic issue, using ideas from DSC.

One of the main objectives of this book is to provide a comprehensive reference for engineers, researchers, and students interested in distributed source coding. Results on this topic have so far appeared in different journals and conferences. We hope that the book will finally provide an integrated view of this active and ever evolving research area.

Edited books would not exist without the enthusiasm and hard work of the contributors. It has been a great pleasure for us to interact with some of the very best researchers in this area who have enthusiastically embarked in this project and have contributed these wonderful chapters. We have learned a lot from them. We would also like to thank the reviewers of the chapters for their time and for their constructive comments. Finally we would like to thank the staff at Academic Press—in particular Tim Pitts, Senior Commissioning Editor, and Melanie Benson—for their continuous help.

Pier Luigi Dragotti, London, UK

Michael Gastpar, Berkeley, California, USA

PART

Theory

1

CHAPTER

Foundations of Distributed Source Coding

1

Krishnan Eswaran and Michael Gastpar

Department of Electrical Engineering and Computer Sciences, University of California, Berkeley, CA

CHAPTER CONTENTS

Distributed Source Coding: Theory, Algorithms, and Applications

1.1 INTRODUCTION

Data compression is one of the oldest and most important signal processing questions. A famous historical example is the Morse code, created in 1838, which gives shorter codes to letters that appear more frequently in English (such as "e" and "t"). A powerful abstraction was introduced by Shannon in 1948 [1]. In his framework, the original source information is represented by a sequence of bits (or, equivalently, by one out of a countable set of prespecified messages). Classically, all the information to be compressed was available in one place, leading to *centralized* encoding problems. However, with the advent of multimedia, sensor, and ad-hoc networks, the most important compression problems are now *distributed*: the source information appears at several separate encoding terminals. Starting with the pioneering work of Slepian and Wolf in 1973, this chapter provides an overview of the main advances of the last three and a half decades as they pertain to the fundamental performance bounds in distributed source coding. A first important distinction is *lossless* versus *lossy* compression, and the chapter provides closed-form formulas wherever possible. A second important distinction is *direct* versus *remote* compression; in the direct compression problem, the encoders have direct access to the information that is of interest to the decoder, while in the remote compression problem, the encoders only access that information indirectly through a noisy observation process (a famous example being the so-called CEO problem). An interesting insight discussed in this chapter concerns the sometimes dramatic (and perhaps somewhat unexpected) performance difference between direct and remote compression. The chapter concludes with a short discussion of the problem of communicating sources across noisy channels, and thus, Shannon's separation theorem.

1.2 CENTRALIZED SOURCE CODING

1.2.1 Lossless Source Coding

The most basic scenario of source coding is that of describing source output sequences with bit strings in such a way that the original source sequence can be recovered *without loss* from the corresponding bit string. One can think about this scenario in two ways. First, one can map source realizations to binary strings of different lengths and strive to minimize the expected length of these codewords. Compression is attained whenever some source output sequences are more likely than others: the likelier sequences will receive shorter bit strings. For centralized source coding (see Figure 1.1), there is a rich theory of such codes (including Huffman codes, Lempel-Ziv codes, and arithmetic codes). However, for distributed source coding, this perspective has not yet been very fruitful. The second approach to lossless source coding is to map L samples of the source output sequence to the set of bit strings of a fixed length N, but to allow a "small" error in the reconstruction. Here, "small" means that the probability of reconstruction error goes to zero as the source sequence length

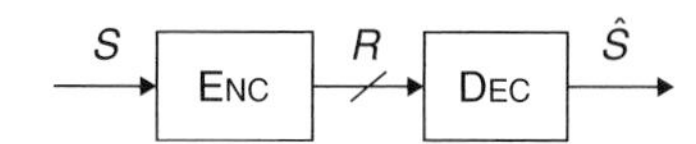

FIGURE 1.1

Centralized source coding.

goes to infinity. The main insight is that it is sufficient to assign bit strings to "typical" source output sequences. One measures the performance of a lossless source code by considering the ratio N/L of the number of bits N of this bit string to the number of source samples L. An *achievable rate* $R = N/L$ is a ratio that allows for an asymptotically small reconstruction error.

Formal definitions of a lossless code and an achievable rate can be found in Appendix A (Definitions A.6 and A.7). The central result of lossless source coding is the following:

Theorem 1.1. Given a discrete information source $\{S(n)\}_{n>0}$, the rate R is achievable via lossless source coding if $R > H_\infty(S)$, where $H_\infty(S)$ is the entropy (or entropy rate) of the source. Conversely, if $R < H_\infty(S)$, R is not achievable via lossless source coding.

A proof of this theorem for the i.i.d. case and Markov sources is due to Shannon [1]. A proof of the general case can be found, for example, in [2, Theorem 3, p. 757].

1.2.2 Lossy Source Coding

In many source coding problems, the available bit rate is not sufficient to describe the information source in a lossless fashion. Moreover, for real-valued sources, lossless reconstruction is not possible for any finite bit rate. For instance, consider a source whose samples are i.i.d. and uniform on the interval [0, 1]. Consider the binary representation of each sample as the sequence. $B_1B_2\ldots$; here, each binary digit is independent and identically distributed (i.i.d.) with probability 1/2 of being 0 or 1. Thus, the entropy of any sample is infinite, and Theorem 1.1 implies that no finite rate can lead to perfect reconstruction.

Instead, we want to use the available rate to describe the source to within the smallest possible average *distortion* D, which in turn is determined by a distortion function $d(\cdot,\cdot)$, a mapping from the source and reconstruction alphabets to the non-negative reals. The precise shape of the distortion function $d(\cdot,\cdot)$ is determined by the application at hand. A widely studied choice is the mean-squared error, that is, $d(s,\hat{s}) = |s-\hat{s}|^2$.

It should be intuitively clear that the larger the available rate R, the smaller the incurred distortion D. In the context of lossy source coding, the goal is thus to study the achievable trade-offs between rate and distortion. Formal definitions of a lossless code and achievable rate can be found in Appendix A (Definitions A.8 and A.9). Perhaps somewhat surprisingly, the optimal trade-offs between rate and distortion can be characterized compactly as a "single-letter" optimization problem usually called the *rate-distortion function.* More formally, we have the following theorem:

Theorem 1.2. Given a discrete memoryless source $\{S(n)\}_{n>0}$ and bounded distortion function $d: \mathcal{S} \times \hat{\mathcal{S}} \rightarrow \mathbb{R}^+$, a rate R is achievable with distortion D for $R > R_S(D)$, where

$$R_S(D) = \min_{\substack{p(\hat{s}|s): \\ E[d(S,\hat{S})] \leqslant D}} I(S; \hat{S}) \tag{1.1}$$

is the rate distortion function. Conversely, for $R < R_S(D)$, the rate R is not achievable with distortion D.

A proof of this theorem can be found in [3, pp. 349–356]. Interestingly, it can also be shown that when $D > 0, R = R_S(D)$ is achievable [4].

Unlike the situation in the lossless case, determining the rate-distortion function requires one to solve an optimization problem. The Blahut–Arimoto algorithm [5, 6] and other techniques (e.g., [7]) have been proposed to make this computation efficient.

While Theorem 1.2 is stated for discrete memoryless sources and a bounded distortion measure, it can be extended to continuous sources under appropriate technical conditions. Furthermore, one can show that these technical conditions are satisfied for memoryless Gaussian sources with a mean-squared error distortion. This is sometimes called the quadratic Gaussian case. Thus, one can use Equation (1.1) in Theorem 1.2 to deduce the following.

Proposition 1.1. *Given a memoryless Gaussian source $\{S(n)\}_{n>0}$ with $S(n) \sim \mathcal{N}(0, \sigma^2)$ and distortion function $d(s, \hat{s}) = (s - \hat{s})^2$,*

$$R_S(D) = \frac{1}{2} \log^+ \frac{\sigma^2}{D}. \tag{1.2}$$

For general continuous sources, the rate-distortion function can be difficult to determine. In lieu of computing the rate-distortion function exactly, an alternative is to find closed-form upper and lower bounds to it. The idea originates with Shannon's work [8], and it has been shown that under appropriate assumptions, Shannon's lower bound for difference distortions ($d(s, \hat{s}) = f(s - \hat{s})$) becomes tight in the high-rate regime [9].

For a quadratic distortion and memoryless source $\{S(n)\}_{n>0}$ with variance σ_S^2 and entropy power Q_S, these upper and lower bounds can be expressed as [10, p. 101]

$$\frac{1}{2} \log \frac{Q_S}{D} \leqslant R_S(D) \leqslant \frac{1}{2} \log \frac{\sigma_S^2}{D}, \tag{1.3}$$

where the entropy power is given in Definition A.4. From Table 1.1, one can see that the bounds in (1.3) are tight for memoryless Gaussian sources.

Table 1.1 Variance and Entropy Power of Common Distributions

Source Name	Probability Density Function	Variance	Entropy Power
Gaussian	$f(x) = \frac{1}{\sqrt{2\pi e \sigma^2}} e^{-(x-\mu)^2/2\sigma^2}$	σ^2	σ^2
Laplacian	$f(x) = \frac{\lambda}{2} e^{-\lambda\lvert x-\mu\rvert}$	$\frac{2}{\lambda^2}$	$\frac{e}{\pi} \cdot \frac{2}{\lambda^2}$
Uniform	$f(x) = \begin{cases} \frac{1}{2a}, & -a \leq x-\mu \leq a \\ 0, & \text{otherwise} \end{cases}$	$\frac{a^2}{3}$	$\frac{6}{\pi e} \cdot \frac{a^2}{3}$

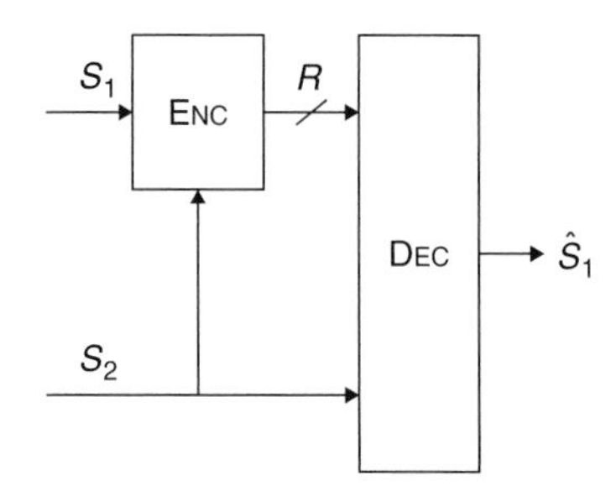

FIGURE 1.2

Conditional rate distortion.

The conditional source coding problem (see Figure 1.2) considers the case in which a correlated source is available to the encoder and decoder to potentially decrease the encoding rate to achieve the same distortion. Definitions A.10 and A.11 formalize the problem.

Theorem 1.3. Given a memoryless source S_1, memoryless source side information S_2 available at the encoder and decoder with the property that $(S_1(k), S_2(k))$ are i.i.d. in k, and distortion function $d: \mathcal{S}_1 \times \hat{\mathcal{S}}_1 \rightarrow \Re^+$, the conditional rate-distortion function is

$$R_{S_1|S_2}(D) = \min_{\substack{p(\hat{s}_1|s_1,s_2): \\ E[d(S_1,\hat{S}_1)] \leq D}} I(S_1; \hat{S}_1|S_2). \tag{1.4}$$

A proof of Theorem 1.3 can be found in [11, Theorem 6, p. 11].

Because the rate-distortion theorem gives an asymptotic result as the blocklength gets large, convergence to the rate-distortion function for any finite blocklength has also been investigated. Pilc [4] as well as Omura [12] considered some initial investigations in this direction. Work by Marton established the notion of a source coding error exponent [13], in which she considered upper and lower bounds to the probability that for memoryless sources, an optimal rate-distortion codebook exceeds distortion D as a function of the blocklength.

1.2.3 Lossy Source Coding for Sources with Memory

We start with an example. Consider a Gaussian source S with $S(i) \sim \mathcal{N}(0, 2)$ where pairs $\vec{Y}(k) = (S(2k-1), S(2k))$ have the covariance matrix

$$\Sigma = \begin{pmatrix} 2 & 1 \\ 1 & 2 \end{pmatrix}, \tag{1.5}$$

and $\vec{Y}(k)$ are i.i.d. over k. The discrete Fourier transform (DFT) of each pair can be written as

$$\tilde{S}(2k-1) = \frac{1}{\sqrt{2}}(S(2k-1) + S(2k)) \tag{1.6}$$

$$\tilde{S}(2k) = \frac{1}{\sqrt{2}}(S(2k-1) - S(2k)), \tag{1.7}$$

which has the covariance matrix

$$\tilde{\Sigma} = \begin{pmatrix} 3 & 0 \\ 0 & 1 \end{pmatrix}, \tag{1.8}$$

and thus the source $\tilde{S}$ is independent, with i.i.d. even and odd entries. For squared error distortion, if C is the codeword sent to the decoder, we can express the distortion as $nD = \sum_{i=1}^{n} E[(S(i) - E[S(i)|C])^2]$. By linearity of expectation, it is possible to rewrite this as

$$nD = \sum_{i=1}^{n} E\left[\left(\tilde{S}(i) - E[\tilde{S}(i)|C]\right)^2\right] \tag{1.9}$$

$$= \sum_{k=1}^{\lceil n/2 \rceil} E\left[\left(\tilde{S}(2k-1) - E[\tilde{S}(2k-1)|C]\right)^2\right] + \sum_{k=1}^{\lfloor n/2 \rfloor} E\left[\left(\tilde{S}(2k) - E[\tilde{S}(2k)|C]\right)^2\right]. \tag{1.10}$$

Thus, this is a rate-distortion problem in which two independent Gaussian sources of different variances have a constraint on the sum of their mean-squared errors. Sometimes known as the parallel Gaussian source problem, it turns out there is a well-known solution to it called reverse water-filling [3, p. 348, Theorem 13.3.3], which in this case, evaluates to the following:

$$R_S(D) = \frac{1}{2}\log\frac{\sigma_1^2}{D_1} + \frac{1}{2}\log\frac{\sigma_2^2}{D_2} \tag{1.11}$$

$$D_i = \begin{cases} \nu, & \nu < \sigma_i^2, \\ \sigma_i^2, & \nu \geq \sigma_i^2, \end{cases} \tag{1.12}$$

where $\sigma_1^2 = 3, \sigma_2^2 = 1$, and ν is chosen so that $D_1 + D_2 = D$.

This diagonalization approach allows one to state the following result for stationary ergodic Gaussian sources.

Proposition 1.2. *Let S be a stationary ergodic Gaussian source with autocorrelation function $E[S_n S_{n-k}] = \phi(k)$ and power spectral density*

$$\Phi(\omega) = \sum_{k=-\infty}^{\infty} \phi_{(k)} e^{-jk\omega}. \tag{1.13}$$

Then the rate-distortion function for S under mean-squared error distortion is given by:

$$R(D_\nu) = \frac{1}{4\pi} \int_{-\pi}^{\pi} \max\left\{0, \log \frac{\Phi(\omega)}{\nu}\right\} d\omega \tag{1.14}$$

$$D_\nu = \frac{1}{2\pi} \int_{-\pi}^{\pi} \min\{\nu, \Phi(\omega)\} d\omega. \tag{1.15}$$

PROOF. See Berger [10, p. 112]. ■

While it can be difficult to evaluate this in general, upper and lower bounds can give a better sense for its behavior. For instance, let σ^2 be the variance of a stationary ergodic Gaussian source. Then a result of Wyner and Ziv [14] shows that the rate-distortion function can be bounded as follows:

$$\frac{1}{2} \log \frac{\sigma^2}{D} - \Delta_S \leqslant R_S(D) \leqslant \frac{1}{2} \log \frac{\sigma^2}{D}, \tag{1.16}$$

where Δ_S is a constant that depends only on the power spectral density of S.

1.2.4 Some Notes on Practical Considerations

The problem formulation considered in this chapter focuses on the existence of codes for cases in which the encoder has access to the entire source noncausally and knows its distribution. However, in many situations of practical interest, some of these assumptions may not hold. For instance, several problems have considered the effects of delay and causal access to a source [15–17]. Some work has also considered cases in which no underlying probabilistic assumptions are made about the source [18–20]. Finally, the work of Gersho and Gray [21] explores how one might actually go about designing implementable vector quantizers.

1.3 DISTRIBUTED SOURCE CODING

The problem of source coding becomes significantly more interesting and challenging in a network context. Several new scenarios arise:

- Different parts of the source information may be available to separate encoding terminals that cannot cooperate.
- Decoders may have access to additional side information about the source information; or they may only obtain a part of the description provided by the encoders.

We start our discussion by an example illustrating the classical problem of source coding with side information at the decoder.

Example 1.1

Let $\{S(n)\}_{n>0}$ be a source where source samples $S(n)$ are uniform over an alphabet of size 8, which we choose to think of as binary vectors of length 3. The decoder has access to a corrupted version of the source $\{\tilde{S}(n)\}_{n>0}$ where each sample $\tilde{S}(n)$ takes values in a set of ternary sequences $\{0, *, 1\}^3$ of length 3 with the property that

$$Pr\left(\tilde{S}(n) = (c_1, c_2, c_3) \middle| S(n) = (b_1, b_2, b_3)\right) = \begin{cases} \frac{1}{4}, & \text{if } (c_1, c_2, c_3) = (b_1, b_2, b_3) \\ \frac{1}{4}, & \text{if } (c_1, c_2, c_3) = (*, b_2, b_3) \\ \frac{1}{4}, & \text{if } (c_1, c_2, c_3) = (b_1, *, b_3) \\ \frac{1}{4}, & \text{if } (c_1, c_2, c_3) = (b_1, b_2, *) \end{cases} \tag{1.17}$$

Thus, the decoder has access to at least two of every three bits per source symbol, but the encoder is unaware of which ones. Consider the partition of the alphabet $\mathcal{S}$ into

$$\mathcal{S}_1 = \left\{\begin{array}{c}(0,0,0), \\ (0,1,1), \\ (1,1,0), \\ (1,0,1)\end{array}\right\} \quad \mathcal{S}_2 = \left\{\begin{array}{c}(1,1,1), \\ (1,0,0), \\ (0,0,1), \\ (0,1,0)\end{array}\right\}. \tag{1.18}$$

If the decoder knows which of these partitions a particular sample $S(n)$ is in, $\tilde{S}(n)$ is sufficient to determine the exact value of $S(n)$. Thus, for each source output, the encoder can use one bit to indicate to which of the two partitions the source sample belongs. Thus, at a rate of 1 bit per source sample, the decoder can perfectly reconstruct the output. However, in the absence of $\{\tilde{S}(n)\}_{n>0}$ at the decoder, Theorem 1.1 implies the best possible rate is $H(S_1) = 3$ bits per source sample.

Example 1.1 illustrates a simple version of a strategy known as *binning*. It turns out that binning can be applied more generally and is used to prove many of the results in distributed source coding.

1.3.1 Lossless Source Coding

More generally, we now consider the scenario illustrated in Figure 1.3, where two separate encoding terminals each observe part of the data. That is, with respect to Figure 1.3, the source streams S_1 and S_2 are dependent on each other. The coding question now involves two separate source codes that appear at rates R_1 and R_2, respectively, and a receiver where the source codes are jointly decoded. Formal definitions are provided in Appendix A (Definitions A.12 and A.13). Since the sources are dependent, the rates R_1 and R_2 constrain one another. That is, if more bits are used to describe one of the sources, typically, the number of bits for the other can be reduced. Specifically, if we

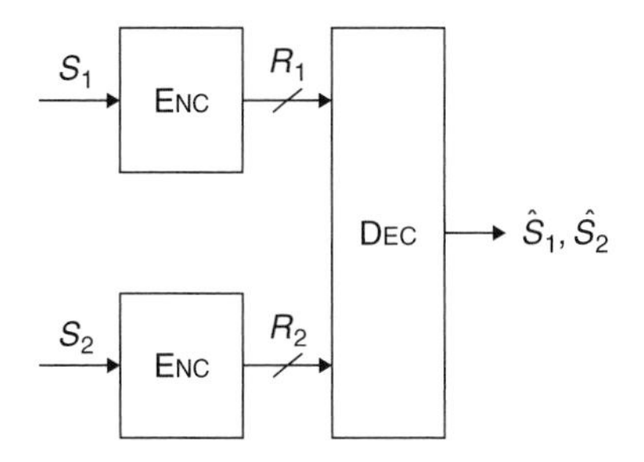

FIGURE 1.3

Distributed source coding problem.

assume $R_2 > \log|\mathcal{S}_2|$, we can assume that the decoder knows S_2 without error, and thus this problem also includes the special case of side information at the decoder.

Theorem 1.4. Given discrete memoryless sources S_1 and S_2, define $\mathcal{R}$ as

$$\mathcal{R} = \Big\{ (R_1, R_2) : R_1 + R_2 \geqslant H(S_1, S_2),$$
$$R_1 \geqslant H(S_1|S_2), R_2 \geqslant H(S_2|S_1) \Big\}. \tag{1.19}$$

Furthermore, let $\mathcal{R}^0$ be the interior of $\mathcal{R}$. Then $(R_1,R_2) \in \mathcal{R}^0$ are achievable for the two-terminal lossless source coding problem, and $(R_1,R_2) \notin \mathcal{R}$ are not.

The proof of this result was shown by Slepian and Wolf [22]; to show achievability involves a random binning argument reminiscent of Example 1.1. However, by contrast to that example, the encoders now bin over the entire vector of source symbols, and they only get probabilistic guarantees that successful decoding is possible.

Variations and extensions of the lossless coding problem have been considered by Ahlswede and Körner [23], who examine a similar setup case in which the decoder is only interested in S_1; Wyner [24], who considers a setup in which one again wants to reconstruct sources S_1 and S_2, but there is now an additional encoding terminal with access to another correlated information source S_3; and Gel'fand and Pinsker [25], who consider perfect reconstruction of a source to which each encoding terminal has only a corrupted version. In all of these cases, a random binning strategy can be used to establish optimality. However, one notable exception to this is a paper by Körner and Marton [26], which shows that when one wants to reconstruct the modulo-2 sum of correlated sources, there exists a strategy that performs better than binning.

1.3.2 Lossy Source Coding

By analogy to the centralized compression problem, it is again natural to study the problem where instead of perfect recovery, the decoder is only required to provide estimates of the original source sequences to within some distortions. Reconsidering Figure 1.3, we now ask that the source S_1 be recovered at distortion D_1, when assessed with distortion measure $d_1(\cdot,\cdot)$, and S_2 at distortion D_2, when assessed with distortion measure $d_2(\cdot,\cdot)$. The question is again to determine the necessary rates, R_1 and R_2, respectively, as well as the coding schemes that permit satisfying the distortion constraints. Formal definitions are provided in Appendix A (Definitions A.14 and A.15). For this problem, a natural achievable strategy arises. One can first quantize the sources at each encoder as in the centralized lossy coding case and then bin the quantized values as in the distributed lossless case. The work of Berger and Tung [27–29] provides an elegant way to combine these two techniques that leads to the following result. The result was independently discovered by Housewright [30].

Theorem 1.5: *"Quantize-and-bin."* Given sources S_1 and S_2 with distortion functions d_k: $\mathcal{S}_k \times \mathcal{U}_k \to \Re^+$, $k \in \{1,2\}$, the achievable rate-distortion region includes the following set:

$$\begin{aligned} \mathcal{R} = \{(\vec{R},\vec{D}) : \exists U_1, U_2 \text{ s.t. } & U_1 - S_1 - S_2 - U_2, \\ E[d_1(S_1,U_1)] \leqslant D_1, & \, E[d_2(S_2,U_2)] \leqslant D_2, \\ R_1 > I(S_1;U_1|U_2), & \, R_2 > I(S_2;U_2|U_1), \\ R_1 + R_2 > & \, I(S_1,S_2;U_1U_2)\}. \end{aligned} \tag{1.20}$$

A proof for the setting of more than two users is given by Han and Kobayashi [31, Theorem 1, pp. 280–284]. A major open question stems from the optimality of the "quantize-and-bin" achievable strategy. While work by Servetto [32] suggests it may be tight for the two-user setting, the only case for which it is known is the quadratic Gaussian setting, which is based on an outer bound developed by Wagner and Anantharam [33, 34].

Theorem 1.6: *Wagner, Tavildar, and Viswanath [35].* Given sources S_1 and S_2 that are jointly Gaussian and i.i.d. in time, that is, $(S_1(k), S_2(k)) \sim \mathcal{N}(0,\Sigma)$, with covariance matrix

$$\Sigma = \begin{pmatrix} \sigma_1^2 & \rho\sigma_1\sigma_2 \\ \rho\sigma_1\sigma_2 & \sigma_2^2 \end{pmatrix} \tag{1.21}$$

for $D_1, D_2 > 0$ define $\mathcal{R}(D_1, D_2)$ as

$$\begin{aligned} \mathcal{R}(D_1,D_2) = \Bigg\{ (R_1,R_2) : R_1 \geqslant \frac{1}{2}\log^+ \frac{(1-\rho^2+\rho^2 2^{-2R_2})\sigma_1^2}{D_1}, \\ R_2 \geqslant \frac{1}{2}\log^+ \frac{(1-\rho^2+\rho^2 2^{-2R_1})\sigma_2^2}{D_2}, \\ R_1 + R_2 \geqslant \frac{1}{2}\log^+ \frac{(1-\rho^2)\sigma_1^2\sigma_2^2\beta(D_1,D_2)}{2D_1D_2} \Bigg\}, \end{aligned} \tag{1.22}$$

where

$$\beta(D_1,D_2) = 1 + \sqrt{1 + \frac{4\rho^2 D_1 D_2}{(1-\rho^2)^2\sigma_1^2\sigma_2^2}}. \tag{1.23}$$

Furthermore, let $\mathcal{R}^0$ be the interior of $\mathcal{R}$. Then for distortions $D_1, D_2 > 0$, $(R_1, R_2) \in \mathcal{R}^0(D_1, D_2)$ are achievable for the two-terminal quadratic Gaussian source coding problem and $(R_1, R_2) \notin \mathcal{R}(D_1, D_2)$ are not.

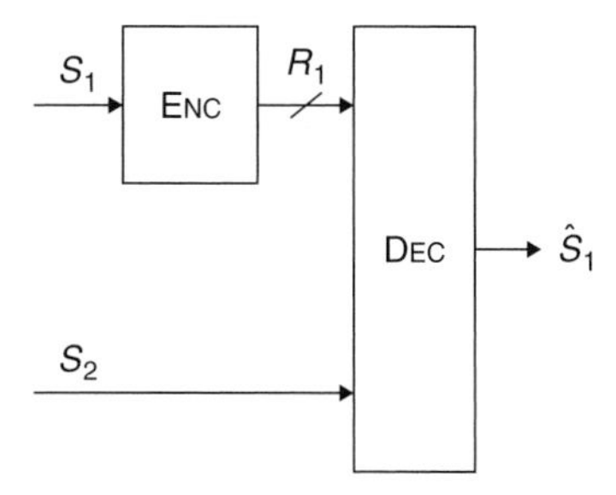

FIGURE 1.4

The Wyner–Ziv source coding problem.

In some settings, the rates given by the "quantize-and-bin" achievable strategy can be shown to be optimal. For instance, consider the setting in which the second encoder has an unconstrained rate link to the decoder, as in Figure 1.4. This configuration is often referred to as the Wyner–Ziv source coding problem.

Theorem 1.7. Given a discrete memoryless source S_1, discrete memoryless side information source S_2 with the property that $(S_1(k), S_2(k))$ are i.i.d. over k, and bounded distortion function $d: \mathcal{S} \times \mathcal{U} \rightarrow \Re^+$, a rate R is achievable with lossy source coding with side information at the decoder and with distortion D if $R > R^{WZ}_{S_1|S_2}(D)$. Here

$$R^{WZ}_{S_1|S_2}(D) = \min_{\substack{p(u|s): \\ U - S_1 - S_2 \\ E[d(S_1, U)] \leqslant D}} I(S_1; U|S_2) \tag{1.24}$$

is the rate distortion function for side information at the decoder. Conversely, for $R < R^{WZ}_{S_1|S_2}(D)$, the rate R is not achievable with distortion D.

Theorem 1.7 was first proved by Wyner and Ziv [36]. An accessible summary of the proof is given in [3, Theorem 14.9.1, pp. 438–443].

The result can be extended to continuous sources and unbounded distortion measures under appropriate regularity conditions [37]. It turns out that for the quadratic Gaussian case, that is, jointly Gaussian source and side information with a mean-squared error distortion function, these regularity conditions hold, and one can characterize the achievable rates as follows. Note the correspondence between this result and Theorem 1.6 as $R_2 \rightarrow \infty$.

Proposition 1.3. *Consider a source S_1 and side information source S_2 such that $(S_1(k), S_2(k)) \sim \mathcal{N}(0, \Sigma)$ are i.i.d. in k, with*

$$\Sigma = \sigma^2 \begin{pmatrix} 1 & \rho \\ \rho & 1 \end{pmatrix} \tag{1.25}$$

Then for distortion function $d(s_1, u) = (s_1 - u)^2$ and for $D > 0$,

$$R^{WZ}_{S_1|S_2}(D) = \frac{1}{2} \log^+ \frac{(1-\rho^2)\sigma^2}{D}. \tag{1.26}$$

1.3.2.1 Rate Loss

Interestingly, in this Gaussian case, even if the encoder in Figure 1.4 *also* had access to the source X_2 as in the conditional rate-distortion problem, the rate-distortion function would still be given by Proposition 1.3. Generally, however, there is a penalty for the absence of the side information at the encoder. A result of Zamir [38] has shown that for memoryless sources with finite variance and a mean-squared error distortion, the rate-distortion function provided in Theorem 1.7 can be no larger than $\frac{1}{2}$ bit/source sample than the rate-distortion function given in Theorem 1.3.

It turns out that this principle holds more generally. For centralized source coding of a length M zero-mean vector Gaussian source $\{\vec{S}(n)\}_{n>0}$ with covariance matrix Σ_S having diagonal entries σ_S^2 and eigenvalues $\lambda_1^{(M)}, \ldots, \lambda_M^{(M)}$ with the property that $\lambda_m^{(M)} \geqslant \varepsilon$ for some $\varepsilon > 0$ and all m, and squared error distortion, the rate-distortion function is given by [10]

$$R_{\vec{S}}(D) = \sum_{m=1}^{M} \frac{1}{2} \log \frac{\lambda_m}{D_m}, \tag{1.27}$$

$$D_m = \begin{cases} \nu & \nu < \lambda_m, \\ \lambda_m & \text{otherwise}, \end{cases} \tag{1.28}$$

where $\sum_{m=1}^{M} D_m = D$. Furthermore, it can be lower bouned by [39]

$$R_{\vec{S}}(D) \geqslant \frac{M}{2} \log \frac{M\varepsilon}{D}. \tag{1.29}$$

Suppose each component $\{S_m(n)\}_{n>0}$ were at a separate encoding terminal. Then it is possible to show that by simply quantizing, without binning, an upper bound on the sum rate for a distortion D is given by [39]

$$\sum_{m=1}^{M} R_m \leqslant \frac{M}{2} \log \left(1 + \frac{M\sigma_S^2}{D}\right). \tag{1.30}$$

Thus, the scaling behavior of (1.27) and (1.30) is the same with respect to both small D and large M.

1.3.2.2 Optimality of "Quantize-and-bin" Strategies

In addition to Theorems 1.6 and 1.7, "quantize-and-bin" strategies have been shown to be optimal for several special cases, some of which are included in the results of Kaspi and Berger [40]; Berger and Yeung [41]; Gastpar [42]; and Oohama [43].

By contrast, "quantize-and-bin" strategies have been shown to be strictly suboptimal. Analogous to Körner and Marton's result in the lossless setting [26], work by Krithivasan and Pradhan [44] has shown that rate points outside those prescribed by the "quantize-and-bin" achievable strategy are achievable by exploiting the structure of the sources for multiterminal Gaussian source coding when there are more than two sources.

1.3.2.3 Multiple Descriptions Problem

Most of the problems discussed so far have assumed a centralized decoder with access to encoded observations from all the encoders. A more general model could also include multiple decoders, each with access to only some subset of the encoded observations. While little is known about the general case, considerable effort has been devoted to studying the multiple descriptions problem. This refers to the specific case of a centralized encoder that can encode the source more than once, with subsets of these different encodings available at different encoders. As in the case of the distributed lossy source coding problem above, results for several special cases have been established [45–55].

1.3.3 Interaction

Consider Figure 1.5, which illustrates a setting in which the decoder has the ability to communicate with the encoder and is interested in reconstructing S. Seminal work by Orlitsky [56, 57] suggests that under an appropriate assumption, the benefits of this kind of interaction can be quite significant. The setup assumes two random variables S and U with a joint distribution, one available at the encoder and the other at the decoder. The decoder wants to determine S with zero probability of error, and the goal is to minimize the total number of bits used over all realizations of S and U with positive probability. The following example illustrates the potential gain.

Example 1.2

(The League Problem [56]) Let S be uniformly distributed among one of 2^m teams in a softball league. S corresponds to the winner of a particular game and is known to the encoder. The decoder knows U, which corresponds to the two teams that played in the tournament. Since the encoder does not know the other competitor in the game, if the decoder cannot communicate with the encoder, the encoder must send m bits in order for the decoder to determine the winner with zero probability of error.

Now suppose the decoder has the ability to communicate with the encoder. It can simply look for the first position at which their binary expansion differs and request that from the encoder. This request costs $\log_2 m$ bits since the decoder simply needs to send one of the m different positions. Finally, the encoder simply sends the value of S at this position, which costs an additional 1 bit. The upshot is that as m gets larger, the noninteractive strategy requires *exponentially more bits* than the interactive one.

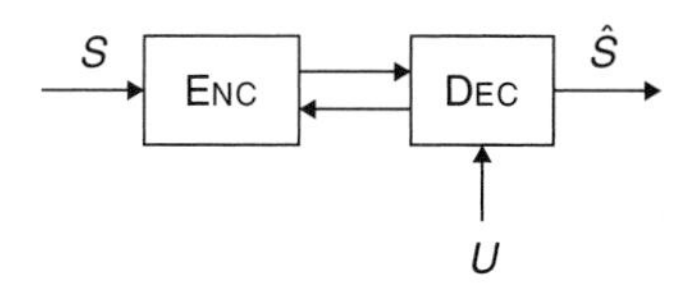

FIGURE 1.5

Interactive source coding.

1.4 REMOTE SOURCE CODING

In many of the most interesting source coding scenarios, the encoders do not get to observe directly the information that is of interest to the decoder. Rather, they may observe a noisy function thereof. This occurs, for example, in camera and sensor networks. We will refer to such source coding problems as *remote source coding*. In this section, we discuss two main insights related to remote source coding:

1. For the centralized setting, direct and remote source coding are the same thing, except with respect to different distortion measures (see Theorem 1.8).
2. For the distributed setting, how severe is the penalty of distributed coding versus centralized? For direct source coding, one can show that the penalty is often small. However, for remote source coding, the penalty can be dramatic (see Equation (1.47)).

The remote source coding problem was initially studied by Dobrushin and Tsybakov [58]. Wolf and Ziv explored a version of the problem with a quadratic distortion, in which the source is corrupted by additive noise, and found an elegant decoupling between estimating the noisy source and compressing the estimate [59]. The problem was also studied by Witsenhausen [60].

We first consider the centralized version of the problem before moving on to the distributed setting. For simplicity, we will focus on the case in which the source and observation processes are memoryless.

1.4.1 Centralized

The remote source coding problem is depicted in Figure 1.6, and Definitions A.16 and A.17 provide a formal description of the problem.

Consider the case in which the source and observation process are jointly memoryless. In this setting, the remote source coding is equivalent to a standard source coding problem with a modified distortion function [10]. For instance, given a distortion function $d: \mathcal{S} \times \hat{\mathcal{S}} \to \Re^+$, one can construct the distortion function $\tilde{d}: \mathcal{U} \times \mathcal{S}$, defined for all $u \in \mathcal{U}$ and $\hat{s} \in \hat{\mathcal{S}}$ as

$$\tilde{d}(u, \hat{s}) = E[d(S, \hat{s}) | U = u], \tag{1.31}$$

where (S, U) share the same distribution as (S_1, U_1). The following result is then straightforward from Theorem 1.2.

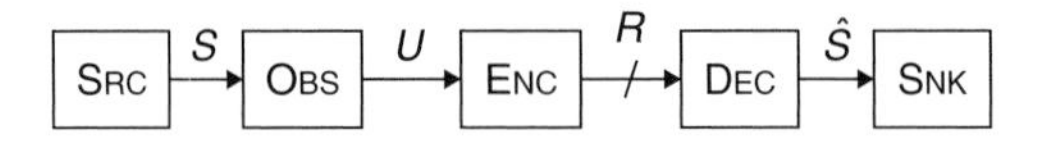

FIGURE 1.6

In the remote source coding problem, one no longer has direct access to the underlying source S but can view a corrupted version U of S through a noisy observation process.

Theorem 1.8: *Remote rate-distortion theorem [10].* Given a discrete memoryless source S, bounded distortion function $d: \mathcal{S} \times \hat{\mathcal{S}} \to \Re^+$, and observations U such that $(S(k), U(k))$ are i.i.d. in k, the remote rate-distortion function is

$$R_S^{\text{remote}}(D) = \min_{\substack{p(\hat{s}|u): \hat{S} - U - S \\ E[d(S,\hat{S})] \leq D}} I(U; \hat{S}). \tag{1.32}$$

This theorem extends to continuous sources under suitable regularity conditions, which are satisfied by finite-variance sources under a squared error distortion.

Theorem 1.9: *Additive remote rate-distortion bounds.* For a memoryless source S, bounded distortion function $d: \mathcal{S} \times \hat{\mathcal{S}} \to \Re^+$, and observations $U_i = S_i + W_i$, where W_i is a sequence of i.i.d. random variables,

$$\frac{1}{2} \log \frac{Q_V}{D - D_0} \leq R_S^{\text{remote}}(D) \leq \frac{1}{2} \log \frac{\sigma_V^2}{D - D_0}, \tag{1.33}$$

where $V = E[S|U]$, and $D_0 = E(S - V)^2$, and where

$$\frac{Q_S Q_W}{Q_U} \leq D_0 \leq \frac{\sigma_S^2 \sigma_W^2}{\sigma_U^2}. \tag{1.34}$$

This theorem does not seem to appear in the literature, but it follows in a relatively straightforward fashion by combining the results of Wolf and Ziv [59] with Shannon's upper and lower bounds. In addition, for the case of a Gaussian source S and Gaussian observation noise W, the bounds in both (1.33) and (1.34) are tight. This can be verified using Table 1.1.

Let us next consider the remote rate-distortion problem in which the encoder makes $M \geq 1$ observations of each source sample. The goal is to illustrate the dependence of the remote rate-distortion function on the number of observations M. To keep things simple, we restrict attention to the scenario shown in Figure 1.7.

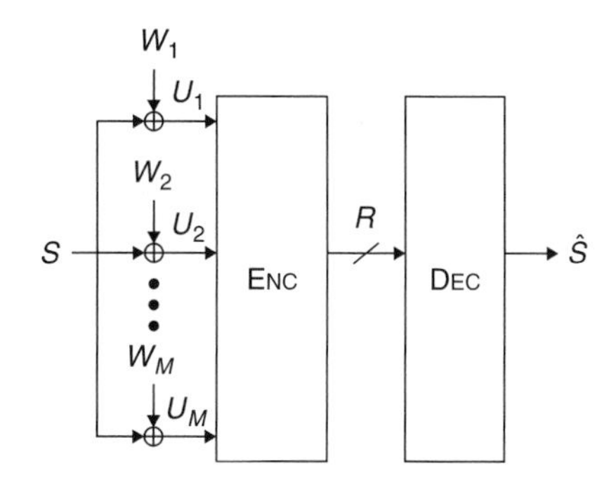

FIGURE 1.7

An additive remote source coding problem with M observations.

More precisely, we suppose that S is a memoryless source and that the observation process is

$$U_m(k) = S(k) + W_m(k), \quad k \geqslant 1, \tag{1.35}$$

where $W_m(k)$ are i.i.d. (both in k and in m) Gaussian random variables of mean zero and variance $\sigma^2_{W_m}$. For this special case, it can be shown [61, Lemma 2] that for any given time k, we can collapse all M observations into a single equivalent observation, characterized by

$$U(k) = \frac{1}{M}\sum_{m=1}^{M}\frac{\sigma_W^2}{\sigma_{W_m}^2}U_m(k) \tag{1.36}$$

$$= S(k) + \frac{1}{M}\sum_{m=1}^{M}\frac{\sigma_W^2}{\sigma_{W_m}^2}W_m(k), \tag{1.37}$$

where $\sigma_W^2 = \left(\frac{1}{M}\sum_{m=1}^{M}\frac{1}{\sigma^2_{W_m}}\right)^{-1}$. This works because $U(k)$ is a sufficient statistic for $S(k)$ given $U_1(k), \ldots, U_M(k)$. However, at this point, we can use Theorem 1.9 to obtain upper and lower bounds on the remote rate-distortion function. For example, using (1.34), we can observe that as long as the source S satisfies $h(S) > -\infty$, D_0 scales linearly with σ_W^2, and thus, inversely proportional to M.

When the source is Gaussian, a precise characterization exists. In particular, Equations (1.33) and (1.34) allow one to conclude that the rate-distortion function is given by the following result.

Proposition 1.4. *Given a memoryless Gaussian source S with $S(i) \sim \mathcal{N}(0, \sigma_S^2)$, squared error distortion, and the M observations corrupted by an additive Gaussian noise model and given by Equation (1.35), the remote rate-distortion function is*

$$R_S^{\text{remote}}(D) = \frac{1}{2}\log\frac{\sigma_S^2}{D} + \frac{1}{2}\log\frac{\sigma_S^2}{\sigma_U^2 - \frac{\sigma_W^2\sigma_S^2}{MD}}, \tag{1.38}$$

where

$$\sigma_U^2 = \sigma_S^2 + \frac{\sigma_W^2}{M} \quad \text{and} \quad \sigma_W^2 = \frac{1}{\frac{1}{M}\sum_{m=1}^{M}\frac{1}{\sigma^2_{W_m}}}. \tag{1.39}$$

As in the case of direct source coding, there is an analogous result for the case of jointly stationary ergodic Gaussian sources and observations.

Proposition 1.5. *Let S be a stationary ergodic Gaussian source S, U an observation process that is jointly stationary ergodic Gaussian with S, and $\Phi_S(\omega)$, $\Phi_U(\omega)$, and*

$\Phi_{S,U}(\omega)$ *be their corresponding power and cross spectral densities. Then for mean-squared error distortion, the remote rate-distortion function is given by*

$$R_S^{\text{remote}}(D_\nu) = \frac{1}{4\pi}\int_{-\pi}^{\pi} \max\left\{0, \frac{|\Phi_{S,U}(\omega)|^2}{\nu\Phi_U(\omega)}\right\} d\omega \tag{1.40}$$

$$D_\nu = \frac{1}{2\pi}\int_{-\pi}^{\pi}\left(\frac{\Phi_S(\omega)\Phi_U(\omega) - |\Phi_{S,U}(\omega)|^2}{\Phi_U(\omega)}\right) d\omega + \frac{1}{2\pi}\int_{-\pi}^{\pi} \min\left\{\nu, \frac{|\Phi_{S,U}(\omega)|^2}{\Phi_U(\omega)}\right\} d\omega. \tag{1.41}$$

PROOF. See Berger [10, pp. 124–129]. ■

Observe that $\frac{1}{2\pi}\int_{-\pi}^{\pi}\left(\frac{\Phi_S(\omega)\Phi_U(\omega)-|\Phi_{S,U}(\omega)|^2}{\Phi_U(\omega)}\right) d\omega$ is simply the mean-squared error resulting from applying a Wiener filter on U to estimate S.

1.4.2 Distributed: The CEO Problem

Let us now turn to the *distributed* version of the remote source coding problem. A particularly appealing special case is illustrated in Figure 1.8. This problem is often motivated with the following scenario. A chief executive officer or chief estimation officer (CEO) is interested in estimating a random process. M agents observe noisy versions of the random process and have noiseless bit pipes with finite rate to the CEO. Under the assumption that the agents cannot communicate with one another, one wants to analyze the fidelity of the CEO's estimate of the random process subject to these rate constraints. Because of the scenario, this is often called the CEO problem [62].

A formal definition of the CEO problem can be given by a slight variation of Definition A.15, namely, by adding an additional encoder with direct access to the underlying source, and considers the rate-distortion region when the rate of this

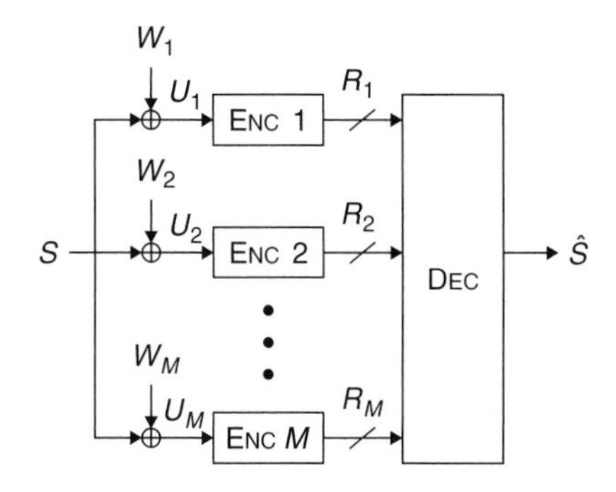

FIGURE 1.8

The additive CEO problem.

encoder is set to zero. The formal relationship is established in Definitions A.18 and A.19.

The CEO problem has been used as a model for sensor networks, where the encoders correspond to the sensors, each with access to a noisy observation of some state of nature. Unfortunately (and perhaps not too surprisingly), the general CEO problem remains open. On the positive side, the so-called *quadratic Gaussian* CEO problem has been completely solved. Here, with respect to Figure 1.8, the source S is a memoryless Gaussian source, and the noises $W_m(k)$ are i.i.d. (both in k and in m) Gaussian random variables of mean zero and variance $\sigma^2_{W_m}$. The setting was first considered by Viswanathan and Berger [63], and later refined (see, e.g. [43, 64, 65] or [66, 67]), leading to the following theorem:

Theorem 1.10. Given a memoryless Gaussian source S with $S(i) \sim \mathcal{N}(0, \sigma_S^2)$, squared error distortion, and the M observations corrupted by an additive Gaussian noise model and given by Equation (1.35), and assume that $\sigma^2_{W_1} \leqslant \sigma^2_{W_2} \leqslant \cdots \leqslant \sigma^2_{W_M}$. Then the sum rate-distortion function for the CEO problem is given by

$$R_S^{\mathrm{CEO}}(D) = \frac{1}{2}\log\frac{\sigma_S^2}{D} - \frac{1}{2}\log\left(\prod_{m=1}^{K}\sigma^2_{W_m}\right) - \frac{K}{2}\log\left(\frac{1}{KD_K} - \frac{1}{KD}\right), \tag{1.42}$$

where K is the largest integer in $\{1, \ldots, M\}$ satisfying

$$\frac{1}{D} - \frac{1}{\sigma_S^2} \geqslant -\frac{K-1}{\sigma^2_{W_K}} + \sum_{m=1}^{K-1}\frac{1}{\sigma^2_{W_m}} \tag{1.43}$$

and

$$\frac{1}{D_K} = \frac{1}{\sigma_S^2} + \sum_{m=1}^{K}\frac{1}{\sigma^2_{W_m}}. \tag{1.44}$$

Furthermore, if $\sigma^2_{W_m} = \sigma^2_W$ for all $m = 1, \ldots, M$,

$$R_S^{\mathrm{CEO}}(D) = \frac{1}{2}\log\frac{\sigma_S^2}{D} + \frac{M}{2}\log\frac{\sigma_S^2}{\sigma_U^2 - \frac{\sigma_W^2\sigma_S^2}{MD}}, \tag{1.45}$$

where

$$\sigma_U^2 = \sigma_S^2 + \frac{\sigma_W^2}{M}. \tag{1.46}$$

Alternate strategies for achieving the sum rate have been explored [68] and achievable strategies for the case of vector-Gaussian sources have been studied [69, 70], but the rate region for this setting remains open.

We conclude this section by discussing the *rate loss* between the *joint* remote source coding problem (illustrated in Figure 1.7) and the *distributed* remote source coding problem (illustrated in Figure 1.8). For the quadratic Gaussian scenario with equal noise variances $\sigma_{W_m}^2 = \sigma_W^2$, this rate loss is exactly characterized by comparing Proposition 1.4 with Theorem 1.10, leading to

$$R_S^{\mathrm{CEO}}(D) - R_S^{\mathrm{remote}}(D) = \frac{M-1}{2} \log \frac{\sigma_S^2}{\sigma_S^2 + \frac{\sigma_W^2}{M}\left(1 - \frac{\sigma_S^2}{D}\right)}. \tag{1.47}$$

By letting $M \to \infty$, this becomes

$$R_S^{\mathrm{CEO}}(D) - R_S^{\mathrm{remote}}(D) = \frac{\sigma_W^2}{2}\left(\frac{1}{D} - \frac{1}{\sigma_S^2}\right), \tag{1.48}$$

so the rate loss scales inversely with the target distortion. Note that this behavior is dramatically different from the one found in Section 1.3.2.1; as we have seen there, for *direct* source coding, both centralized and distributed encoding have the same scaling law behavior with respect to large M and small D (at least under the assumptions stated in Section 1.3.2.1). By contrast, as the quadratic Gaussian example reveals, for *remote* source coding, centralized and distributed encoding can have very different scaling behaviors.

It turns out that the quadratic Gaussian example has the worst case rate loss behavior among all problems that involve additive Gaussian observation noise and mean-squared error distortion (but arbitrary statistics of the source S). More precisely, we have the following proposition, proved in [61]:

Proposition 1.6. *Given a real-valued zero-mean memoryless source S with variance σ_S^2, squared error distortion, and the M observations corrupted by an additive Gaussian noise model and given by Equation (1.35) with $\sigma_{W_i}^2 = \sigma_W^2$, the difference between the sum rate-distortion function for the CEO problem and the remote rate-distortion function is no more than*

$$\begin{aligned} R_S^{\mathrm{CEO}}(D) - R_S^{\mathrm{remote}}(D) \leqslant{} & \frac{M-1}{2} \log \frac{\sigma_S^2}{\sigma_S^2 + \frac{\sigma_W^2}{M}\left(1 - \frac{\sigma_S^2}{D}\right)} \\ & + \frac{1}{2} \log \frac{(M-1)D\sigma_W^2 + \sigma_S^2\left(M\left(2D + 2\sqrt{D\sigma_W^2}\right) - MD\left(D + 2\sqrt{D\sigma_W^2}\right)\right)}{D\left(M\sigma_S^2 + \sigma_W^2\right) - \sigma_W^2\sigma_S^2}. \end{aligned} \tag{1.49}$$

By letting $M \to \infty$, this becomes

$$R_S^{\mathrm{CEO}}(D) - R_S^{\mathrm{remote}}(D) \leqslant \frac{\sigma_W^2}{2}\left(\frac{1}{D} - \frac{1}{\sigma_S^2}\right) + \frac{1}{2}\log\left(1 + \left(\frac{1}{D} - \frac{1}{\sigma_S^2}\right)\left(D + 2\sqrt{D\sigma_W^2}\right)\right). \tag{1.50}$$

Thus, the rate loss cannot scale more than inversely with the distortion D even for non-Gaussian sources. Bounds for the rate loss have been used to give tighter achievable bounds for the sum rate when the source is non-Gaussian [71].

Although the gap between distributed and centralized source coding of a remote source can be significant, cooperative interaction among the encoders can potentially decrease the gap. While this has been studied when the source is Gaussian [72], the question remains open for general sources.

1.5 JOINT SOURCE-CHANNEL CODING

When information is to be stored on a perfect (error-free) discrete (e.g., binary) medium, then the source coding problem surveyed in this chapter is automatically relevant. However, in many tasks, the source information is compressed because it must be communicated across a noisy channel. There, it is not evident that the formalizations discussed in this chapter are relevant. For the stationary ergodic point-to-point problem, Shannon's *separation theorem* proves that, indeed, an optimal architecture is to first compress the source information down to the capacity of the channel, and then communicate the resulting encoded bit stream across the noisy channel in a reliable fashion.

To make this precise, we discuss the scenario illustrated in Figure 1.9. Formally, consider the following class of noisy channels:

Definition 1.1. Given a random vector X^n, a discrete memoryless channel (used without feedback) is characterized by a conditional distribution $p(y|x)$ such that for a given channel input x^n, the channel output Y^n satisfies

$$Pr(Y^n = y^n | X^n = x^n) = \prod_{i=1}^{n} Pr(Y(i) = y(i) | X(i) = x(i)), \tag{1.51}$$

where $Pr(Y(i) = y | X(i) = x) = p(y|x)$.

Moreover, by reliable communication, we mean the following:

Definition 1.2. Given a source S and discrete memoryless channel $p(y|x)$, an (N, α, δ) *lossless joint source-channel code* consists of an encoding function $f: \mathcal{S}^N \to \mathcal{X}^K$ and decoding function $g: \mathcal{Y}^K \to \mathcal{S}^N$ such that $X^K = f(S^N)$ and

$$Pr(S^K \neq g(Y^K)) \leq \delta, \tag{1.52}$$

where $\alpha K = N$. Thus, α is sometimes called the *mismatch* of the source to the channel.

Definition 1.3. A source S is *recoverable* over a discrete memoryless channel $p(y|x)$ with mismatch α if for all $\delta > 0$, there exists N^* such that for all $N \geq N^*$, there exists an (N, α, δ) lossless joint source-channel code.

Then, the so-called separation theorem can be expressed as follows:

Theorem 1.11: *Joint Source-Channel Coding Theorem.* A discrete memoryless source S is recoverable over a discrete memoryless channel $p(y|x)$ with mismatch α if $H(S) < \alpha \max_{p(x)} I(X; Y)$. Conversely, if $H(S) > \alpha \max_{p(x)} I(X; Y)$, it is not recoverable.

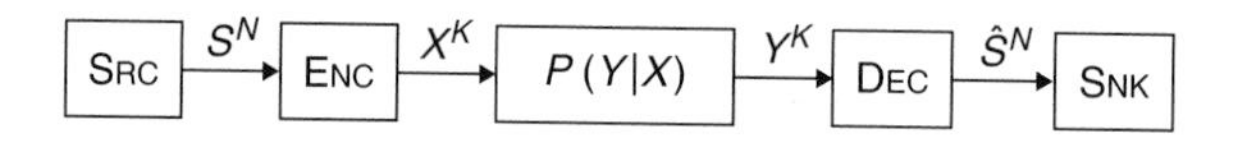

FIGURE 1.9

In many situations, one might need to transmit a source over a noisy channel.

A proof of this theorem can be found, for example, in [3]. One should note that Theorem 1.11 can be stated more generally for both a larger class of sources and for channels.

The crux, however, is that there is no equivalent of the separation theorem in communication *networks*. One of the earliest examples of this fact was given by [73]: when correlated sources are transmitted in a lossless fashion across a multiple-access channel, it is *not* generally optimal to first compress the sources with the Slepian–Wolf code discussed in Section 1.3, thus obtaining smaller and essentially independent source descriptions, and sending these descriptions across the multiple-access channel using a capacity-approaching code. Rather, the source correlation can be exploited to access dependent input distributions on the multiple-access channels, thus *enlarging* the region of attainable performance. A more dramatic example was found in [74–76]. In that example, modeling a simple "sensor" network, a single underlying source is observed by M terminals subject to independent noises. These terminals communicate over an additive multiple-access channel to a "fusion center" whose only goal is to recover the single underlying source to within the smallest squared error. For this problem, one may use the CEO source code discussed in Section 1.4.2, followed by a regular capacity-approaching code for the multiple-access channel. Then, the smallest squared error distortion D decreases like $1/\log M$. However, it can be shown that the optimal performance for a joint source-channel code decreases like $1/M$, that is, *exponentially* better. For some simple instances of the problem, the optimal joint source-channel code is simply uncoded transmission of the noisy observations [77]. Nevertheless, for certain classes of networks, one can show that a distributed source code followed by a regular channel code is good enough, thus establishing partial and approximate versions of network separation theorems. An account of this can be found in [78].

ACKNOWLEDGMENTS

The authors would like to acknowledge anonymous reviewers and others who went through earlier versions of this manuscript. In particular, Mohammad Ali Maddah-Ali pointed out and corrected an error in the statement of Theorem 1.10.

APPENDIX A: Formal Definitions and Notations

This appendix contains technical definitions that will be useful in understanding the results in this chapter.

A.1 NOTATION

Unless otherwise stated, capital letters X, Y, Z denote random variables, lower-case letters x, y, z instances of these random variables, and calligraphic letters $\mathcal{X}, \mathcal{Y}, \mathcal{Z}$ sets.

The shorthand $p(x,y), p(x)$, and $p(y)$ will be used for the probability distributions $Pr(X=x, Y=y), Pr(X=x)$, and $Pr(Y=y)$. X^N will be used to denote the random vector $X_1, X_2, \ldots, X_N$. The alternate vector notation $\vec{R}$ and $\vec{D}$ will sometimes be used to represent achievable rate-distortion regions. Finally, we will use the notation $X - Y - Z$ to denote that the random variables X, Y, Z form a Markov chain. That is, X and Z are independent given Y.

We start by defining the notion of entropy, which will be useful in the sequel.

Definition A.1. Let X, Y be discrete random variables with joint distribution $p(x,y)$. Then the *entropy* $H(X)$ of X is a real number given by the expression

$$H(X) = -E[\log p(X)] = -\sum_x p(x) \log p(x), \tag{A.53}$$

the *joint entropy* $H(X,Y)$ of X and Y is a real number given by the expression

$$H(X,Y) = -E[\log p(X,Y)] = -\sum_{x,y} p(x,y) \log p(x,y), \tag{A.54}$$

and the *conditional entropy* $H(X|Y)$ of X given Y is a real number given by the expression

$$H(X|Y) = -E[\log p(X|Y)] = H(X,Y) - H(Y). \tag{A.55}$$

With a definition for entropy, we can now define mutual information.

Definition A.2. Let X, Y be discrete random variables with joint distribution $p(x,y)$. Then the *mutual information* $I(X;Y)$ between X and Y is a real number given by the expression

$$I(X;Y) = H(X) - H(X|Y) = H(Y) - H(Y|X). \tag{A.56}$$

Similar definitions exist in the continuous setting.

Definition A.3. Let X, Y be continuous random variables with joint density $f(x,y)$. Then we define the *differential entropy* $h(X) = -E[\log f(X)]$, *joint differential entropy* $h(X,Y) = -E[\log f(X,Y)]$, *conditional differential entropy* $h(X|Y) = -E[\log f(X|Y)]$, and *mutual information* $I(X;Y) = h(X) - h(X|Y) = h(Y) - h(Y|X)$.

Definition A.4. Let X be a continuous random variable with density $f(x)$. Then its *entropy power* Q_X is given by

$$Q_X = \frac{1}{2\pi e} \exp\{2h(X)\}. \tag{A.57}$$

The central concept in this chapter is the *information source*, which is merely a discrete-time random process $\{S(n)\}_{n>0}$. For the purposes of this chapter, we assume that all information sources are stationary and ergodic. When the random variables $\{S(n)\}_{n>0}$ take values in a *discrete* set, we will refer to the source as a *discrete information source*. For simplicity, we will refer to a source $\{S(n)\}_{n>0}$ with the abbreviated notation S.

Definition A.5. The entropy rate of a stationary and ergodic discrete information source S is

$$H_\infty(S) = \lim_{n\to\infty} \frac{1}{n} H(S(1), S(2), \ldots, S(n)) = \lim_{n\to\infty} H(S(n)|S(n-1), S(n-2), \ldots, S(1)). \tag{A.58}$$

In many of the concrete examples discussed in this chapter, S will be assumed to be a sequence of independent and identically distributed (i.i.d.) random variables. This is usually referred to as a *memoryless* source, and one can show that $H_\infty(S) = H(S(1))$. For notational convenience, we will denote this simply as $H(S)$.

A.1.1 Centralized Source Coding

Definition A.6. Given a source S, an (N, R, δ) *lossless source code* consists of an encoding function $f: \mathcal{S}^N \to \{1, \ldots, M\}$ and decoding function $g: \{1, \ldots, M\} \to \mathcal{S}^N$ such that

$$Pr\left(S^N \neq g\left(f(S^N)\right)\right) \leqslant \delta, \tag{A.59}$$

where $M = 2^{NR}$.

In the above definition, R has units in bits per source sample. Thus, R is often referred to as the *rate* of the source code.

Definition A.7. Given a source S, a rate R is *achievable* via lossless source coding if for all $\delta > 0$, there exists N^* such that for $N \geqslant N^*$, there exists an $(N, R+\delta, \delta)$ lossless source code.

Similarly, definitions can be given in the lossy case.

Definition A.8. Given a source S and distortion function $d: \mathcal{S} \times \hat{\mathcal{S}} \to \Re^+$, an (N, R, D) *centralized lossy source code* consists of an encoding function $f: \mathcal{S}^N \to \{1, \ldots, M\}$ and decoding function $g: \{1, \ldots, M\} \to \hat{\mathcal{S}}^N$ such that for $\hat{S}^N = g(f(S^N))$

$$\sum_{i=1}^{N} E[d(S(i), \hat{S}(i))] \leqslant ND, \tag{A.60}$$

where $M = 2^{NR}$.

Definition A.9. Given a source S and distortion function $d: \mathcal{S} \times \hat{\mathcal{S}} \to \mathbb{R}^+$, a rate R is *achievable* with distortion D if for all $\delta > 0$, there exists a sequence of (N, R_N, D_N) centralized lossy source codes and N^* such that for $N \geqslant N^*$, $R_N \geqslant R + \delta$, $D_N \leqslant D + \delta$.

Definition A.10. Given a source S_1, source side information S_2 available at the encoder and decoder, and distortion function $d: \mathcal{S}_1 \times \hat{\mathcal{S}}_1 \to \Re^+$, an (N, R, D) *conditional lossy source code* consists of an encoding function $f: \mathcal{S}_1^N \times \mathcal{S}_2^N \to \{1, \ldots, M\}$ and decoding function $g: \{1, \ldots, M\} \times \mathcal{S}_2^N \to \hat{\mathcal{S}}_1^N$ such that for $\hat{S}_1^N = g(f(S_1^N, S_2^N), S_2^N)$,

$$\sum_{i=1}^{N} E[d(S(i), \hat{S}(i))] \leqslant ND, \tag{A.61}$$

where $M = 2^{NR}$.

Definition A.11. Given a source S_1, source side information S_2 available at the encoder and decoder, and distortion function $d: \mathcal{S}_1 \times \hat{\mathcal{S}}_1 \to \Re^+$, a rate R is *achievable* with distortion D if for all $\delta > 0$, there exists a sequence of (N, R_N, D_N) conditional lossy source codes and N^* such that for $N \geqslant N^*$, $R_N \geqslant R + \delta$, $D_N \leqslant D + \delta$. The supremum over all achievable R is the *conditional rate-distortion function*, denoted as $R_{S_1|S_2}(D)$.

A.1.2 Distributed Source Coding

We first consider the lossless case.

Definition A.12. Given discrete sources S_1 and S_2, an (N, R_1, R_2, δ) *two-terminal lossless source code* consists of two encoding functions $f_1: \mathcal{S}_1^N \to \{1, \ldots, M_1\}$ and $f_2: \mathcal{S}_2^N \to \{1, \ldots, M_2\}$ as well as a decoding function $g: \{1, \ldots, M_1\} \times \{1, \ldots, M_2\} \to \mathcal{S}_1^N \times \mathcal{S}_2^N$ such that

$$Pr\left((S_1^N, S_2^N) \neq g\left(f_1(S_1^N), f_2(S_2^N)\right)\right) \leqslant \delta, \tag{A.62}$$

where $M_i = 2^{NR_i}$.

Definition A.13. Given discrete sources S_1 and S_2, a rate pair (R_1, R_2) is *achievable* for the two-terminal lossless source coding problem if for all $\delta > 0$, there exists a sequence of $(N, R_{1,N}, R_{2,N}, \delta_N)$ lossless source codes and N^* such that for $N \geqslant N^*$, $R_{1,N} \geqslant R_1 + \delta$, $R_{2,N} \geqslant R_2 + \delta$, and $\delta_N \leqslant \delta$.

Analogous definitions can be provided in the lossy setting. Here, we consider the case of M users with access to side information, as depicted in Figure A.10.

Definition A.14. Given M sources S_m, $m \in \{1, \ldots, M\}$ and K distortion functions $d_k: \prod_{m=1}^{M} \mathcal{S}_m \times \mathcal{U}_k \to \Re^+$, $k \in \{1, \ldots, K\}$, an $(N, \vec{R}, \vec{D})$ *distributed lossy source code* consists of M encoding functions $f_m: \mathcal{S}_m^N \to \{1, \ldots, 2^{NR_m}\}$ and decoding functions $g_k: \prod_{m=1}^{M} \{1, \ldots, 2^{NR_m}\} \to \mathcal{U}_k^N$ such that for $U_k^N = g_k(f_1(S_1^N), f_2(S_2^N), \ldots, f_M(S_M^N))$

$$\sum_{i=1}^{N} E\left[d_k\left(S_1(i), S_2(i), \ldots, S_{M+1}(i), U_k(i)\right)\right] \leqslant ND_k. \tag{A.63}$$

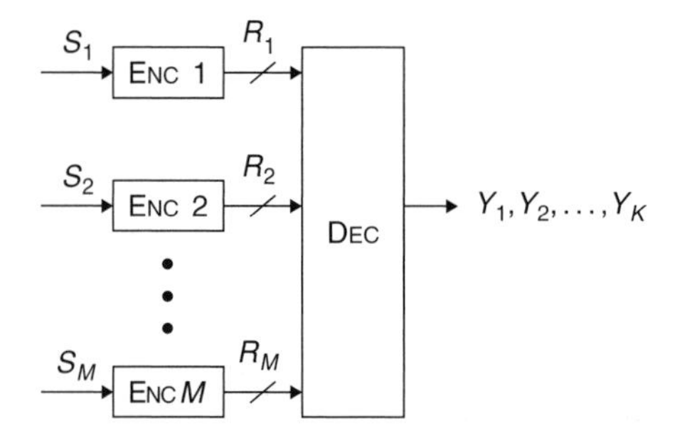

FIGURE A.10

Distributed source coding problem.

Definition A.15. Given M sources S_m, $m \in \{1,\ldots,M\}$ and K distortion functions d_k: $\prod_{m=1}^{M} \mathcal{S}_m \times \mathcal{U}_k \to \Re^+$, $k \in \{1,\ldots,K\}$, a rate-distortion vector $(\vec{R}, \vec{D})$ is *achievable* with a distributed lossy source code if for all $\delta > 0$, there exists a sequence of $(N, \vec{R}_N, \vec{D}_N)$ lossy source codes with side information at the decoder and N^* such that for all $N \geqslant N^*$, $R_{m,N} \geqslant R_m + \delta$, $D_{k,N} \leqslant D_k + \delta$, $m \in \{1,\ldots,M\}$, $k \in \{1,\ldots,K\}$.

A.1.3 Remote Source Coding

Definition A.16. Given a source S, distortion function $d: \mathcal{S} \times \hat{\mathcal{S}} \to \Re^+$, and observation process U, an (N,R,D) *remote source code* consists of an encoding function $f: \mathcal{U}^N \to \{1,\ldots,M\}$ and decoding function $g: \{1,\ldots,M\} \to \hat{\mathcal{S}}^N$ such that for $\hat{S}^N = g(f(U^N))$

$$\sum_{i=1}^{N} E[d(S(i), \hat{S}(i))] \leqslant ND, \tag{A.64}$$

where $M = 2^{NR}$.

Definition A.17. Given a source S, distortion function $d: \mathcal{S} \times \mathcal{S} \to \Re^+$, and observation process U, a rate R is *achievable* with distortion D if for all $\delta > 0$, there exists a sequence of (N, R_N, D_N) remote source codes and N^* such that for all $N \geqslant N^*$, $R_N \geqslant R + \delta$, $D_N \leqslant D + \delta$. The supremum over all achievable R is the *remote rate-distortion function*, denoted as $R_S^{\text{remote}}(D)$.

The CEO problem is a special case of the distributed source coding problem, as depicted in Figure A.11.

Definition A.18. Given $M+1$ sources, S and U_m, $m \in \{1,\ldots,M\}$ and a distortion function $d: \mathcal{S} \times \hat{\mathcal{S}} \to \Re^+$, an $(N, \vec{R}, D)$ *CEO source code*—where $\vec{R}$ is a length M vector on the nonnegative reals—is an $M+1$ user, single-distortion function, $(N, \vec{\tilde{R}}, D)$ distributed lossy source code, where $\tilde{R}_1 = 0$, $\tilde{R}_m = R_{m-1}$ for $m \in \{2,\ldots,M+1\}$ $S = S_1$, $U_m = S_{m+1}$, with the distortion function $d_1: \prod_{m=1}^{M+1} \mathcal{S}_m \times \hat{\mathcal{S}}_1 \to \Re^+$ having the property that

$$d_1(s, u_1, \ldots, u_M, \hat{s}_1) = d(s, \hat{s}_1). \tag{A.65}$$

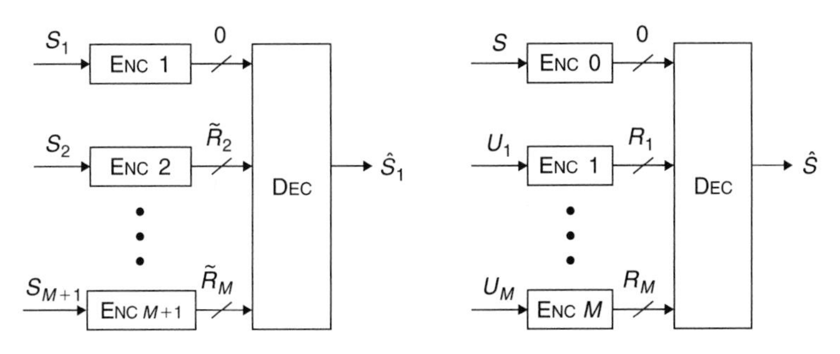

FIGURE A.11

CEO problem as special case of distributed source coding problem.

Definition A.19. Given $M+1$ sources, S and U_m, $m \in \{1,\ldots,M\}$ and a distortion function $d: \mathcal{S} \times \hat{\mathcal{S}} \rightarrow \Re^+$, a sum rate-distortion pair (R,D) is *achievable* with a CEO source code if for all $\delta > 0$, there exists a sequence of $(N, \vec{R}_N, D_N)$ CEO source codes and N^* such that for all $N \geqslant N^*$, $\sum_{m=1}^{M} R_{m,N} \geqslant R+\delta$, $D_N \leqslant D+\delta$. For a given distortion D, the supremum over R for all achievable sum rate-distortion pairs (R,D) specifies the sum rate-distortion function for the CEO problem, denoted by $R_S^{\mathrm{CEO}}(D)$.

REFERENCES

[1] C. E. Shannon, "A mathematical theory of communication," *Bell System Technical Journal*, vol. 27, pp. 379–423, 623–656, 1948.

[2] T. S. Han and S. Verdú, "Approximation theory of output statistics," *IEEE Transactions on Information Theory*, vol. 39, no. 3, pp. 752–772, May 1993.

[3] T. Cover and J. Thomas, *Elements of Information Theory*. New York: John Wiley and Sons, 1991.

[4] R. Pilc, "The transmission distortion of a source as a function of the length encoding block length," *Bell System Technical Journal*, vol. 47, no. 6, pp. 827–885, 1968.

[5] S. Arimoto, "An algorithm for calculating the capacity and rate-distortion functions," *IEEE Transactions on Information Theory*, vol. 18, pp. 14–20, 1972.

[6] R. Blahut, "Computation of channel capacity and rate-distortion functions," *IEEE Transactions on Information Theory*, vol. 18, pp. 460–473, 1972.

[7] K. Rose, "A mapping approach to rate-distortion computation and analysis," *IEEE Transactions on Information Theory*, vol. 40, pp. 1939–1952, 1994.

[8] C. Shannon, "Coding theorems for a discrete source with a fidelity criterion," *IRE Convention Rec.*, vol. 7, pp. 142–163, 1959.

[9] T. Linder and R. Zamir, "On the asymptotic tightness of the Shannon lower bound," *IEEE Transactions on Information Theory*, pp. 2026–2031, November 1994.

[10] T. Berger, *Rate Distortion Theory: A Mathematical Basis for Data Compression*, ser. Information and System Sciences Series. Englewood Cliffs, NJ: Prentice-Hall, 1971.

[11] R. Gray, "Conditional rate-distortion theory," Stanford University, Tech. Rep., October 1972.

[12] J. Omura, "A coding theorem for discrete-time sources," *IEEE Transactions on Information Theory*, vol. 19, pp. 490–498, 1973.

[13] K. Marton, "Error exponent for source coding with a fidelity criterion," *IEEE Transactions on Information Theory*, vol. 20, no. 2, pp. 197–199, March 1974.

[14] A. Wyner and J. Ziv, "Bounds on the rate-distortion function for sources with memory," *IEEE Transactions on Information Theory*, vol. 17, no. 5, pp. 508–513, September 1971.

[15] R. Gray, "Sliding-block source coding," *IEEE Transactions on Information Theory*, vol. 21, no. 4, pp. 357–368, July 1975.

[16] H. Witsenhausen, "On the structure of real-time source coders," *Bell System Technical Journal*, vol. 58, no. 6, pp. 1437–1451, 1979.

[17] D. Neuhoff and R. Gilbert, "Causal source codes," *IEEE Transactions on Information Theory*, vol. 28, no. 5, pp. 710–713, September 1982.

[18] A. Lapidoth, "On the role of mismatch in rate distortion theory," *IEEE Transactions on Information Theory*, vol. 43, pp. 38–47, January 1997.

[19] A. Dembo and T. Weissman, "The minimax distortion redundancy in noisy source coding," *IEEE Transactions on Information Theory*, vol. 49, pp. 3020–3030, 2003.

[20] T. Weissman, "Universally attainable error-exponents for rate-distortion coding of noisy sources," *IEEE Transactions on Information Theory*, vol. 50, no. 6, pp. 1229–1246, 2004.

[21] A. Gersho and R. Gray, *Vector Quantization and Signal Compression*. Boston: Kluwer Academic Publishers, 1992.

[22] D. Slepian and J. Wolf, "Noiseless coding of correlated information sources," *IEEE Transactions on Information Theory*, vol. 19, pp. 471–480, 1973.

[23] R. Ahlswede and J. Korner, "Source coding with side information and a converse for degraded broadcast channels," *IEEE Transactions on Information Theory*, vol. 21, no. 6, pp. 629–637, November 1975.

[24] A. Wyner, "On source coding with side information at the decoder," *IEEE Transactions on Information Theory*, vol. 21, no. 3, pp. 294–300, May 1975.

[25] S. Gel'fand and M. Pinsker, "Coding of sources on the basis of observations with incomplete information (in Russian)," *Problemy Peredachi Informatsii (Problems of Information Transmission)*, vol. 25, no. 2, pp. 219–221, April–June 1979.

[26] J. Korner and K. Marton, "How to encode the modulo-two sum of binary sources (Corresp.)," *IEEE Transactions on Information Theory*, vol. 25, no. 2, pp. 219–221, March 1979.

[27] T. Berger and S. Tung, "Encoding of correlated analog sources," in *1975 IEEE-USSR Joint Workshop on Information Theory*. Moscow: IEEE Press, 1975, pp. 7–10.

[28] S. Tung, "Multiterminal source coding," Ph.D. dissertation, Cornell University, 1977.

[29] T. Berger, "Multiterminal source coding," in *Lecture Notes presented at CISM Summer School on the Information Theory Approach to Communications*, 1977.

[30] K. B. Housewright, "Source coding studies for multiterminal systems," Ph.D. dissertation, University of California, Los Angeles, 1977.

[31] T. S. Han and K. Kobayashi, "A unified achievable rate region for a general class of multiterminal source coding systems," *IEEE Transactions on Information Theory*, pp. 277–288, May 1980.

[32] S. D. Servetto, "Multiterminal source coding with two encoders—I: A computable outer bound," Arxiv, November 2006. [Online]. Available: http://arxiv.org/abs/cs/0604005v3

[33] A. Wagner and V. Anantharam, "An improved outer bound for the multiterminal source coding problem," in *Proc. IEEE International Symposium on Information Theory*, Adelaide, Australia, 2005.

[34] ———, "An infeasibility result for the multiterminal source-coding problem," *IEEE Transactions on Information Theory*, 2008. [Online]. Available: http://arxiv.org/abs/cs.IT/0511103

[35] A. Wagner, S. Tavildar, and P. Viswanath, "Rate region of the quadratic Gaussian two-encoder source-coding problem," *IEEE Transactions on Information Theory*, vol. 54, no. 5, pp. 1938–1961, May 2008.

[36] A. Wyner and J. Ziv, "The rate-distortion function for source coding with side information at the decoder," *IEEE Transactions on Information Theory*, vol. 22, pp. 1–10, January 1976.

[37] A. Wyner, "The rate-distortion function for source coding with side information at the decoder—II: General sources," *Information and Control*, vol. 38, pp. 60–80, 1978.

[38] R. Zamir, "The rate loss in the Wyner-Ziv problem," *IEEE Transactions on Information Theory*, vol. 42, pp. 2073–2084, November 1996.

[39] K. Eswaran and M. Gastpar, "On the significance of binning in a scaling-law sense," in *Information Theory Workshop*, Punta del Este, Uruguay, March 2006.

[40] A. H. Kaspi and T. Berger, "Rate-distortion for correlated sources with partially separated encoders," *IEEE Transactions on Information Theory*, vol. IT-28, no. 6, pp. 828–840, 1982.

[41] T. Berger and R. Yeung, "Multiterminal source encoding with one distortion criterion," *IEEE Transactions on Information Theory*, vol. IT-35, pp. 228–236, 1989.

[42] M. Gastpar, "The Wyner-Ziv problem with multiple sources," *IEEE Transactions on Information Theory*, vol. 50, no. 11, pp. 2762–2768, November 2004.

[43] Y. Oohama, "Rate-distortion theory for Gaussian multiterminal source coding systems with several side informations at the decoder," *IEEE Transactions on Information Theory*, vol. 51, pp. 2577–2593, 2005.

[44] D. Krithivasan and S. S. Pradhan, "Lattices for distributed source coding: Jointly Gaussian sources and reconstruction of a linear function," Arxiv, July 2007. [Online]. Available: http://arxiv.org/abs/0707.3461

[45] H. Witsenhausen, "On source networks with minimal breakdown degradation," *Bell System Technical Journal*, vol. 59, no. 6, pp. 1083–1087, July–August 1980.

[46] ———, "On team guessing with independent informations," *Math. Oper. Res.*, vol. 6, no. 2, pp. 293–304, May 1981.

[47] ———, "Team guessing with lacunary information," *Math. Oper. Res.*, vol. 8, no. 1, pp. 110–121, February 1983.

[48] H. Witsenhausen and A. Wyner, "Source coding for multiple descriptions II: A binary source," *Bell System Technical Journal*, vol. 60, no. 10, pp. 2281–2292, December 1981.

[49] A. El Gamal and T. Cover, "Achievable rates for multiple descriptions," *IEEE Transactions on Information Theory*, vol. 28, no. 6, pp. 851–857, November 1982.

[50] J. Wolf, A. Wyner, and J. Ziv, "Source coding for multiple descriptions," *Bell System Technical Journal*, vol. 59, no. 8, pp. 1417–1426, October 1980.

[51] T. Berger and Z. Zhang, "Minimum breakdown degradation in binary source coding," *IEEE Transactions on Information Theory*, vol. 29, no. 6, pp. 807–814, November 1983.

[52] L. Ozarow, "On a source-coding problem with two channels and three receivers," *Bell System Technical Journal*, vol. 59, no. 10, pp. 1909–1921, 1980.

[53] R. Ahlswede, "The rate-distortion region for multiple descriptions without excess rate," *IEEE Transactions on Information Theory*, vol. 31, no. 6, pp. 721–726, November 1985.

[54] V. Goyal, "Multiple description coding: Compression meets the network," *IEEE Signal Processing Magazine*, vol. 18, no. 5, pp. 74–93, September 2001.

[55] R. Venkataramani, G. Kramer, and V. Goyal, "Multiple description coding with many channels," *IEEE Transactions on Information Theory*, vol. 49, no. 9, pp. 1111–1126, September 2003.

[56] A. Orlitsky, "Worst-case interactive communication I: Two messages are almost optimal," *IEEE Transactions on Information Theory*, vol. 36, no. 5, pp. 2106–2114, September 1990.

[57] A. Orlitsky, "Worst-case interactive communication II: Two messages are not optimal," *IEEE Transactions on Information Theory*, vol. 37, no. 4, pp. 995–1005, July 1991.

[58] R. Dobrushin and B. Tsybakov, "Information transmission with additional noise," *IEEE Transactions on Information Theory*, vol. 8, pp. 293–304, September 1962.

[59] J. Wolf and J. Ziv, "Transmission of noisy information to a noisy receiver with minimum distortion," *IEEE Transactions on Information Theory*, vol. 16, pp. 406–411, 1970.

[60] H. Witsenhausen, "Indirect rate distortion problems," *IEEE Transactions on Information Theory*, vol. 26, pp. 518–521, September 1962.

[61] K. Eswaran and M. Gastpar, "Rate loss in the CEO problem," in *39th Annual Conference on Information Sciences and Systems*, Baltimore, MD, 2005.

[62] T. Berger, Z. Zhang, and H. Viswanathan, "The CEO problem," *IEEE Transactions on Information Theory*, vol. 42, pp. 887–902, May 1996.

[63] H. Viswanathan and T. Berger, "The quadratic Gaussian CEO problem," *IEEE Transactions on Information Theory*, vol. 43, pp. 1549–1559, 1997.

[64] Y. Oohama, "The rate-distortion function for the quadratic Gaussian CEO problem," *IEEE Transactions on Information Theory*, vol. 44, pp. 1057–1070, May 1998.

[65] ———, "Multiterminal source coding for correlated memoryless Gaussian sources with several side informations at the decoder," in *Information Theory and Communications Workshop, Proceedings of the 1999 IEEE*, South Africa, June 1999.

[66] J. Chen, X. Zhang, T. Berger, and S. B. Wicker, "An upper bound on the sum-rate distortion function and its corresponding rate allocation schemes for the CEO problem," *IEEE Journal on Selected Areas in Communications: Special Issue on Sensor Networks*, pp. 1–10, 2003.

[67] V. Prabhakaran, D. Tse, and K. Ramchandran, "Rate region of the quadratic Gaussian CEO problem," in *Proceedings of ISIT*, pp. 119-, 2004.

[68] S. C. Draper and G. W. Wornell, "Successively structured CEO problems," in *Proceedings of ISIT*, pp. 65-, 2002.

[69] X. Zhang and S. Wicker, "On the rate region of the vector Gaussian CEO problem," in *39th Annual Conference on Information Sciences and Systems*, Baltimore, MD, 2005.

[70] S. Tavildar and P. Viswanath, "The sum rate for the vector Gaussian CEO problem," in *2005 Asilomar Conference*, pp. 3–7, 2005.

[71] K. Eswaran and M. Gastpar, "On the quadratic AWGN CEO problem and non-Gaussian source," in *Proceedings of ISIT*, pp. 219–223, 2005.

[72] V. Prabhakaran, K. Ramchandran, and D. Tse, "On the role of interaction between sensors in the CEO problem," in *Annual Allerton Conference on Communication, Control, and Computing*, Monticello, IL, October 2004.

[73] T. Cover, A. El Gamal, and M. Salehi, "Multiple access channels with arbitrarily correlated sources," *IEEE Transactions on Information Theory*, vol. 26, no. 6, pp. 648–657, November 1980.

[74] M. Gastpar and M. Vetterli, "On the capacity of wireless networks: The relay case," in *Proc. IEEE Infocom 2002*, vol. 3, New York, June 2002, pp. 1577–1586.

[75] ———, "Source-channel communication in sensor networks," *Information Processing in Sensor Networks*, vol. 2, pp. 162–177, 2003.

[76] ———, "Power, spatio-temporal bandwidth, and distortion in large sensor networks," *IEEE Journal on Selected Areas in Communications (Special Issue on Self-Organizing Distributive Collaborative Sensor Networks)*, vol. 23, no. 4, pp. 745–754, April 2005.

[77] M. Gastpar, "Uncoded transmission is exactly optimal for a simple Gaussian sensor network," in *Proc. 2007 Information Theory and Applications Workshop*, San Diego, CA, February 2007.

[78] M. Gastpar, M. Vetterli, and P. L. Dragotti, "Sensing reality and communicating bits: A dangerous liaison," *IEEE Signal Processing Magazine*, vol. 23, no. 4, pp. 70–83, July 2006.

CHAPTER

Distributed Transform Coding

2

Varit Chaisinthop and Pier Luigi Dragotti
Department of Electrical and Electronic Engineering,
Imperial College London, UK

CHAPTER CONTENTS

2.1 INTRODUCTION

Compression or approximation of an observed source is a central problem in signal processing and communications, and transform coding has over the years emerged as the dominating compression strategy. This is because transform coders are normally

Distributed Source Coding: Theory, Algorithms, and Applications

very efficient and computationally simple. Transforms are used in most of the compression standards, including image compression standards such as JPEG and JPEG 2000 ([20, 34]) and video compression standards such as H.26x [31].

Transform coding has been widely studied in recent years, and many important results and optimality conditions have been derived. For example, it is well known that, in some particular settings, the Karhunen–Loève transform (KLT) is the optimal transform for compression or linear approximation [15, 18]. More recently, the analysis of popular transform coders has led to new insights into the general problem of source approximation and to new interesting connections between compression and nonlinear approximation. This analysis has also clarified why the wavelet transform (WT) is the best transform in some particular situations [4, 21, 32].

In distributed source coding, however, the source is not available at a single location and thus it is not possible to apply a single transform to the entire source. Instead, the source is partially observed by separate encoders that have to devise a local compression strategy. It is thus natural to study how the classical centralized transform coding paradigm changes in this new context. For example, if each encoder performs transform coding locally, should the local transform be different from the one used in a centralized compression? And how are the other modules of a transform coder going to change? Is quantization, bit allocation, and entropy coding going to be different?

Distributed transform coding has been analyzed in recent years, and some answers can be provided. When the source is Gaussian, the KLT emerges again as the best (and in some cases optimal) transform, even though changes to the classical structure of the transform have to be implemented. This has been studied in [9–12], and the asymptotic behavior of the distributed KLT has been considered in [28, 29]. A similar problem has also been studied in [22, 37]. In the high bit-rate regime, some optimality conditions for transforms have also been proved [8, 27].

The problem of distributed transform coding remains widely open, however, when the Gaussianity of the source and the high bit-rate assumptions are relaxed. The design of distributed transform coders in this particular case is normally based on heuristics, and usually the local transform is not different from the one used in the centralized scenario. Most of the modifications are done in the quantization and entropy coding stages. For example, in the context of distributed video coding, the structure of the discrete cosine transform (DCT) is the same as in classical video coding. However, the DCT coefficients are quantized in a different way in order to exploit the fact that correlated information is available at the decoder. This strategy, though effective, is not necessarily optimal.

In the next section, we review the main results on centralized transform coding. Section 2.3 provides an overview of the main results that appeared in [12] where extensions of the KLT to the distributed scenario were considered. Sections 2.4 and 2.5 address the case where the input source is not Gaussian. The high bit-rate regime is discussed, and extensions of DCT and WT are presented. We conclude in Section 2.6.

2.2 FOUNDATIONS OF CENTRALIZED TRANSFORM CODING

2.2.1 Transform Coding Overview

The problem of compression or source coding can be divided into two types: unconstrained and constrained. In an unconstrained source coding problem, the encoder has access to the entire source vector $\mathbf{x} \in \mathbb{R}^N$. It is therefore possible to devise an optimal source code for a particular source vector. In practice, however, it is hardly the case that the entire source vector can be observed. In addition, even though we can achieve good compression results, the high complexity of an unconstrained source coder makes it impractical to implement.

A typical compression scheme is shown in Figure 2.1. A standard compressor consists of three independent blocks: a block implementing linear transform, a quantizer, and a lossless entropy encoder. This type of structure is called transform coding. A transform code is an example of a constrained source code. Constrained source codes are, loosely speaking, suboptimal but low in complexity, which arises from the modularization of the encoding process. This approach allows "simple coding" to be used with high efficiency. Simple coding refers to the use of scalar quantizer and scalar entropy coding. This is one of the main reasons transform code is the most widely used source code today. Given a random vector $\mathbf{x}$ of size N, the simplicity of the transform code allows $\mathbf{x}$ with a large value of N to be encoded.

The task of an encoder is to map the source vector $\mathbf{x} \in \mathbb{R}^N$ to a bitstream of finite length. An invertible linear transform produces decorrelated transform coefficients $\mathbf{y} = \mathbf{Tx}$ where $\mathbf{y} \in \mathbb{R}^N$. The quantizer then maps $\mathbb{R}^N$ to a discrete set of indices I, which produce estimates of the coefficients $\widehat{\mathbf{y}} \in \mathbb{R}^N$. This is then followed by a lossless entropy encoder that performs a reversible mapping from I to a bit stream.

The decoder essentially reverses the operation of the encoder to reconstruct the approximation $\widehat{\mathbf{x}}$ of the source. Since entropy coding is a lossless process, the quantizer indices can be recovered exactly from the bitstream to give the estimates of transform

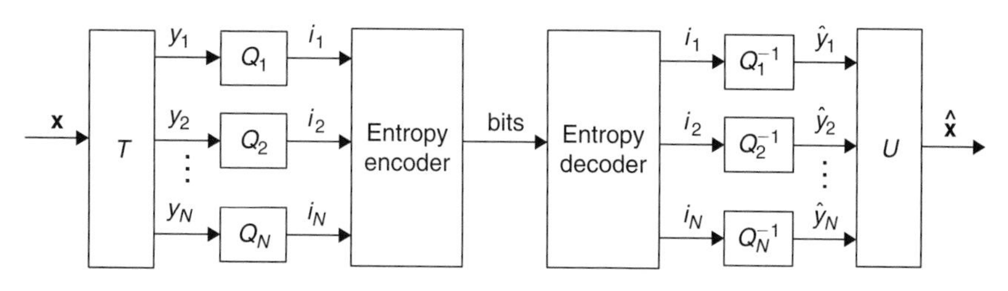

FIGURE 2.1

Transform coding. A typical compression scheme consists of three elements: linear transform, quantization, and lossless compression (entropy coding).

coefficients $\widehat{\mathbf{y}}$. A linear transform, generally denoted as U, is then applied to $\widehat{\mathbf{y}}$ to produce the approximation of the source $\widehat{\mathbf{x}} = U\widehat{\mathbf{y}}$ where, as often is the case, $U = T^{-1}$.

The quality of a lossy encoder is normally measured by the mean-squared error (MSE) distortion of the approximation $\widehat{\mathbf{x}}$. The MSE distortion is denoted by $D = E[d(\mathbf{x}, \widehat{\mathbf{x}})]$ where

$$D = E\left[d(\mathbf{x}, \widehat{\mathbf{x}})\right] = E\left[\|\mathbf{x} - \widehat{\mathbf{x}}\|^2\right] = E\left[\sum_{i=1}^{N} (x_i - \widehat{x}_i)^2\right]$$

and $E[\cdot]$ is the expectation operator. In order to gauge the performance of a source code, the distortion is normally measured against the rate R, which is the expected number of bits produced by the encoder divided by the length N of the encoded source vector $\mathbf{x}$. This is normally termed the distortion-rate performance. Thus the performance of one source code is said to be better than the other in a rate-distortion sense if, at the same given rate R, the earlier can achieve a lower distortion than the latter.

We will now look at each building block of a transform coder in more detail.

2.2.2 Lossless Compression

Entropy code is a form of lossless compression, which is reversible and therefore can only be applied to discrete sources. Consider a discrete source X that produces I different values a_i and the probability of occurrence of each value is denoted by $p(a_i)$. An entropy code assigns a unique binary representation $b(a_i)$ or a codeword to each a_i. The goal is to find a codeword $b(a_i)$ for each symbol so that the expected length of the binary representation of X is minimized. That is, we want to minimize:

$$E[l(X)] = \sum_{i=0}^{I-1} p(a_i) l_i \tag{2.1}$$

where l_i is the length of the binary codeword associated with the symbol a_i.

Since entropy coding is a lossless process, the binary representation has to be invertible. Each codeword also has to be uniquely decodable. This means that no codeword can be a prefix of another codeword. Therefore, the entire message sequences can be encoded without adding any "punctuation" to tell the decoder where each codeword ends and begins.

An entropy code is said to be optimal if it is a prefix code that minimizes $E[l(X)]$. Huffman codes [19] and arithmetic codes [35] are examples of optimal codes. The lower bound on the expected length (2.1) of a prefix code is given by the entropy $H(X)$ of the source where

$$H(X) = -\sum_{i=0}^{I-1} p(a_i) \log_2(a_i)$$

and the performance of an optimal code is bounded by

$$H(X) \leq L(\gamma) \leq H(X) + 1. \tag{2.2}$$

2.2.3 Quantizers

The amplitude values of a discrete-time signal are normally continuous. Quantization is a lossy process that maps the continuous values of the source $\mathbb{R}^{\mathrm{N}}$ into a finite set of alphabet or a reproduction codebook $C = \{\widehat{x}_i\}_{i \in I} \subset \mathbb{R}^{\mathrm{N}}$, where I is a discrete set. Usually, each component or sample of the observed signal $\mathbf{x} \in \mathbb{R}^{\mathrm{N}}$ is quantized individually ($N = 1$), and the quantizer is called scalar quantizer. A more sophisticated form of quantization that operates on a group of samples of the source vector ($N > 1$) is called vector quantization. For a detailed treatment of vector quantization, we refer the reader to [13].

Most of today's compressors use a simple uniform scalar quantizer together with prediction or transforms. In the case of a uniform quantizer, the input is divided into intervals of equal size. The distance between each interval is the *step size* Δ. The output y is derived from the nearest integer multiple of Δ that is closest to the input x. Formally, this can be written as $y = round(x/\Delta)$. If the quantizer has the index $i = \{0, 1, \ldots, I-1\}$ that outputs I different symbols y_i, we would need $R = \lceil \log_2(I) \rceil$ bits to represent each symbol with a fixed-length binary code. The quantizer is then said to have rate R. What we have here is, in fact, a fixed-rate quantizer. However, adding an entropy code to this quantizer gives a variable-rate quantizer specified by the mapping where the rate R is given by the expected code length divided by N. Furthermore, from (2.2), we know that the performance of an optimal entropy code is bounded by the entropy. Thus, the variable-rate quantizer with an optimal entropy code is also called entropy constrained.

While the optimal design of a quantizer ([7, 15, 16]) is beyond the scope of this chapter we can say that an optimal quantizer is the one that minimizes the distortion for a given rate or minimizes the rate for a target distortion and the rate-distortion curve of the considered source gives us the performance bound. Unfortunately, in practice, the rate-distortion function is known only for a few cases. However, one can obtain a remarkable result for a Gaussian source where the distortion is measured as the MSE. Assuming that the source produces independent and identically distributed (i.i.d.) Gaussian random variables with variance σ^2, the distortion-rate bound of the source subject to MSE is

$$D(R) = \sigma^2 2^{-2R}. \tag{2.3}$$

Given the bound in Equation (2.3), we now want to examine how close the performance of a scalar quantizer is to this bound. A precise answer has been obtained for a high bit-rate analysis (i.e., for large R) [15, 16]. For a fixed-rate quantizer, the optimal quantizer for a Gaussian source is nonuniform and achieves the distortion-rate function of

$$D(R) = \frac{\sqrt{3}\pi}{2} \sigma^2 2^{-2R}. \tag{2.4}$$

In comparison to the bound in (2.3), the distortion of this nonuniform quantizer is higher by $\sim 4.35d\dot{B}$, or the rate loss is ~ 0.72 bits per symbol. This means that this quantizer needs 0.72 more bits per symbol to achieve the same distortion of the best possible lossy encoder.

One can improve the performance given in (2.4) by adding the entropy encoder to obtain a variable-rate quantizer. Interestingly, at high rate, the optimal quantizer is uniform. In this case, the achieved distortion-rate performance is given by

$$D(R) = \frac{\pi e}{6}\sigma^2 2^{-2R}.$$

The distortion is now $\sim 1.53d\dot{B}$ higher than the bound, and the redundancy is ~ 0.255 bits per symbol, which is a significant improvement when compared to (2.4).

Interestingly, at high rate, for non-Gaussian sources, it is still optimal to use a uniform quantizer followed by an entropy encoder, and the operational distortion-rate function is

$$D(R) \approx c\sigma^2 2^{-2R} \tag{2.5}$$

where c is a constant whose value depends on the probability density function of the source. In summary, at high bit rates, one can conclude that the best quantization strategy is to use a uniform quantizer followed by an entropy encoder, which would result in a fairly simple lossy compression scheme whose performance is very close to that of the best possible performance bound.

2.2.4 Bit Allocation

Since each transform coefficient is quantized and entropy coded separately, the total number of bits (or the bit budget) is split among the coefficients in some way. The question of how these bits should be allocated to each coefficient is referred to as the bit allocation problem. Given a set of quantizers whose performances are described by the distortion-rate functions

$$D_i = g_i(\sigma_i, R_i), \quad R_i \in \Re_i, \quad i = 1, 2, \ldots, N,$$

where $\sigma_i \in \mathbb{R}$ and a set of available rates $\Re_i$ is a subset of nonnegative real numbers, the problem is then to minimize the average distortion $D = \sum_{i=1}^{N} D_i$ given a maximum rate $R = \sum_{i=1}^{N} R_i$. This is therefore a constrained optimization problem, which can be solved using Lagrange multipliers. See [23] for a detailed treatment of this topic.

Intuitively, the initial bit allocation is not optimal if the average distortion can be reduced by taking bits away from one coefficient and giving them to another. Therefore, we can say that one necessary condition for an optimal bit allocation is that the slope of each g_i at R_i must be equal. From the high-resolution analysis of the performance of a quantizer given in (2.5), we have that

$$g_i(\sigma_i, R_i) = c_i\sigma_i^2 2^{-2R_i}, \quad \Re_i = [0, \infty), \quad i = 1, 2, \ldots, N. \tag{2.6}$$

Ignoring the fact that all the rates must be nonnegative, an equal-slope condition leads to an optimal bit allocation as

$$R_i = \frac{R}{N} + \frac{1}{2}\log_2 \frac{c_i}{\left(\prod_{i=1}^{N} c_i\right)^{1/N}} + \frac{1}{2}\log_2 \frac{\sigma_i^2}{\left(\prod_{i=1}^{N} \sigma_i^2\right)^{1/N}}.$$

With the above bit allocation, all the distortions D_i are equal and the average distortion is

$$D = \left(\prod_{i=1}^{N} c_i\right)^{1/N} \left(\prod_{i=1}^{N} \sigma_i^2\right)^{1/N} 2^{-2R}. \tag{2.7}$$

Each R_i must be nonnegative for the above solution to be valid. At lower rates, the components with smallest $c_i \cdot \sigma_i^2$ are given no bits and the remaining components are given correspondingly higher allocations. For uniform quantizers, the bit allocation determines the step size Δ_i for each component. The equal-distortion property means that optimality can be achieved when all the step sizes are equal.

2.2.5 Transforms

Earlier we said that transform coding allows simple coding of scalar quantization and entropy code to be used efficiently. Let us now consider the problem of compressing sources with block memory. Given a source vector $\mathbf{x} \in \mathbb{R}^{\mathbf{N}}$ of length N that consists of statistically dependent samples, clearly, it would be inefficient to scalar quantize each element independently since we would not exploit the dependency of the samples. An alternative is to use a vector quantizer to perform a joint compression. This is not a practical option, however, as the complexity normally increases exponentially as N increases. Instead, we want to devise a transform coder as shown in Figure 2.1. We denote with T the transform applied to the source vector $\mathbf{x}$ leading to the transform coefficients $\mathbf{y} = \mathbf{Tx}$, where each coefficient y_k is then scalar quantized. The reconstructed source vector is $\widehat{\mathbf{x}} = U\,\widehat{\mathbf{y}}$, where $\widehat{\mathbf{y}}$ is the scalar quantized transform coefficients and usually $U = T^{-1}$. Our goal is then to devise the transform T that minimizes the MSE $D = E\left[\|\mathbf{x} - \widehat{\mathbf{x}}\|^2\right]$ for a given rate R.

2.2.5.1 *Optimal Transforms for Gaussian Sources*

We start by assuming that $\mathbf{x}$ is a jointly Gaussian zero-mean vector with covariance matrix $\Sigma_{\mathbf{x}} = E\left[\mathbf{x}\mathbf{x}^{\mathbf{T}}\right]$. In addition, we assume that T is orthogonal. Since T is orthogonal, the Euclidean lengths are preserved, which gives $D = E\left[\|\mathbf{x} - \widehat{\mathbf{x}}\|^2\right] = E\left[\|\mathbf{y} - \widehat{\mathbf{y}}\|^2\right]$.

A Karhunen-Loève transform (KLT) is an orthogonal transform that diagonalizes $\Sigma_{\mathbf{x}}$ leading to $\Sigma_{\mathbf{y}} = E\left[\mathbf{y}\mathbf{y}^{\mathbf{T}}\right] = T\Sigma_{\mathbf{x}}T^T = \Lambda$. Since the diagonal matrix $\Sigma_{\mathbf{y}} = T\Sigma_{\mathbf{x}}T^T$ is the covariance matrix of $\mathbf{y} = \mathbf{Tx}$, the KLT transform gives uncorrelated coefficients. Our Gaussian assumption means that the coefficients $y_k, k = 1, 2, \ldots, N$, are independent Gaussian variables with variance λ_k^2. Here λ_k^2 is the kth diagonal element of $\Sigma_{\mathbf{y}}$.

Therefore, because of this independence, we can compress each coefficient y_k independently. From (2.5) and (2.6), we have that the kth component of $\mathbf{y}$ contributes a distortion

$$D_k(R_k) = c\lambda_k^2 2^{-2R_k}$$

where R_k is the rate allocated to y_k and $c = \frac{\pi e}{6}$ for variable-rate entropy constrained quantization or $c = \frac{\sqrt{3}\pi}{2}$ for a fixed-rate quantization. The overall distortion-rate function is

$$D(R) = E\left[\|\mathbf{x} - \widehat{\mathbf{x}}\|^2\right] = E\left[\|\mathbf{y} - \widehat{\mathbf{y}}\|^2\right] = \frac{1}{N}\sum_{k=1}^{N} D_k, \tag{2.8}$$

with $R = \sum_k R_k$. The goal then is to minimize (2.8) subject to $R = \sum_k R_k$. This is then the problem of bit allocation discussed in the previous section. With these assumptions, with any rate and bit allocation, one can show that the KLT is an optimal transform.

Theorem 2.1. [15] Consider a transform coder with orthogonal analysis transform T and synthesis transform $U = T^{-1} = T^T$. If there is a single function g as defined by Equation (2.6) that describes the quantization of each transform coefficient as

$$D_k(R_k) = E\left[y_k - \widehat{y_k}^2\right] = g(\lambda_k, R_k),\ \ k = 1, 2, \ldots, N$$

where λ_k^2 is the variance of y_k and R_k is the rate allocated to y_k, then for any rate allocation $(R_1, R_2, \ldots, R_N)$ there is a KLT that minimizes the distortion. In the typical case where g is nonincreasing, a KLT that gives $(\lambda_1^2, \lambda_2^2, \ldots, \lambda_N^2)$ sorted in the same order as bit allocation minimizes the distortion.

Recall that from the high-resolution analysis of the performance of the quantizer described by (2.6), with optimal bit allocation, the average distortion is given by (2.7). For our independent Gaussian transform coefficients, the distortion simplifies to

$$D(R) = E\left[\|\mathbf{y} - \widehat{\mathbf{y}}\|^2\right] = c\left(\prod_{k=1}^{N}\lambda_k^2\right)^{1/N} 2^{-2R}. \tag{2.9}$$

Therefore, the optimal orthogonal transform is the one that minimizes the geometric mean of the transform coefficient variances given by $\left(\prod_{k=1}^{N}\lambda_k^2\right)^{1/N}$. Applying Hadamard's inequality to $\Sigma_{\mathbf{y}}$ gives

$$(\det T)(\det \Sigma_{\mathbf{x}})(\det T^T) = \det \Sigma_{\mathbf{y}} \leqslant \prod_{k=1}^{N}\lambda_k^2.$$

Since $\det T = 1$, the left-hand side of this inequality is invariant to the choice of T. Equality is achieved when KLT is used. Thus KLTs minimize the distortion.

One can measure the factor at which the distortion is reduced by the use of transform with a *coding gain*. Assuming high rate and optimal bit allocation, the coding gain of a transform is given by

$$\text{coding gain} = \frac{\left(\prod_{i=1}^{N} (\Sigma_{\mathbf{x}})_{ii}\right)^{1/N}}{\left(\prod_{k=1}^{N} \lambda_k^2\right)^{1/N}},$$

which is a function of variance vector, $(\lambda_1^2, \lambda_2^2, \ldots, \lambda_N^2)$, of the transform coefficients and the variance vector of the diagonal elements of $\Sigma_{\mathbf{x}}$. Our previous discussion shows that the KLTs maximize coding gain.

If the input source is stationary but not Gaussian, we can still use the same scheme. In this case, the KLT decorrelates the components of $\mathbf{x}$ but does not provide independent components. We can still perform the same bit allocation analysis as shown before. Although the approach tends to give good results, we are not guaranteed that the performance is optimal.

2.2.6 Linear Approximation

Linear approximation is to some extent a "simplified" form of compression where one stores or transmits only a subset of the coefficients of $\mathbf{y}$ and the retained coefficients are not quantized. More formally, consider a function in a certain space and an orthogonal basis $\{g_n\}_{n\in\mathbb{N}}$. We want to approximate this function using only M elements of $\{g_n\}_{n\in\mathbb{N}}$. In linear approximation problems, the choice of these M elements is fixed a priori and is *independent* of the f we are trying to approximate. By choosing the first M elements of the basis, we have the following linear approximation of f:

$$f_M = \sum_{n=0}^{M-1} \langle f, g_n \rangle g_n.$$

Since the basis is orthogonal, the MSE of this approximation is

$$\varepsilon_M = \|f - f_M\|^2 = \sum_{n=M}^{\infty} |\langle f, g_n \rangle|^2. \tag{2.10}$$

Therefore, given a class of signals and a choice of possible bases, the best basis in this context is the one that gives the smallest MSE as given in (2.10). Interestingly enough, one can show that if the signals are realizations of a jointly Gaussian process, which generates jointly Gaussian vectors, the best basis to approximate such signals is again the KLT.

In summary, in this section we have seen that when the source is Gaussian the KLT is the best transform in both the compression and linear approximation scenarios. In the case of compression, moreover, a strategy that involves the KLT, uniform scalar quantizers, and entropy encoders achieves optimal performance if the rate is allocated correctly.

In the next sections, we will study how these main building blocks of centralized transform coding are changed when the compression is performed in a distributed rather than centralized fashion.

2.3 THE DISTRIBUTED KARHUNEN–LOÈVE TRANSFORM

The problem we consider now is depicted in Figure 2.2: there are L terminals, each observing a part of a vector $\mathbf{x}$. If the encoding of $\mathbf{x}$ is performed jointly, then we are back to the centralized transform coding scenario discussed in the previous section. The situation encountered in this and the following sections is more complicated, however. Terminals cannot communicate among themselves. Instead, each terminal individually provides an approximation of its observation to the central decoder. The decoder receives the different approximations sent by each terminal and performs a joint reconstruction. The fundamental issue then is to devise the best local transform coding strategy knowing that the encoding is independent but the decoding is performed jointly.

In this section, we discuss the case when $\mathbf{x}$ is a jointly Gaussian vector. In this scenario, some precise answers can be provided, and extensions of the classical KLT are presented. Sections 2.4 and 2.5 instead study the case when $\mathbf{x}$ is not Gaussian, and an overview of existing distributed transform coding strategies is provided. In

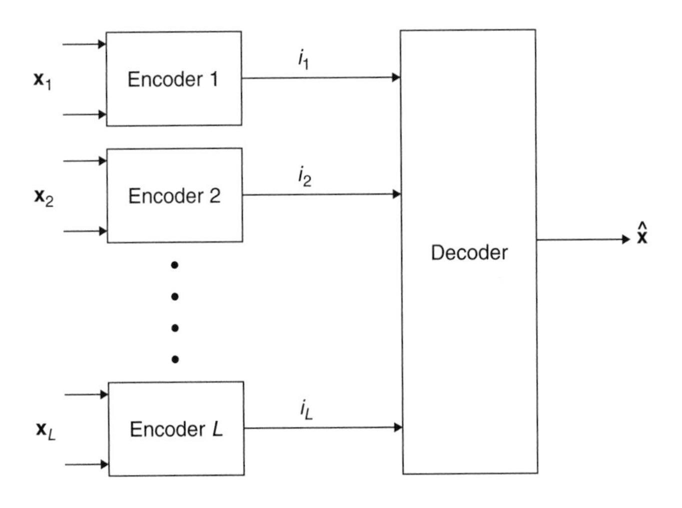

FIGURE 2.2

The distributed transform coding problem. Encoder l has access to the subvector $\mathbf{x}_l$ and has to provide the best description i_l of such vector knowing that the reconstruction of all the descriptions is performed jointly.

particular, other transforms such as the wavelet transform and the discrete cosine transform are considered.

2.3.1 Problem Statement and Notation

We now assume that the L terminals depicted in Figure 2.2 are observing a part of an N-dimensional, jointly Gaussian real-valued vector $\mathbf{x}$. We further assume that $\mathbf{x}$ has zero mean and covariance matrix Σ_x. If the encoding of $\mathbf{x}$ is performed jointly, the best transform is the KLT, and this is true in both the linear approximation and compression scenarios. In the distributed scenario, however, encoders cannot communicate with each other, and each terminal devises the best local transform coding strategy based on its partial observation of $\mathbf{x}$ and on the knowledge of the global covariance matrix Σ_x. To be more specific, we assume that the first terminal observes the first M_1 components of $\mathbf{x}$ denoted by $\mathbf{x}_1 = (x_1, x_2, \ldots, x_{M_1})$, the second encoder observes the next M_2 components $\mathbf{x}_2 = (x_{M_1+1}, x_{M_1+2}, \ldots, x_{M_1+M_2})$, and so on. Each terminal provides an approximation of the observed partial vector to the central decoder whose goal is to produce the best possible estimate of the entire vector $\mathbf{x}$ from the received approximations. More precisely, the goal is to find a vector $\hat{x}$ that minimizes the mean-squared error (MSE) $E\left[\|\mathbf{x}-\hat{\mathbf{x}}\|^2\right]$.

Let us consider for the time being the linear approximation problem. That is, each terminal produces a k_l-dimensional approximation of the observed components, which is equivalent to saying that terminal l applies a $k_l \times M_l$ matrix T_l to its local observation. Hence, stacking all the transforms together, we can say that the central receiver observes the vector $\mathbf{y}$ given by

$$\mathbf{y} = T\mathbf{x} = \begin{pmatrix} T_1 & 0 & \ldots & 0 \\ 0 & T_2 & \ldots & 0 \\ \vdots & \vdots & \ddots & \vdots \\ 0 & 0 & \ldots & T_L \end{pmatrix} \mathbf{x}. \tag{2.11}$$

The decoder then has to estimate $\mathbf{x}$ from $\mathbf{y}$. This situation is clearly more constrained than the one encountered in the centralized transform coding scenario since the transform T is block-diagonal in the present distributed scenario, whereas it is unconstrained in the centralized case.

Under the MSE criterion, the optimal estimate $\hat{\mathbf{x}}$ of $\mathbf{x}$ is given by the conditional expectation of $\mathbf{x}$ given $\mathbf{y}$. Moreover, in the Gaussian case, the estimator is linear and is given by

$$\hat{\mathbf{x}} = E[\mathbf{x}|\mathbf{y}] = E[\mathbf{x}|T\mathbf{x}] = \Sigma_x T^T (T\Sigma_x T^T)^{-1} T\mathbf{x}.$$

The corresponding MSE can be written as

$$D = E[\|\mathbf{x}-\hat{\mathbf{x}}\|^2] = trace(\Sigma_x - \Sigma_x T^T (T\Sigma_x T^T)^{-1} T\Sigma_x). \tag{2.12}$$

Therefore, for fixed k_l's, the distributed approximation problem can be stated as the minimization of (2.12) over all block-diagonal matrices T as given in (2.11). A simple solution to this problem does not seem to exist. However, in some particular settings, some precise answers can be provided [12]. Moreover, in [12] an iterative algorithm that finds (locally) optimal solutions is also proposed.

The compression problem is to some extent even more involved since the transformed vectors $\mathbf{y}_l = T_l\mathbf{x}_l$ need to be quantized and the binary representation of the quantized version of $\mathbf{y}_l$ is transmitted to the receiver. The objective now is to obtain the best possible estimate of $\mathbf{x}$ from the quantized versions of the approximations $\mathbf{y}_l$, $l = 1, 2, \ldots, L$ under an overall bit budget R. Thus the problem now is not only to find the right block-diagonal transform T but also the right allocation of the bits among the encoders and the correct quantization strategy.

In the next subsection, we consider the two-terminal scenario and review the linear approximation and compression results presented in [12]. One special case leading to the *Conditional KLT* is also reviewed, and examples are given. The following subsection then discusses the multiterminal scenario.

2.3.2 The Two-terminal Scenario

We now consider the two-terminal scenario depicted in Figure 2.3 where Terminal 1 observes the vector $\mathbf{x}_1$ given by the first M components of $\mathbf{x}$ and the second terminal observes the vector $\mathbf{x}_2$ given by the last $N - M$ components of $\mathbf{x}$. The covariance matrix of $\mathbf{x}_1$ (i.e., $E[\mathbf{x}_1\mathbf{x}_1^T]$) is denoted with Σ_1. Similarly, we have $\Sigma_2 = E[\mathbf{x}_2\mathbf{x}_2^T]$ and $\Sigma_{12} = E[\mathbf{x}_1\mathbf{x}_2^T]$.

The transform T_2 applied by the second terminal to $\mathbf{x}_2$ is assumed to be fixed and known at both terminals. The decoder, however, does not have access to $T_2\mathbf{x}_2$ but to the noisy version $\mathbf{y}_2 = T_2\mathbf{x}_2 + \mathbf{z}_2$, where $\mathbf{z}_2$ is a zero-mean jointly Gaussian vector

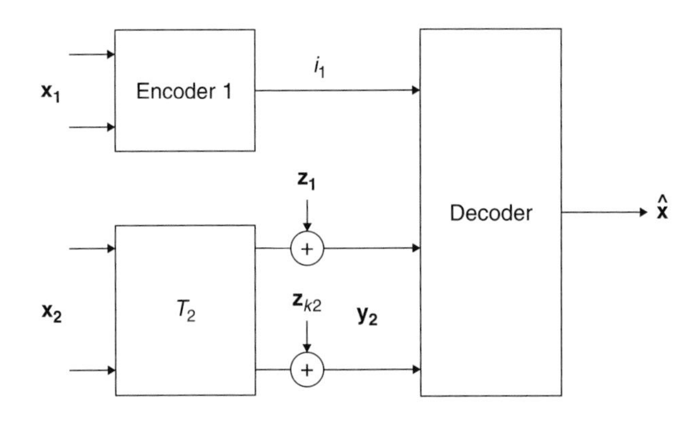

FIGURE 2.3

The two-terminals scenario: Encoder 2 applies a transform T_2 to the observed vector $\mathbf{x}_2$. The transform is fixed and known at both encoders. The decoder receives a noisy version $\mathbf{y}_2 = T_2\mathbf{x}_2 + \mathbf{z}$; the noise may model the quantization error or is due to noise in the communication channel. The open issue is to find the best transform coding strategy at Encoder 1 under these circumstances.

independent of $\mathbf{x_2}$. Notice that T_2 has dimension $k_2 \times (N-M)$ and $\mathbf{z_2}$ has dimension $k_2 \times 1$. The noise might be due to the transmission channel or may model the effect of the compression of the transformed coefficients $T_2\mathbf{x_2}$.

In line with the problem statement of the previous section, two perspectives are considered. In the first one, Terminal 1 has to provide the best k_1-dimensional approximation of the observed vector $\mathbf{x_1}$ given that $\mathbf{y_2}$ is available at the decoder (but not at the Encoder 1). In the second scenario, $\mathbf{x_1}$ has to be compressed using an available bit budget R, and the task is again to devise the best compression strategy given that $\mathbf{y_2}$ is available at the decoder. In both scenarios, the goal is the minimization of

$$E[\|\mathbf{x}-\hat{\mathbf{x}}\|^2|\mathbf{y_2}], \tag{2.13}$$

where $\hat{\mathbf{x}}$ is the reconstructed vector.

Let us focus on the approximation problem first. Because of the assumption that $\mathbf{x}$ and $\mathbf{z_2}$ are Gaussian, there exist constant matrices A_1 and A_2 such that

$$\mathbf{x_2} = A_1\mathbf{x_1} + A_2\mathbf{y_2} + \mathbf{v}, \tag{2.14}$$

where $\mathbf{v}$ is a Gaussian random vector independent of $\mathbf{x_1}$ and $\mathbf{y_2}$. Fundamentally, the term $A_1\mathbf{x_1} + A_2\mathbf{y_2}$ represents the linear estimation of $\mathbf{x_2}$ from $\mathbf{x_1}$ and $\mathbf{y_2}$, and $\mathbf{v}$ is the uncertainty. Namely, $\mathbf{v}$ represents what cannot be estimated of $\mathbf{x_2}$ from $\mathbf{x_1}$ and $\mathbf{y_2}$.

Using the same arguments, we can also write

$$\begin{pmatrix} I_M \\ A_1 \end{pmatrix}\mathbf{x_1} = B_2\mathbf{y_2} + \mathbf{w}, \tag{2.15}$$

where I_M is the M-dimensional identity matrix and $\mathbf{w}$ is a Gaussian random vector independent of $\mathbf{y_2}$ and with correlation matrix Σ_w. It is worth mentioning at this stage that, because of the peculiar structure of the vector $\binom{I_M}{A_1}\mathbf{x_1}$, the matrix Σ_w has rank $\leqslant M$. It is also possible to show that Σ_w is given by [12]:

$$\Sigma_w = \begin{pmatrix} I_M \\ A_1 \end{pmatrix} \left(\Sigma_1 - \Sigma_{12}T_2^T(T_2\Sigma_2T_2^T + \Sigma_z)^{-1}T_2\Sigma_{12}^T\right)\left(I_M \; A_1^T\right).$$

In the linear approximation problem, we need to find the matrix T_1 of dimension $k_1 \times M$ that minimizes (2.13). To determine such matrix, we rewrite the distortion (2.13) as follows [12]:

$$\begin{aligned}
E[\|\mathbf{x}-\hat{\mathbf{x}}\|^2|\mathbf{y_2}] &= E[\|\mathbf{x_1}-\hat{\mathbf{x}}_1\|^2|\mathbf{y_2}] + E[\|\mathbf{x_2}-\hat{\mathbf{x}}_2\|^2|\mathbf{y_2}] \\
&\overset{(a)}{=} E[\|\mathbf{x_1}-\hat{\mathbf{x}}_1\|^2|\mathbf{y_2}] + E[\|A_1\mathbf{x_1}+A_2\mathbf{y_2}+\mathbf{v}-\hat{\mathbf{x}}_2\|^2|\mathbf{y_2}] \\
&\overset{(b)}{=} E[\|\mathbf{x_1}-\hat{\mathbf{x}}_1\|^2|\mathbf{y_2}] + E[\|A_1\mathbf{x_1}-A_1\hat{\mathbf{x}}_1\|^2|\mathbf{y_2}] + E[\|\mathbf{v}\|^2] \\
&= E\left[\left\|\begin{pmatrix} I_M \\ A_1 \end{pmatrix}\mathbf{x_1} - \begin{pmatrix} I_M \\ A_1 \end{pmatrix}\hat{\mathbf{x}}_1\right\|^2 \middle| \mathbf{y_2}\right] + E[\|\mathbf{v}\|^2],
\end{aligned}$$

where in (a) we have used (2.14) and in (b) we have used the fact that the optimal estimation of $\mathbf{x_2}$ from $(\hat{\mathbf{x}}_1, \mathbf{y_2})$ is $\hat{\mathbf{x}}_2 = A_1\hat{\mathbf{x}}_1 + A_2\mathbf{y_2}$. Also recall that $\mathbf{v}$ is independent of

$\mathbf{x}_1$ and $\mathbf{y}_2$. We now apply an orthogonal transform Q^T to $\binom{I_M}{A_1}\mathbf{x}_1$ so that the components of the resulting vector $\mathbf{y}_1$ are conditionally independent given $\mathbf{y}_2$. Since we are focusing on the Gaussian case only, from (2.15) we obtain that Q^T is the $N \times N$ matrix that diagonalizes Σ_w. Denote the eigendecomposition of Σ_w as follows:

$$\Sigma_w = Q\text{diag}(\lambda_1^2, \lambda_2^2, \ldots, \lambda_N^2)Q^T, \tag{2.16}$$

where the eigenvalues $\lambda_i^2, i = 1, 2, \ldots, N$ are in nonincreasing order. Then denote with T_1 the matrix

$$T_1 = Q^T \begin{pmatrix} I_M \\ A \end{pmatrix}. \tag{2.17}$$

By using the fact that Q is unitary and the fact that the components of $\mathbf{y}_1$ are conditionally independent, we arrive at this simplified expression for the MSE:

$$E[\|\mathbf{x} - \hat{\mathbf{x}}\|^2 | \mathbf{y}_2] = \sum_{m=1}^{N} E[\|y_m - \hat{y}_m\|^2 | \mathbf{y}_2] + E[\|\mathbf{v}\|^2].$$

The error $E[\|y_m - \hat{y}_m\|^2 | \mathbf{y}_2]$ is zero if the component y_m is transmitted and is equal to λ_m^2 if y_m is discarded. We therefore deduce that the best k_1-dimensional approximation of $\mathbf{x}_1$ is obtained by keeping the k_1 components of $\mathbf{y}_1$ related to the largest eigenvalues of Σ_w. Since the eigenvalues in (2.16) are in nonincreasing order, the optimal k_1-dimensional approximation of $\mathbf{x}_1$ is given by the first k_1 components of the vector $\mathbf{y}_1 = T_1\mathbf{x}_1$ and the reconstruction error is

$$E[\|\mathbf{x} - \hat{\mathbf{x}}\|^2 | \mathbf{y}_2] = \sum_{m=k_1+1}^{N} \lambda_m^2 + E[\|\mathbf{v}\|^2.$$

The matrix T_1 is known as the *local KLT* [12].

We now provide intuition for the shape of the optimal transform T_1. If Terminal 1 were to provide a linear approximation of $\mathbf{x}_1$ without considering the side information and the fact that the entire vector $\mathbf{x}$ needs to be recovered, it would simply compute the eigendecomposition of Σ_1 and send the k_1 components related to the largest eigenvalues of Σ_1. However, since some information about $\mathbf{x}_1$ can be estimated from $\mathbf{y}_2$ and since the received approximation is also used to estimate $\mathbf{x}_2$, the relative importance of the components of $\mathbf{x}_1$ may change. This is clearly reflected in Equation (2.15). The *local KLT* provides the optimal solution because it takes these issues into account and shows that the correct covariance matrix to diagonalize is $\Sigma_{\mathbf{w}}$.

In the compression scenario, Terminal 1 can use only R bits to encode $\mathbf{x}_1$, and the aim is again the minimization of (2.13). The interesting finding of [12] is that the optimal encoding strategy consists in applying the local KLT (i.e., the transform T_1) to $\mathbf{x}_1$, followed by independent compression of the components of $\mathbf{y}_1 = T_1\mathbf{x}_1$. The rate allocation among the components depends on the eigenvalue distribution of Σ_w. Namely, the components of $\mathbf{y}_1$ related to the largest eigenvalues get more rates.

This optimal encoding strategy is remarkably similar to the compression strategy used in the classical centralized scenario. However, the similarities terminate here. It is in fact also important to point out some differences. First, contrary to the centralized case, the rate allocation in the distributed case depends on the side information available at the decoder (indeed, the side information has an effect on the structure of the matrix Σ_w). Second, the constructive compression schemes involved in the distributed case are normally very different from those used in the classical scenario. More specifically, in the classical scenario the components of $\mathbf{y}_1$ can be compressed using, for example, standard scalar quantizers. In contrast, in the distributed scenario, a standard scalar quantizer might not be optimal, and sophisticated Wyner-Ziv coding schemes might be required. For an overview of existing Wyner-Ziv constructive codes, we refer the reader to [17].

A case of particular interest is the one where $\mathbf{y}_2 = \mathbf{x}_2$. This means that the exact observation of Terminal 2 is available at the decoder. In this scenario $A_1 = 0$ and Σ_w becomes:

$$\Sigma_w = \Sigma_1 - \Sigma_{12}\Sigma_2^{-1}\Sigma_{12}^T.$$

Notice that Σ_w now has dimension $M \times M$. The local KLT is the $M \times M$ matrix T_c that diagonalizes Σ_w. Such a matrix is called *Conditional KLT* in [12].

To conclude this section, we consider the following example: Assume a three-dimensional vector $\mathbf{x}$ with covariance matrix

$$\Sigma_x = \begin{pmatrix} 1.2 & 0 & 0 \\ 0 & 1 & 0.5 \\ 0 & 0.5 & 1 \end{pmatrix}.$$

Encoder 1 observes the first two components of this vector: (x_1, x_2), while Encoder 2 has access to the third one. Consider, as usual, the approximation problem first. We assume Terminal 1 has to provide a one-dimensional approximation. If Encoder 1 were to use the classical centralized strategy, it would transmit x_1 only. In fact, since the first two components are uncorrelated, the classical KLT is in this case the identity matrix and the first component has higher variance. However, this solution is in general suboptimal since x_2 is correlated with x_3.

Assume, for example, that Encoder 2 is switched off, which means that the component x_3 is lost. In this case, x_2 is more important since it is the only component that can provide some information about x_3. In this context, the local KLT is still diagonal; however, the nonzero eigenvalues of Σ_w are $\lambda_1^2 = 1.2$ and $\lambda_2^2 = 1.25$, which shows that x_2 represents the components that will be transmitted. One can easily see that the MSE using the classical KLT is $D = 2$, whereas the local KLT would lead to $D_{lKLT} = 1.95$. If we were to do compression, then the rate R that Encoder 1 can use is allocated according to the eigenvalues λ_1^2 and λ_2^2. This means that x_2 will get more rate than x_1. Since x_3 is lost, the compression scheme used by Terminal 1 can be, for example, a classical scalar quantizer, and classical high-rate quantization theory can be used to evaluate the operational rate-distortion function.

Consider now the dual situation, namely, assume that x_3 is available at the decoder. The component x_2 is less important as it can be estimated from x_3. In this context, the local KLT is again diagonal, but the nonzero eigenvalues of Σ_w are $\lambda_1^2 = 1.2$ and $\lambda_2^2 = 0.75$. Thus Terminal 1 transmits x_1. In the compression scenario, the rate R is again allocated according to λ_1^2 and λ_2^2; however, the compression schemes involved in this case are different. The component x_1 is independent of x_3 and thus can be compressed using an independent coder such as a scalar quantizer. The component x_2 instead depends on x_3; thus the compression scheme that needs to be used is a Wynez-Ziv encoder. The distributed transform coding strategy of this second case is shown in Figure 2.4(b) and compared against the independent compression strategy of Figure 2.4(a).

The more general scenario considered in this section assumes that a noisy version of x_3 is available at the decoder. We now understand that the variance of this noise is going to determine which component is sent by Encoder 1. Strong noise is equivalent to assuming that x_3 is lost and so x_2 should be sent. On the other hand, weak noise indicates that x_3 is mostly available at the decoder, and that therefore x_1 should be transmitted. In the same way, the allocation of the rate R between x_1 and x_2 is influenced by the noise. In the case of strong noise, more rate is given to x_2. With weak noise, more rate is allocated to x_1.

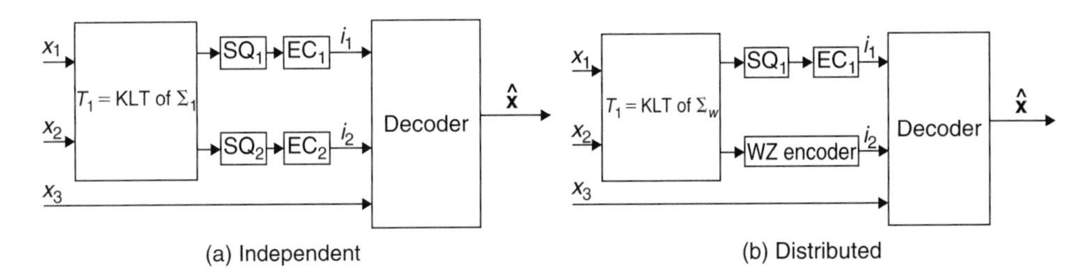

FIGURE 2.4

Independent versus distributed transform coding for the example given in Section 2.3.2. In Figure 2.4(a), Encoder 1 performs an independent encoding of x_1, x_2; that is, it does not take into account the fact that x_3 is available at the decoder. In this case, the correct transform is the KLT of Σ_1, and the rate allocated to the first component is $R_1 \sim R/2 + \log_2 \sigma_1 - \frac{1}{2}\log_2 \sigma_1\sigma_2$, where σ_1^2, σ_2^2 are the eigenvalues of Σ_1 and $R_2 \sim R/2 + \log_2 \sigma_2 - \frac{1}{2}\log_2 \sigma_1\sigma_2$. Moreover, the transformed components can be compressed using standard scalar quantizers and entropy encoders as described in Section 2.2. The optimal distributed transform coding strategy is depicted in Figure 2.4(b) and is remarkably different from the one shown in part (a). The optimal transform is the local KLT, namely, the KLT of Σ_w. This leads to a different bit allocation where $R_1 \sim R/2 + \log_2 \lambda_1 - \frac{1}{2}\log_2 \lambda_1\lambda_2$ and $R_2 \sim R/2 + \log_2 \lambda_2 - \frac{1}{2}\log_2 \lambda_1\lambda_2$. Finally, the second transformed component is correlated with x_3; therefore, it has to be compressed using a Wyner–Ziv encoder, which is normally based on channel coding principles and is very different from a standard scalar quantizer.

Finally, we would also like to point out that, because of the very peculiar structure of the covariance matrix of this example, the local KLT turns out to be always diagonal. However, in most situations the local KLT is not diagonal and is very different from the centralized one. We refer the reader to [12] for examples showing these differences.

2.3.3 The Multiterminal Scenario and the Distributed KLT Algorithm

The general situation in which L terminals have to optimize their local transform simultaneously is more complicated, and, as pointed out previously, an optimal solution does not appear to exist. However, the local perspective discussed in the previous section can be used to devise an iterative algorithm to find the best distributed KLT. The basic intuition is that if all but one terminal have decided their approximation strategy, the remaining terminal can optimize its local transformation using the local KLT discussed in the previous section. Then in turn, each encoder can optimize its local transform given that the other transforms are fixed. This is a sort of round robin strategy, but it is implemented off-line, and encoders do not need to interact since they all have access to the global covariance matrix Σ_x. The algorithm is not guaranteed to converge to the global minimum but may get stuck in a local one. The convergence of such algorithm has been studied in [12].

The distributed KLT algorithm for compression operates along the same principles. Now we have a total rate R available. At each iteration only a fraction of the rate is allocated. Let us denote with $R(i)$ the rate used in the ith iteration. In turn, each terminal is allowed to use the $R(i)$ bits and optimizes its local compression scheme, while all other encoders are kept fixed. The encoder that achieves the best overall reduction of the distortion is allocated the rate $R(i)$. The process is then iterated on a new rate $R(i+1)$, and the process stops when the total bit budget R has been achieved. Notice that the approach is in spirit very similar to a standard greedy bit allocation strategy used in traditional compression algorithms.

2.4 ALTERNATIVE TRANSFORMS

In many practical situations, the assumption that the source is Gaussian might not be correct, and in fact the problem of properly modeling real-life signals such as images and video remains open. In these cases, the KLT is rarely used and is often replaced by the DCT or WT.

In this section, we discuss how transform coding is used in existing distributed compression schemes of sources such as images and video. We briefly describe how the concept of source coding with side information is combined with the standard model of transform coding to form a distributed transform coding scheme. The main results of high bit-rate analysis of such a scheme will also be presented. Lastly, in Section 2.5, we will look at an alternative approach to distributed coding, which uses the results of the new sampling theory of signal with finite rate of innovation [33].

2.4.1 Practical Distributed Transform Coding with Side Information

Existing practical distributed coding schemes are the result of combining transform coding with Wyner–Ziv coding (also known as source coding with side information at the decoder). Wyner–Ziv coding is the counterpart of the Slepian and Wolfs theorem [30], which considers the lossless distributed compression of two correlated discrete sources X and Y. The Wyner–Ziv theorem [36] is an extension of the lossy distributed compression of X with the assumption that the lossless version of Y is available at the decoder as side information. A Slepian–Wolf coder for distributed source coding with side information employs channel coding techniques. A Wyner–Ziv encoder can be thought of as a quantizer followed by a Slepian–Wolf encoder.

In cases of images and video, existing distributed coding schemes add the Wyner–Ziv encoder into the standard transform coding structure. As with the centralized case, a linear transform is independently applied to each image or video frame. Each transform coefficient is still treated independently, but it is fed into a Wyner–Ziv coder instead of a scalar quantizer and an entropy coder. We refer the reader to [24] for an overview of distributed video coding.

Examples of these coding schemes are the Wyner–Ziv video codec [1, 14], and PRISM [25] where both schemes employ the block-based discrete cosine transform (DCT) in the same way as the centralized case. In the Wyner–Ziv video codec, the quantized transform coefficients are grouped into bit-planes and fed into a rate-compatible punctured turbo coder (RCPT). Syndrome encoding by a trellis channel coder was used to code the scalar quantized coefficients in [25].

In the work of [8], a distributed coding scheme for multiview video signals was proposed where each camera observes and encodes a dynamic scene from a different viewpoint. Motion-compensated spatiotemporal wavelet transform was used to explore the dependencies in each video signal. As with other distributed source coding schemes, the proposed scheme employs the coding structure of Wyner–Ziv coding with side information. Here, one video signal was encoded with the conventional source coding principle and was used as side information at the decoder. Syndrome coding was then applied to encode the transform coefficients of the other video signals. Interestingly, it was also shown that, at high rates, for such coding structure the optimal motion-compensated spatiotemporal transform is the Haar wavelet. Another wavelet-based distributed coding scheme was presented in [3], where the wavelet-based Slepian–Wolf coding was used to encode hyperspectral images that are highly correlated within and across neighboring frequency bands. A method to estimate the correlation statistics of the wavelet coefficients was also presented.

2.4.2 High-rate Analysis of Source Coding with Side Information at Decoder

In this section we briefly describe some of the main results of applying high-rate quantization theory to distributed source coding. We refer the reader to [26, 27] for a detailed explanation and derivation of these results.

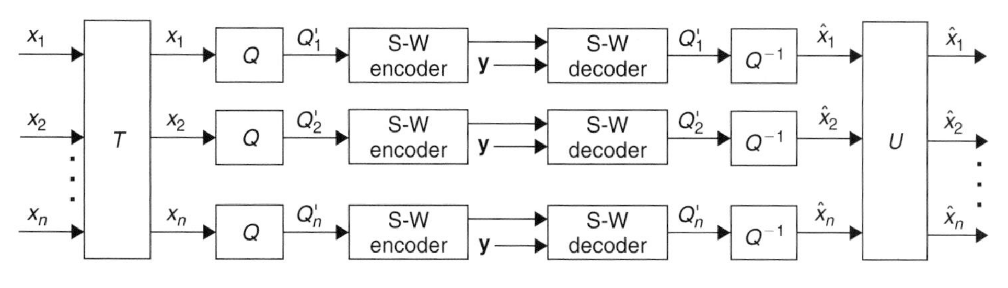

FIGURE 2.5

Distributed transform coding scheme.

Let us consider a transform coding setting in Figure 2.5. We denote with $\mathbf{x} = (x_1, \ldots, x_n)$ the continuous random vector of finite dimension n and let the random vector $\mathbf{y}$ be the side information available at the decoder. We assume that the linear transform T is orthogonal and that $U = T^{-1} = T^T$. Each component of the transform coefficients $\mathbf{x}' = T\mathbf{x}$ is individually scalar quantized. The quantizer index is then assumed to be coded with an ideal Slepian-Wolf coder. The decoder uses $\mathbf{y}$ to decode and reconstruct the approximation of the transform coefficients $\mathbf{x}'$. The estimate of the original source vector is then obtained by applying an inverse transform $\widehat{\mathbf{x}} = U\widehat{\mathbf{x}}'$.

The high-rate analysis in [26,27] assumes the following: the expected distortion D_i of each component of $\mathbf{x}$ is measured as $D_i = E[(x_i' - \widehat{x}_i')]$; the rate required to code the quantization index Q_i' is the conditional entropy $R_i = H(Q_i'|\mathbf{y})$; the total expected distortion per sample is measured as the MSE where $D = \frac{1}{n}E\left[\|\mathbf{x} - \widehat{\mathbf{x}}\|^2\right]$; and the total expected rate per sample is $R = \frac{1}{n}\sum_i R_i$. The goal was then to minimize the Lagrangian cost function $J = D + \lambda R$.

With the above assumptions and assuming large R_i, it was shown in [27] that the expected distortion in each component of $\mathbf{x}$ is

$$D_i \simeq \frac{1}{12} 2^{h(x_i'|\mathbf{y})} 2^{-2R_i},$$

where $h(x_i'|\mathbf{y})$ is the differential entropy. The authors then gave four necessary conditions for the minimization of the Lagrangian cost function. In particular, it was stated that an optimal choice of U is one that diagonalizes the expected conditional covariance matrix $\overline{\Sigma}_{\mathbf{x}|\mathbf{y}} = E_{\mathbf{y}}[\Sigma_{\mathbf{x}|\mathbf{y}((y))}]$. Thus U is the KLT for $\overline{\Sigma}_{\mathbf{x}|\mathbf{y}}$.

2.5 NEW APPROACHES TO DISTRIBUTED COMPRESSION WITH FRI

In [2] a new approach to distributed video compression was proposed. The uniqueness of this approach lies in the combined use of discrete wavelet transform (DWT) and the concept of sampling of signals with finite rate of innovation (FRI), which shifts the task of disparity compensation or motion estimation to the decoder side. The proposed scheme does not employ any traditional channel coding technique.

In this section, we will discuss the basic concepts behind this approach and show the analysis of the rate-distortion bound for a simple example. For simplicity we will only consider a simplified case of a translating bi-level polygon. We refer to [2] for results of more general schemes with realistic cases. We start by presenting the main results of sampling FRI signals before moving on to the actual coding scheme.

2.5.1 Background on Sampling of 2D FRI Signals

The definition and sampling schemes of FRI signals are given in detail in [5, 6, and 33]. We will only discuss the main results that will be used in the sequel. Let a 2D continuous signal be $f(x,y)$ with $x, y \in \mathbb{R}$ and let the 2D sampling kernel be $\varphi(x,y)$. In a typical sampling setup, the samples obtained by sampling $f(x,y)$ with $\varphi(x,y)$ are given by

$$S_{m,n} = \langle f(x,y), \varphi(x/T - m, y/T - n) \rangle,$$

where $\langle \cdot \rangle$ denotes the inner product and $m, n \in \mathbb{Z}$. Assume that the sampling kernel satisfies the polynomial reproduction property that is,

$$\sum_{m \in \mathbb{Z}} \sum_{n \in \mathbb{Z}} c_{m,n}^{(p,q)} \varphi(x-m, y-n) = x^p y^q \tag{2.18}$$

with $p, q \in \mathbb{Z}$ and a proper set of coefficients $c_{m,n}^{(p,q)}$. It follows that, with $T = 1$, the continuous geometric moment $m_{p,q}$ of order $(p+q)$ of the signal $f(x,y)$ is given by

$$m_{p,q} = \sum_{m \in \mathbb{Z}} \sum_{n \in \mathbb{Z}} c_{m,n}^{(p,q)} S_{m,n}. \tag{2.19}$$

Therefore, given a set of coefficients $c_{m,n}^{(p,q)}$, one can retrieve the continuous moments from an arbitrarily low-resolution set of samples $\widehat{S}_{m,n}$ provided that $f(x,y)$ lies in the region where Equation (2.19) is satisfied.

The polynomial reproduction property is satisfied by any valid scaling function $\varphi_j(x,y)$ of the WT, where j represents the number of levels of the wavelet decomposition. Thus we can change the resolution of $S_{m,n}$ by altering j with the corresponding sampling period of $T = 2^j$. It can be shown that the required coefficients $c_{m,n}^{(p,q)}$ are given by

$$c_{m,n}^{(p,q)} = \langle x^p y^q, \widetilde{\varphi}_j(x-m, y-n) \rangle \tag{2.20}$$

where $\widetilde{\varphi}_j(x,y)$ is the dual of $\varphi_j(x,y)$. In the next section, we show that the motion parameters can be extracted from the set of continuous moments $m_{p,q}$ obtained using Equation (2.19).

2.5.2 Detailed Example: Coding Scheme for Translating a Bi-level Polygon

In order to gain some intuition, we will consider a simplified example of a sequence of a bi-level equilateral polygon, moving by translation, in a uniform background. Let us define a block of N video frames or N multiview images to be $f_i(x,y)$, $i=1,2,\ldots,N, x,y\in\mathbb{R}$. We set $f_1(x,y)$ to be the reference frame. It then follows that $f_i(x,y)=f_1(x-x_i,y-y_i), i=2,3,\ldots,N$ where $\underline{t_i}=[x_i,y_i]$ is the translation vector. Using the definition of the moment of the ith frame, $m^i_{p,q}$, and let $x'=x-x_i$, we have that

$$m^i_{1,0}=\int\int f_1(x',y')(x'+x_i)dx'dy'=m^1_{1,0}+m^1_{0,0}x_i. \tag{2.21}$$

Therefore $\underline{t_i}$ can be calculated using first and zero-th order moments as $x_i=\frac{(m^i_{1,0}-m^1_{1,0})}{m^1_{0,0}}$ and similarly $y_i=\frac{(m^i_{0,1}-m^1_{0,1})}{m^1_{0,0}}$. Note that $(\overline{x},\overline{y})=\left(\frac{m_{1,0}}{m_{0,0}},\frac{m_{0,1}}{m_{0,0}}\right)$ is defined as the barycenter of the signal. In terms of video coding, the reference frame $f_1(x,y)$ is usually denoted as the "key frame." For the rest of this section, we refer to $f_i(x,y), i=2,3,\ldots,N$ as the "nonkey frames."

The coding scheme for this sequence is as follows: Each corner point of the polygon in the key frame $f_1(x,y)$ is quantized and transmitted to the decoder, and the moments of $f_1(x,y)$ are computed directly. The nonkey frames $f_i(x,y), i=2,3,\ldots,N$ are then sampled with a kernel $\varphi(x,y)$ that satisfies the condition given in (2.18). The samples $S_i(m,n), i=2,3,\ldots,N$ are quantized before being transmitted. The decoder retrieves the zero-th and first-order moments of each frame using Equations (2.19) and (2.20), and the translation vectors $\underline{t_i}, i=2,3,\ldots,N$ are retrieved as shown in (2.21). Nonkey frames are then reconstructed as $f_i(x,y)=f_1(x-x_i,y-y_i)$. This scheme is summarized in Figure 2.6.

Note that the encoder only performs standard, independent, transform coding on each frame, and since only the samples of the nonkey frames are transmitted along with the key frame, the overall bit rate is reduced. The dependency between each

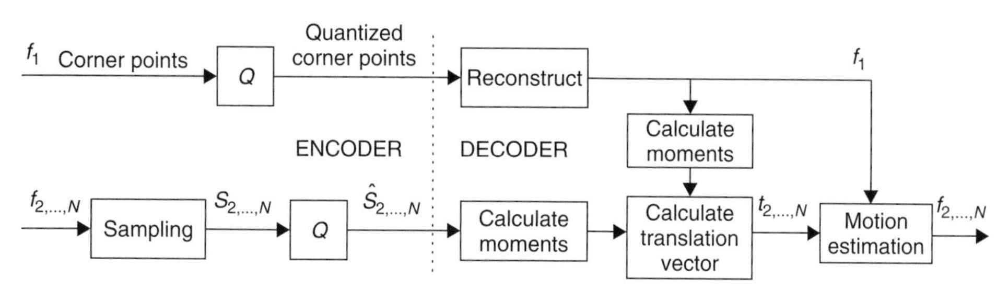

FIGURE 2.6

Distributed coding scheme for a bi-level, translating, polygon sequence based on FRI.

frame is exploited at the decoder via joint decoding. Thus, this coding scheme has a low-complexity encoder.

In the following section, we will examine the distortion-rate behavior of this distributed coding scheme and compare the rate-distortion performance with that of an ideal joint (interframe) coder and an ideal independent (intraframe) coder.

2.5.2.1 Rate-distortion Analysis: Joint Encoder and Independent Encoder

Figure 2.7 presents an example of a bi-level polygon in a uniform background and a visualization of the reconstruction error as a result of a mismatch between the values of a translation vector at the encoder and the decoder.

For the ideal joint encoder, we assume that the location of each corner point of the polygon and the frame-by-frame translation vectors $\underline{t_i}$ are quantized and transmitted to the decoder directly. The independent encoder on the other hand transmits the quantized corner points of every video frame to the decoder.

Let $f_i(x, y)$, $i = 1, 2, \ldots, N$ be an N-frame video sequence with size $A \times A$ that contains a bi-level, equilateral polygon of C corner points with amplitude B. First we want to find the theoretical distortion-rate bound of an ideal joint encoder. The distortion D is always measured as the mean-squared error (MSE) in the following analysis.

The joint encoder has to quantize two sets of parameters: the corner points and the translation vectors. Thus we have two types of distortion that are independent and additive, which can then be analyzed separately. Furthermore, at high bit rate, we can also assume that, for each parameter, the quantization errors along the x-direction and y-direction in the Cartesian coordinate are independent and, hence, also additive. Lastly, we assume that the locations of the corner points are independent.

Let R_C be the number of bits allocated to represent each corner point and let R_T denote the number of bits allocated to represent each translation vector. With the above assumption, one can show that the optimal bit allocation that minimizes D for this joint encoder is given by

$$R_C = R_T + R_E,$$

FIGURE 2.7

(Left) A bi-level square in a uniform background. (Right) A visualization of the reconstruction error resulting from the mismatch between the values of a translation vector at the encoder and the decoder.

where the constant $R_E = 2\log_2\left(\frac{N}{2(N-1)}\right)$. At a high rate, the distortion-rate bound $D_{INTER}(R)$ is

$$D_{INTER}(R) \leq 2A^2B^2C_{INTER}\left(2^{-\frac{R}{2(C+N-1)}}\right),$$

where R is the total number of bits and C_{INTER} is a constant.

We also want to find the distortion-rate bound of the independent coder. In this case the total number of bits used to encode the sequence is $R = NCR_C$. With the same set of assumptions as above, one can also show that, at a high rate, the distortion-rate $D_{INTRA}(R)$ bound of the independent coder is given by

$$D_{INTRA}(R) \leq 2A^2B^2\left(2^{-\frac{R}{2(NC)}}\right).$$

Note that the decay rate of $D_{INTER}(R)$, which equals $2^{-\frac{R}{2(C+N-1)}}$, is greater than that of $D_{INTRA}(R)$ given by $2^{-\frac{R}{2(NC)}}$ for $C > 1$. This is as expected since the joint encoder exploits the interframe dependency of the video sequence and achieves a better compression efficiency. In the next section, we will examine the distortion-rate behavior of the proposed FRI-based distributed coding scheme.

2.5.2.2 *Rate-distortion Analysis: Distributed Coding Using Sampling of FRI Signals*

In the proposed scheme shown in Figure 2.6, there are still two sets of parameters to be quantized: the corner points and the samples. The quantization errors in the samples lead to errors in the retrieved translation vectors. Thus, we have two types of distortion: the distortion due to the error in the retrieved translation vector as a result of quantization of the transmitted samples and the distortion due to quantization of corner points. As with the previous analysis, we assume that the two types of distortion are independent.

Given the observed samples, $S_{m,n}$, of size $L_m \times L_n$, from the sampling scheme of FRI signals, clearly, the errors introduced by quantizing $S_{m,n}$ lead to an error in the retrieved moment $M_{p,q}$. In the proposed coding scheme, the frame-to-frame translation vector can be obtained by calculating the barycenters of the two video frames. The barycenter in the Cartesian coordinate is defined as $\left(\frac{M_{1,0}}{M_{0,0}}, \frac{M_{0,1}}{M_{0,0}}\right)$, where $\frac{M_{1,0}}{M_{0,0}}$ and $\frac{M_{0,1}}{M_{0,0}}$ are in fact normalized first-order moments. Let us denote the normalized moment of order $(p+q)$ as $\overline{M}_{p,q} = \frac{M_{p,q}}{M_{0,0}}$ and let $S_{m,n} = \widehat{S}_{m,n} + e_{m,n}$, where $\widehat{S}_{m,n}$ denotes the quantized samples and $e_{m,n}$ represents the quantization error; we have that

$$\overline{M}_{p,q} = \sum_{m}^{L_m}\sum_{n}^{L_n}\frac{C_{m,n}^{p,q}}{M_{0,0}}\widehat{S}_{m,n} + \sum_{m}^{L_m}\sum_{n}^{L_n}\frac{C_{m,n}^{p,q}}{M_{0,0}}e_{m,n} = \widehat{\overline{M}}_{p,q} + \overline{w}_{p,q},$$

where $\overline{C}_{m,n}^{p,q}$ are the normalized polynomial reproduction coefficients.

Let us denote with $\overline{M}_{p,q}^{i} = \widehat{\overline{M}}_{p,q}^{i} + \overline{w}_{p,q}^{i}$, the normalized moment of the ith frame. It is easy to show that the error in the frame-to-frame translation vector between the

ith and jth frame is then given by (q_x^S, q_y^S) where $q_x^S = (\overline{w}_{1,0}^i + \overline{w}_{1,0}^j)$ represents the error in the x-component of the translation vector. Similarly, $q_y^S = (\overline{w}_{0,1}^i + \overline{w}_{0,1}^j)$ is the error in the y-component.

Since each sample is independently scalar quantized, one can assume that the probability density function (PDF) of $\overline{w}_{p,q}$ is Gaussian whose variance depends on the set of polynomial reproduction coefficients $\overline{C}_{m,n}^{p,q}$. We also assume that the errors in each direction of the retrieved translation vector are independent, that is, q_x^S and q_y^S are independent. With these assumptions, it is possible to show that the distortion due to the quantization of the samples, as denoted by D_S, is given by

$$D_S \leqslant \frac{16AB^2}{\sqrt{2\pi}} \left(\sigma_0(\overline{C}_{m,n}^{1,0}) + \sigma_0(\overline{C}_{m,n}^{0,1})\right) 2^{-n_s},$$

where n_s is the number of bits allocated to represent each sample and σ_0 is a constant.

It can then be shown that the optimal bit allocation is

$$\frac{R_C}{2} = n_S + R_S,$$

with the constant $R_S = \log_2\left(\frac{A\sqrt{2\pi}}{16}\frac{N}{(N-1)}\left(\sigma_0(\overline{C}_{m,n}^{1,0}) + \sigma_0(\overline{C}_{m,n}^{0,1})\right)^{-1}\right)$. Note that R_C is the number of bits allocated to represent each corner point. Finally, it follows that the theoretical distortion-rate bound, $D_{DSC\_FRI}(R)$, of this proposed coding scheme is given by

$$D_{DSC\_FRI}(R) \leqslant 2AB^2 C_{DSC\_FRI} 2^{-\frac{R}{2C+\frac{(N-1)L_m L_n}{2}}}, \tag{2.22}$$

where C_{DSC_FRI} is a constant.

Equation (2.22) shows us that the decay rate of $D_{DSC\_FRI}(R)$ is equal to $2^{-\frac{R}{2C+\frac{(N-1)L_m L_n}{2}}}$, which is always lower than the rate of $2^{-\frac{R}{2(C+N-1)}}$ of the joint encoder. However, in comparison with the independent encoder whose decay rate equals $2^{-\frac{R}{2(NC)}}$, the decay rate of the distortion of the proposed FRI-based coding scheme can be higher depending on the value of C and $L_m L_n$. The value of $L_m L_n$ decreases as the level of wavelet decomposition increases, which results in a better compression performance. Thus, we ideally want the level of wavelet decomposition to be as high as possible as there will be fewer coefficients to be transmitted.

Another point to be noted is that as the number of corner points C becomes larger, our proposed coding method performs relatively better than the independent encoder. The key point here is that as the complexity of the object in the video frame increases, this FRI-based distributed coding scheme can achieve better compression efficiency than the independent encoder. This is because, with this scheme, we only need to encode the complex object once in the key frame. However, when C is small, we are better off by just transmitting the simple object in every video frame rather than transmitting a number of samples.

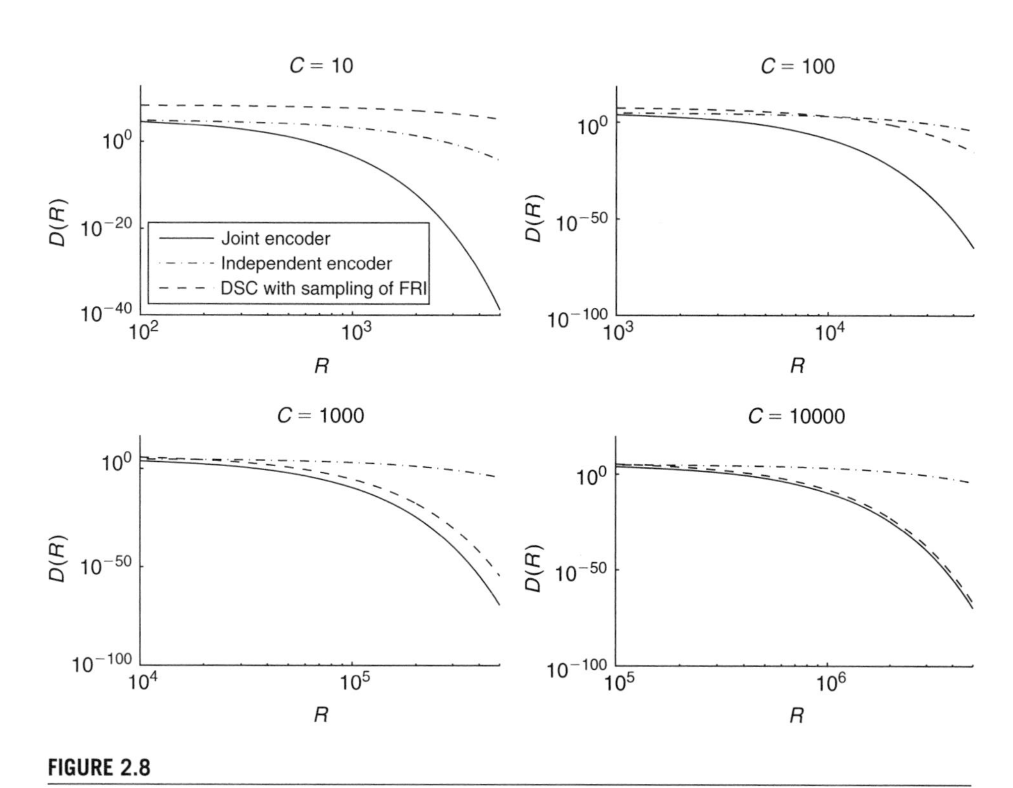

FIGURE 2.8

Plots of $D_{INTER}(R)$ $D_{INTRA}(R)$ and $D_{DSC\_FRI}(R)$. The complexity, C, of the sequence was set to 10, 100, 1000, and 10,000. A Daubechies db2 filter was used and the parameters were: $N=8$, $L_m=8$, and $L_n=8$.

Figure 2.8 shows the plots of the distortion-rate bound for the three coding schemes. A Daubechies db2 filter was used with eight levels of wavelet decomposition. The size of the observed samples was set to 8×8. From the plots, at $C=10$, the independent (intraframe) coder achieves higher compression efficiency than the FRI-based distributed coding scheme. However, the proposed scheme outperforms the independent coder as the number of corner points increases. The performance gap between the joint (interframe) encoder and the proposed scheme also decreases as the complexity of the object in the sequence increases.

The plots shown in Figure 2.8 have been validated in [2] with synthesized video sequences of translating bi-level polygons with a different number of corner points. The scheme has also been extended to the case of a real object whose motion can be described by an affine transform (see [2]). Even with a real object and a more complex motion model, the same intuition remains: it makes sense to encode a complex object once and then transmit only the samples (i.e., the low-pass coefficients of the WT) of the object for the rest of the sequence.

2.6 CONCLUSIONS

In this chapter, we started by reviewing the main known results on transform coding in the centralized compression scenario, we then looked at the distributed transform coding problem that arises when the source is not available at a single location but is distributed at multiple sites. Extensions of the KLT to the distributed case were first presented, and then possible variations of DCT and WT were discussed.

Although there are scenarios where precise answers to the correct design of a distributed transform coder can be provided, in most situations the solutions are based on heuristics and optimal distributed transforms are not known yet. This clearly indicates that the distributed transform coding problem is still widely open and that a vast space of unexplored solutions remain to be discovered.

REFERENCES

[1] E. Setton, A. Aaron, S. Rane, and B. Girod. Transform-domain Wyner–Ziv codec for video. In *Proc. Visual Communications and Image Processing, VCIP-2004*, San Jose, CA, January 2004.

[2] V. Chaisinthop and P. L. Dragotti. A new approach to distributed coding using sampling of signals with finite rate of innovation. In *Proc. Picture Coding Symposium 2007*, Lisbon, Portugal, November 2007.

[3] N.-M. Cheung, C. Tang, A. Ortega, and C. S. Raghavendra. Efficient wavelet-based predictive Slepian-Wolf coding for hyperspectral imagery. *Signal Process.*, 86(11): 3180–3195, 2006.

[4] A. Cohen, I. Daubechies, O. Guleryuz, and M. T. Orchard. On the importance of combining wavelet-based non-linear approximation with coding strategies. *IEEE Trans. on Information Theory*, 48(7):1895–1921, July 2002.

[5] P. L. Dragotti, M. Vetterli, and T. Blu. Exact sampling results for signals with finite rate of innovation using Strang–Fix conditions and local kernels. In *Proc. of IEEE Int. Conf. on Acoustic, Speech and Signal Processing (ICASSP)*, Philadelphia, PA, March 2005.

[6] P. L. Dragotti, M. Vetterli, and T. Blu. Sampling moments and reconstructing signals of finite rate of innovation: Shannon meets Strang–Fix. *IEEE Trans. on Signal Processing*, 55(5):1741–1757, May 2007.

[7] N. Farvardin and J. Modestino. Optimum quantizer performance for a class of non-Gaussian memoryless sources. *Information Theory, IEEE Transactions on*, 30(3):485–497, May 1984.

[8] M. Flierl and P. Vandergheynst. Distributed coding of highly correlated image sequences with motion-compensated temporal wavelets. *Eurasip Journal Applied Signal Processing*, 2006, September 2006.

[9] M. Gastpar, P. L. Dragotti, and M. Vetterli. The distributed Karhunen–Loève transform. In *Multimedia and Signal Processing Workshop*, Virgin Islands, December 2002.

[10] M. Gastpar, P. L. Dragotti, and M. Vetterli. The distributed, partial and conditional Karhunen–Loève transforms. In *Data Compression Conference*, Snowbird, Utah, March 2003.

[11] M. Gastpar, P. L. Dragotti, and M. Vetterli. On compression using the distributed Karhunen–Loève transform. In *Proc. of IEEE Int. Conf. on Acoustic, Speech and Signal Processing, invited paper*, Montreal (CA), May 2004.

[12] M. Gastpar, P. L. Dragotti, and M. Vetterli. The distributed Karhunen-Loève transform. *IEEE Trans. on Information Theory*, 52 (12):5177-5196, December 2006.

[13] A. Gersho and R. M. Gray. *Vector Quantization and Signal Compression*. Kluwer Academic Publishers, Norwell, MA, 1991.

[14] B. Girod, A. M. Aaron, S. Rane, and D. Rebollo-Monedero. Distributed video coding. *Proceedings of the IEEE*, 93(1):71-83, 2005.

[15] V. K Goyal. Theoretical foundations of transform coding. *IEEE Signal Processing Magazine*, 18(5):9-21, September 2001.

[16] R. M. Gray and D. L. Neuhoff. Quantization. *Information Theory, IEEE Transactions on*, 44(6):2325-2383, October 1998.

[17] C. Guillemot and A. Roumy. Towards constructive Slepian-Wolf coding schemes. In *Distributed Source Coding: Theory, Algorithms and Applications*, eds. P. L. Dragotti and M. Gastpar, Academic Press, Elsevier, San Diego, CA, 2009.

[18] J. Huang and P. Schultheiss. Block quantization of correlated Gaussian random variables. *Communications, IEEE Transactions on [legacy, pre-1988]*, 11(3):289-296, September 1963.

[19] D. A. Huffman. A method for the construction of minimum-redundancy codes. *Proceedings of the IRE*, 40(9):1098-1101, 1952.

[20] D. T. Lee. JPEG2000: Retrospective and new developments. *Proceedings of the IEEE*, 93(1): 32-41, 2005.

[21] S. Mallat. *A Wavelet Tour of Signal Processing*. Academic Press, Boston, MA, 1998.

[22] H. Nurdin, R. R. Mazumdar, and A. Bagchi. On estimation and compression of distributed correlated signals with incomplete observations. In *Proc. Conf. Mathematical Theory of Networks and Systems*, Leuven (Belgium), July 2004.

[23] A. Ortega and K. Ramchandran. Rate-distortion methods for image and video compression. *Signal Processing Magazine, IEEE*, 15(6):23-50, 1998.

[24] F. Pereira, C. Brites, and J. Ascenso. Distributed video coding: Basics, codecs and performance. In *Distributed Source Coding: Theory, Algorithms and Applications*, eds. P. L. Dragotti and M. Gastpar, Academic Press, Elsevier, San Diego, CA, 2009.

[25] R. Puri and K. Ramchandran. PRISM: A video coding architecture based on distributed compression principles. In *40th Allerton Conference on Communication, Control and Computing*, Allerton, IL, October 2002.

[26] D. Rebollo-Monedero and B. Girod. Quantization for distributed source coding. In *Distributed Source Coding: Theory, Algorithms and Applications*, eds. P. L. Dragotti and M. Gastpar, Academic Press, Elsevier, San Diego, CA, 2009.

[27] D. Rebollo-Monedero, S. Rane, A. Aaron, and B. Girod. High-rate quantization and transform coding with side information at the decoder. *Eurasip Signal Processing Journal*, 86(11): 3160-3179, November 2006. Special Issue on Distributed Source Coding.

[28] O. Roy and M. Vetterli. On the asymptotic distortion behavior of the distributed Karhunen-Loève transform. In *Proc. of Allerton Conf. Communication, Control and Computing*, Monticello (IL), September 2005.

[29] O. Roy and M. Vetterli. Dimensionality reduction for distributed estimation in the infinite dimensional regime. *IEEE Trans. on Information Theory*, 54(4):1655-1669, April 2008.

[30] D. Slepian and J. Wolf. Noiseless coding of correlated information sources. *IEEE Transactions on Information Theory*, 19(4):471-480, 1973.

[31] G. J. Sullivan and T. Wiegand. Video compression—from concepts to the H.264/AVC standard. *Proceedings of the IEEE*, 93(1):18-31, 2005.

[32] M. Vetterli. Wavelets, approximation and compression. *IEEE Signal Processing Magazine*, 18:59-73, September 2001.

[33] M. Vetterli, P. Marziliano, and T. Blu. Sampling signals with finite rate of innovation. *IEEE Transactions on Signal Processing*, 50(6):1417–1428, 2002.

[34] G. K. Wallace. The JPEG still picture compression standard. *Commun. ACM*, 34(4):30–44, 1991.

[35] I. H. Witten, R. M. Neal, and J. G Cleary. Arithmetic coding for data compression. *Commun. ACM*, 30(6):520–540, 1987.

[36] A. Wyner and J. Ziv. The rate-distortion function for source coding with side information at the decoder. *IEEE Transactions on Information Theory*, 22(1):1–10, 1976.

[37] K. S. Zhang, X. R. Li, P. Zhang, and H. F. Li. Optimal linear estimation fusion part IV: Sensor data compression. In *Proc. of Int. Conf. Fusion*, Queensland (Australia), July 2003.

CHAPTER

Quantization for Distributed Source Coding

3

David Rebollo-Monedero
Department of Telematics Engineering, Universitat Politècnica de Catalunya, Barcelona, Spain

Bernd Girod
Department of Electrical Engineering, Stanford University, Palo Alto, CA

CHAPTER CONTENTS

Distributed Source Coding: Theory, Algorithms, and Applications

3.1 INTRODUCTION

Consider the sensor network depicted in Figure 3.1, where sensors obtain noisy readings of some unseen data of interest that must be transmitted to a central unit. The central unit has access to side information, for instance, archived data or readings from local sensors. At each sensor, neither the noisy observations of the other sensors nor the side information is available. Nevertheless, the statistical dependence among the unseen data, the noisy readings, and the side information may be exploited in the design of each individual sensor encoder and the joint decoder at the central unit to optimize the rate-distortion performance. Clearly, if all the noisy readings and the side information were available at a single location, traditional joint denoising and encoding techniques could be used to reduce the transmission rate as much as possible, for a given distortion. However, since each of the noisy readings must be individually encoded without access to the side information, practical design methods for efficient distributed coders of noisy sources are needed.

The first attempts to design quantizers for Wyner–Ziv (WZ) coding—that is, lossy source coding of a single instance of directly observed data with side information at the decoder—were inspired by the information-theoretic proofs. Zamir and Shamai [50, 51] proved that under certain circumstances, linear codes and high-dimensional nested lattices approach the WZ rate-distortion function, in particular if the source data and side information are jointly Gaussian. This idea was further developed and applied by Pradhan et al. [21, 29, 30] and Servetto [38], who published heuristic designs and

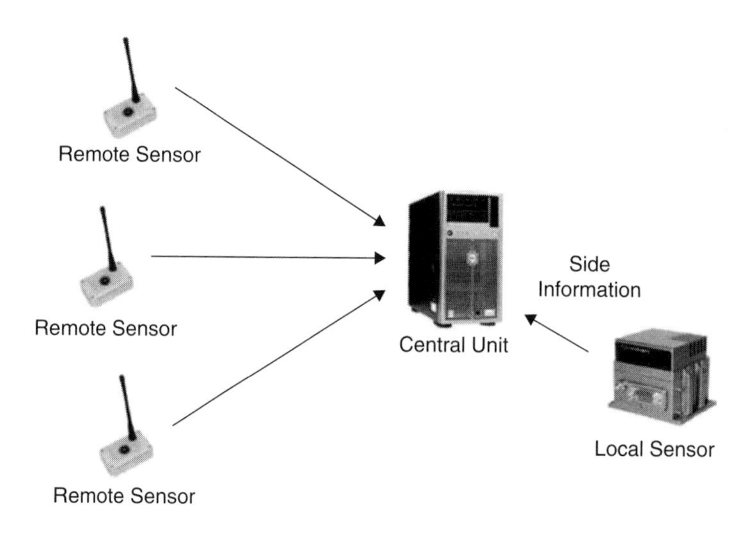

FIGURE 3.1

Distributed source coding in a sensor network.

performance analysis focusing on the Gaussian case, based on nested lattices, with either fixed-rate coding or entropy coding of the quantization indices.

A different approach was followed by Fleming and Effros [11], who generalized the Lloyd algorithm [26] for locally optimal, fixed-rate WZ quantization design. Later, Fleming, Zhao, and Effros [12] included rate-distortion optimized quantizers in which the rate measure is a function of the quantization index, for example, a codeword length. Unfortunately, vector quantizer dimensionality and entropy code blocklength are identical in their formulation; thus the resulting quantizers either lack in performance or are prohibitively complex. An efficient algorithm for finding globally optimal quantizers among those with contiguous code cells was provided in [27], although, regrettably, it has been shown that code cell contiguity precludes optimality in general distributed settings [9]. Cardinal and Van Asche [5] considered Lloyd quantization for ideal symmetric Slepian–Wolf (SW) coding without side information. Cardinal [3, 4] has also studied the problem of quantization of the side information itself for lossless distributed coding of discrete source data (which is not quantized).

It may be concluded from the proof of the converse to the WZ rate-distortion theorem [42] that there is no asymptotic loss in performance by considering block codes of sufficiently large length, which may be seen as vector quantizers, followed by fixed-length coders. This suggests a convenient implementation of WZ coders as quantizers, possibly preceded by transforms, followed by SW coders, analogous to the implementation of nondistributed coders. This implementation is represented in Figure 3.2. The quantizer divides the signal space into cells, which, however, may consist of noncontiguous subcells mapped into the same quantizer index. Xiong et al. [25, 43] implemented a WZ coder as a nested lattice quantizer followed by a SW coder, and in [45], a trellis-coded quantizer was used instead (see also [44]).

An extension of the Lloyd algorithm, more general than Fleming and Effros's, appeared in our own work [37], which considered a variety of coding settings in a unified framework, including the important case of ideal SW distributed coding of one or several sources with side information, that is, when the rate is the joint conditional entropy of the quantization indices given the side information.

An upper bound on the rate loss due to the unavailability of the side information at the encoder was found by Zamir [47], who also proved that for power-difference distortion measures and smooth source probability distributions, this rate loss vanishes

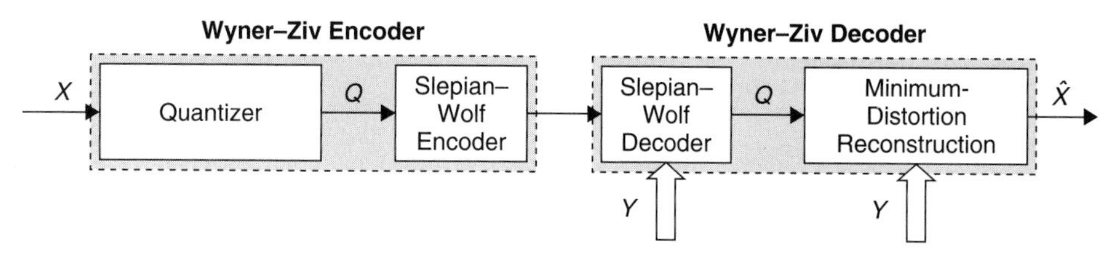

FIGURE 3.2

A practical WZ coder obtained by cascading a quantizer and a SW coder.

in the limit of small distortion. A similar high-resolution result was obtained by Zamir and Berger [48] for distributed coding of several sources *without* side information, also from an information-theoretic perspective, that is, for arbitrarily large dimension. In [40] (unpublished), it was shown that tessellating quantizers followed by SW coders are asymptotically optimal in the limit of small distortion *and* large dimension. The first theoretic characterization of high-rate WZ quantizers with SW coding was presented in our own work [32]. It was inspired by the traditional developments of high-rate quantization theory, specifically Bennett's distortion approximation [1, 28], Gish and Pierce's rate approximation [17], and Gersho's conjecture [15].

As for quantization of a noisy observation of an unseen source, the nondistributed case was studied by Dobrushin, Tsybakov, Wolf, Ziv, Ephraim, Gray, and others [8, 10, 41]. Most of the operational work on distributed coding of noisy sources, that is, for a fixed dimension, deals with quantizer design for a variety of settings, as in the work by Lam and Reibman [22, 23], and Gubner [20], but does not consider entropy coding or the characterization of such quantizers at high rates or transforms. The problem of optimal noisy WZ quantizer design with side information at the decoder and ideal SW coding has only been studied under the assumptions of high rates and particular statistical conditions by our research group [36].

This chapter investigates the design of rate-distortion optimal quantizers for distributed compression in a network with multiple senders and receivers. In such a network, several noisy observations of one or more unseen sources are separately encoded by each sender and the quantization indices transmitted to a number of receivers, which jointly decode all available transmissions with the help of side information locally available. We present a unified framework in which a variety of coding settings is allowed, including ideal SW coding, along with an extension of the Lloyd algorithm for "locally" optimal design. In addition, we characterize optimal quantizers in the limit of high rates and approximate their performance. This work summarizes part of the research detailed in [31] and a number of papers, mainly [34, 35], where additional mathematical technicalities and proofs can be found.

Section 3.2 contains the formulation of the problem studied. A theoretic analysis for optimal quantizer design is presented in Section 3.3. In Section 3.5, optimal quantizers and their performance are characterized asymptotically at high rates. Experimental results for distributed coding of jointly Gaussian data are shown in Section 3.4 and revisited in Section 3.6.

3.2 FORMULATION OF THE PROBLEM

3.2.1 Conventions

Throughout the chapter, the measurable space in which a random variable (r.v.) takes on values will be called an alphabet. All alphabets are assumed to be Polish spaces to ensure the existence of regular conditional probabilities. We shall follow the convention of using upper-case letters for r.v.'s, and lower-case letters for particular values they take on. Probability density functions (PDFs) and probability mass functions (PMFs) are denoted by p and subindexed by the corresponding r.v.

3.2.2 Network Distributed Source Coding

We study the design of optimal quantizers for network distributed coding of noisy sources with side information. Figure 3.3 depicts a network with several lossy encoders communicating with several lossy decoders. Let $m, n \in \mathbb{Z}^+$ represent the number of encoders and decoders, respectively. Let $X, Y = (Y_j)_{j=1}^n$ and $Z = (Z_i)_{i=1}^m$ be r.v.'s defined on a common probability space, statistically ***dependent*** in general, taking values in arbitrary, possibly different alphabets, respectively, $\mathcal{X}, \mathcal{Y}$, and $\mathcal{Z}$. For each $i = 1, \ldots, m$, Z_i represents an *observation* statistically related to some *source data* X of interest, available only at encoder i. For instance, Z_i might be an image corrupted by noise, a feature extracted from the source data such as a projection or a norm, the pair consisting of the source data itself together with some additional encoder side information, or any type of correlated r.v., and X might also be of the form $(X_i)_{i=1}^m$, where X_i could play the role of the source data from which Z_i is originated. For each $j = 1, \ldots, n$, Y_j represents some ***side information***, for example, previously decoded data, or an additional, local noisy observation, available at decoder j only. For each i, a quantizer

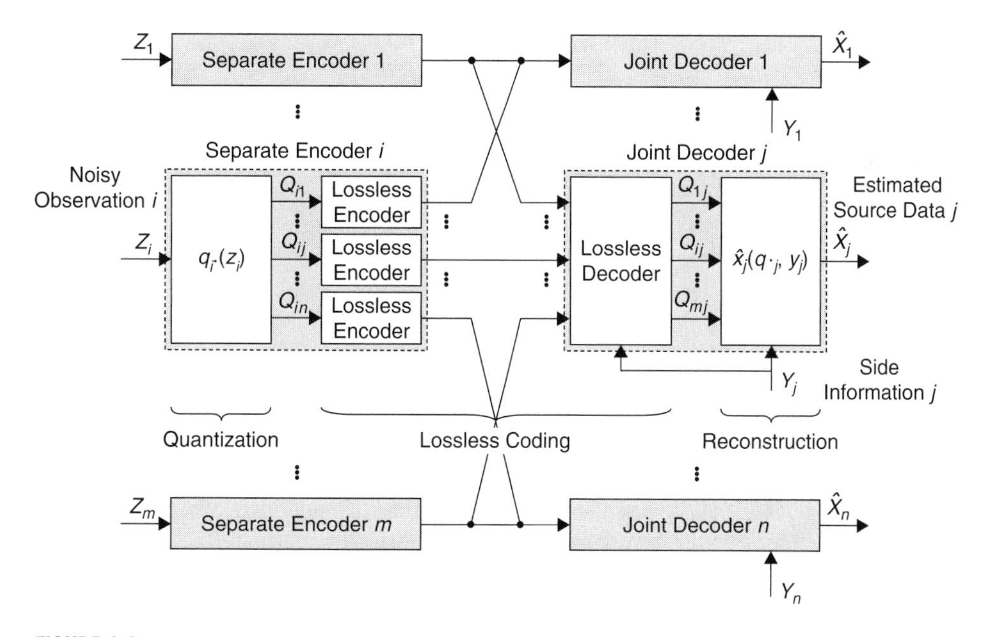

FIGURE 3.3

Distributed quantization of noisy sources with side information in a network with m encoders and n decoders (m and n need not be equal). $\hat{X}$ is the reconstruction of a r.v. X jointly distributed with Y and Z.

$q_{i\cdot}(z_i) = (q_{ij}(z_i))_{j=1}^n$, which can also be regarded as a family of quantizers, is applied to the observation Z_i, obtaining the quantization indices $Q_{i\cdot} = (Q_{ij})_{j=1}^n$. Each quantization index Q_{ij} is losslessly encoded at encoder i and transmitted to decoder j, where it is losslessly decoded. We shall see that both encoding and decoding of quantization indices may be joint or separate. For each j, all quantization indices received by decoder j, $Q_{\cdot j} = (Q_{ij})_{i=1}^m$, and the side information Y_j are used jointly to estimate the unseen source data X. Let $\hat{X}_j$ represent this estimate, obtained with a measurable function $\hat{x}_j(q_{\cdot j}, y_j)$, called *reconstruction function*, in an alphabet $\hat{\mathcal{X}}_j$ possibly different from $\mathcal{X}$. Define $Q = (Q_{ij})_{i=1,j=1}^{m,n}$, $\hat{X} = (\hat{X}_j)_{j=1}^n$, and $\hat{x}(q, y) = (\hat{x}_j(q_{\cdot j}, y_j))_{j=1}^n$. $\mathcal{Q}$ and $\hat{\mathcal{X}}$ will denote the alphabets of Q and $\hat{X}$, respectively. *Partially connected networks* in which encoder i does *not* communicate with decoder j can easily be handled by redefining $Q_{i\cdot}$, $Q_{\cdot j}$ and Q not to include Q_{ij}.

3.2.3 Cost, Distortion, and Rate Measures

We shall now extend the concept of rate measure presented in [37] by means of a much more general definition, accompanied by a characteristic property that will play a major role in the extension of the Lloyd algorithm for optimized quantizer design, called update property. Even though the definition of rate measure and its update property may seem rather abstract and complex initially, we shall see that it possesses great generality and is applicable to a wide range of problems. The terms *cost measure*, *rate measure*, and *distortion measure*, formally equivalent by definition, will be used to emphasize different connotations in the context of applications, usually a Lagrangian cost, a transmission rate, and the distortion introduced by lossy decoding, respectively.

A *cost measure* is a nonnegative measurable function of the form $c(q, x, \hat{x}, y, z)$, possibly defined in terms of the joint probability distribution of (Q, X, Y, Z).[1] Its *associated expected cost* is defined as $C = \mathrm{E}\, c(Q, X, \hat{X}, Y, Z)$. Furthermore, a cost measure will be required to satisfy the following *update property*. For any modification of the joint probability distribution of (Q, X, Y, Z) preserving the marginal distribution of (X, Y, Z), there must exist an induced cost measure c', consistent with the original definition almost surely (a.s.) but expressed in terms of the modified distribution, satisfying $\mathrm{E}\, c'(Q, X, Y, Z) \geqslant C$, where the expectation is taken with respect to the original distribution.

An important example of rate measure is that corresponding to asymmetric SW coding of a quantization index Q given side information Y. The achievable rate in this case has been shown to be $\mathrm{H}(Q|Y)$, when the two alphabets of the random variables involved are finite [39], in the sense that any rate greater than $\mathrm{H}(Q|Y)$ would allow arbitrarily low probability of decoding error, but any rate less than $\mathrm{H}(Q|Y)$ would not.

[1] Rigorously, a cost measure may take as arguments probability distributions (measures) and probability functions; that is, it is a function of a function. We shall become immediately less formal and call the evaluation of the cost measure for a particular probability distribution or function, cost measure as well.

In [6], the validity of this result was generalized to countable alphabets, but it was still assumed that $\mathrm{H}(Y) < \infty$. We show in [31, App. A] (also [35]) that the asymmetric SW result remains true under the assumptions in this chapter, namely, for any Q in a countable alphabet and any Y in an arbitrary alphabet, possibly continuous, regardless of the finiteness of $\mathrm{H}(Y)$.

Concordantly, we define the rate measure $r(q, y) = -\log p_{Q|Y}(q|y)$ to model the use of an *ideal* SW coder, in the sense that both the decoding error probability and the rate redundancy are negligible. Note that the very same rate measure would also model a conditional coder with access to the side information at the encoder. We check that the SW rate measure defined satisfies the update property. The rate measure corresponding to any other PMF $p'_{Q|Y}$ would be $r'(q, y) = -\log p'_{Q|Y}(q, y)$. The modified associated rate would then be $\mathcal{R}' = \mathrm{E}\, r'(Q, Y)$, where r' is defined in terms of the new PMF but the expectation is taken with respect to the original one. Clearly, $\mathcal{R}' - \mathcal{R} = \mathrm{D}(p_{Q|Y} \| p'_{Q|Y}) \geqslant 0$. It can be seen that all rate measures defined in [37] satisfy the update property, thereby making the corresponding coding settings applicable to this framework.

If the alphabets of X, Y, Z, and $\hat{X}$ are equal to some common normed vector space, then an example of a distortion measure is $d(x, \hat{x}, y, z) = \alpha \|x - \hat{x}\|^2 + \beta \|y - \hat{x}\|^2 + \gamma \|z - \hat{x}\|^2$, for any $\alpha, \beta, \gamma \in [0, \infty)$. An example of a cost measure suitable for distributed source coding, the main focus of this work, is $c(q, x, \hat{x}, y) = d(x, \hat{x}) + \lambda\, r(q, y)$, where λ is a nonnegative real number determining the rate-distortion trade-off in the Lagrangian cost $\mathcal{C} = \mathcal{D} + \lambda\, \mathcal{R}$.

3.2.4 Optimal Quantizers and Reconstruction Functions

Given a suitable cost measure $c(q, x, \hat{x}, y, z)$, we address the problem of finding quantizers $q_{i\cdot}(z_i)$ and reconstruction functions $\hat{x}_j(q_{\cdot j}, y_j)$ to minimize the associated expected cost $\mathcal{C}$. The choice of the cost measure leads to a particular noisy distributed source coding system, including a model for lossless coding.

3.2.5 Example: Quantization of Side Information

Even though the focus of this work is the application represented in Figure 3.3, the generality of this formulation allows many others. For example, consider the coding problem represented in Figure 3.4, proposed in [3]. A r. v. Z is quantized. The quantization index Q is coded at rate $\mathcal{R}_1 = \mathrm{H}(Q)$ and used as side information for a SW coder of a discrete random vector X. Hence, the additional rate required is $\mathcal{R}_2 = \mathrm{H}(X|Q)$. We wish to find the quantizer $q(z)$ minimizing $\mathcal{C} = \mathcal{R}_2 + \lambda\, \mathcal{R}_1$. It can be shown that $r_1(q) = -\log p_Q(q)$ and $r_2(x, q) = -\log p_{X|Q}(x|q)$ are well-defined rate measures, using an argument similar to that for $-\log p_{Q|Y}(q|y)$. Therefore, this problem is a particular case of our formulation. In fact, this is a noisy WZ, or a statistical inference problem in which $\mathcal{R}_2$ plays the role of distortion, since minimizing $\mathrm{H}(X|Q)$ is equivalent to minimizing $\mathrm{I}(Z; X) - \mathrm{I}(Q; X)$, nonnegative by the data processing inequality, zero if and only if Q is a sufficient statistic.

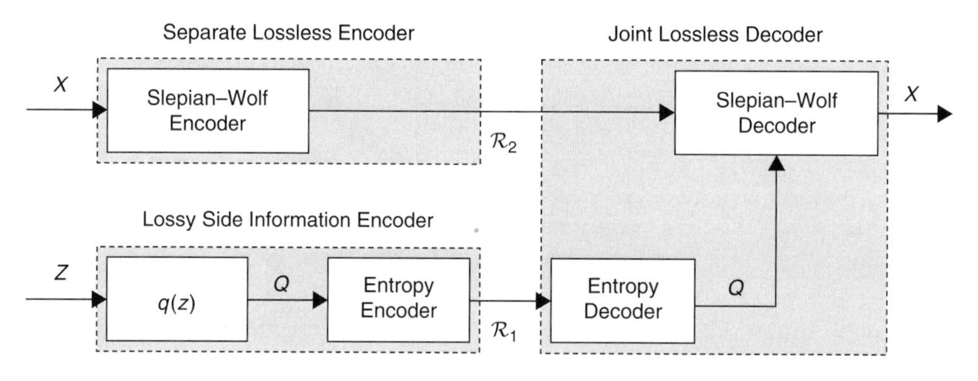

FIGURE 3.4

Quantization of side information.

Aside from particularizations and variations of distributed coding problems such as quantization of side information, quantization with side information at the encoder, broadcast with side information and an extension of the Blahut-Arimoto algorithm to noisy WZ coding, we find that our theoretical framework also unifies apparently unrelated problems such as the bottleneck method and Gauss mixture modeling. An extensive list of examples appears in [31].

3.3 OPTIMAL QUANTIZER DESIGN

In this section we establish necessary conditions for the optimal quantization and reconstruction functions, analogous to the nearest neighbor and centroid condition found in conventional, nondistributed quantization. Each necessary condition may be interpreted as the solution to a Bayesian decision problem. They will be expressed in terms of conditional cost measures, defined below, which play the role of conditional risks in Bayesian decision theory. These conditions will be used later to develop an extension of the Lloyd algorithm suitable for distributed quantization. A detailed, rigorous analysis of these conditions and the derived algorithm is presented in [31].

3.3.1 Optimality Conditions

The *conditional cost measure for encoder i* is defined as

$$\tilde{c}_i(q_{i\cdot}, z_i) = \mathrm{E}\left[\left[c(Q, X, \hat{x}(Q, Y), Y, Z)\right]_{Q_{i\cdot}=q_{i\cdot}} | z_i\right],$$

where $[\text{expression}]_{\text{substitution}}$ denotes substitution in an expression, that is, the expression of the definition is evaluated at $q_{i\cdot}$ and conditioned on the event $\{Z_i = z_i\}$. Observe that the conditional cost measure is completely determined by the joint distribution of X, Y, and Z, the cost measure $c(q,x,\hat{x},y,z)$, the quantization at other encoders $p_{Q_{i'\cdot}|Z_{i'}}(q_{i'\cdot},z_{i'})$, and all the reconstruction functions, grouped as $\hat{x}(q,y)$. Using the fact that $Q_i = q_i(Z_i)$ and iterated expectation, it is easy to see that for *each* sender i, $\mathrm{E}\,\tilde{c}_i(Q_{i\cdot}, Z_i) = \mathcal{C}$. In terms of Bayesian decision theory, the expectation of conditional risks gives the overall Bayes risk. This key property leads to the necessary optimality condition for the quantizer at sender i, which may be informally expressed as:

$$q^*_{i\cdot}(Z_i) \overset{\text{a.s.}}{=} \arg\min_{q_{i\cdot}} \tilde{c}_i(q_{i\cdot}, Z_i), \tag{3.1}$$

provided that a minimum exists.

Similarly to the sender cost measures, the *conditional cost measure for decoder j* is

$$\tilde{\tilde{c}}_j(q_{\cdot j},\hat{x}_j,y_j) = \mathrm{E}[[c(Q,X,\hat{X},Y,Z)]_{\hat{X}_j=\hat{x}_j}|q_{\cdot j},y_j].$$

Observe that the defining expression is evaluated at $\hat{x}_j$ and conditioned on the joint event $\{Q_{\cdot j}=q_{\cdot j},\ Y_j=y_j\}$, and it is completely determined by the joint distribution of X, Y, and Z, the cost measure $c(q,x,\hat{x},y,z)$, all quantizers, and the reconstruction functions at other decoders $\hat{x}_{j'}(q_{\cdot j'},y_{j'})$. As in the case of the sender cost measures, it can be shown that for each j, $\mathrm{E}\,\tilde{\tilde{c}}_j(Q_{\cdot j},\hat{X}_j,Y_j) = \mathcal{C}$. From this key property it follows that an optimal reconstruction function must satisfy

$$\hat{x}^*_j(Q_{\cdot j},Y_j) \overset{\text{a.s.}}{=} \arg\min_{\hat{x}_j} \tilde{\tilde{c}}_j(Q_{\cdot j},\hat{x}_j,Y_j), \tag{3.2}$$

provided that a minimum exists, for each pair $(q_{\cdot j},y_j)$ satisfying $p_{Q_{\cdot j}|Y_j}(q_{\cdot j}|y_j)>0$ ($\hat{x}^*_j(q_{\cdot j},y_j)$ can be arbitrarily defined elsewhere).

3.3.2 Lloyd Algorithm for Distributed Quantization

The necessary optimality conditions (3.1) and (3.2), together with the rate update property defining rate measures, suggest an alternating optimization algorithm that extends the Lloyd algorithm to the quantizer design problem considered in this work:

1. Initialization: For each $i=1,\ldots,m$, $j=1,\ldots,n$, choose initial quantizers $(q^{(1)}_{i\cdot}(z_i))_i$, and initial reconstruction functions $(\hat{x}^{(1)}_j(q_{\cdot j},y_j))_j$. Set $k=1$ and $\mathcal{C}^{(0)}=\infty$.

2. Cost measure update: Update the cost measure $c^{(k)}(q,x,\hat{x},y,z)$, completely determined by probability distributions involving $Q^{(k)}$, X, $\hat{X}$, Y, and Z.

3. Convergence check: Compute the expected cost $\mathcal{C}^{(k)}$ associated with the current quantizers $(q^{(k)}_{i\cdot}(z_i))_i$, reconstruction functions $\hat{x}^{(k)}_j(q_{\cdot j},y_j)$, and cost measure $c^{(k)}(q,x,\hat{x},y,z)$. Depending on its value with respect to $\mathcal{C}^{(k-1)}$, continue or stop.

4. Quantization update: For each i, obtain the next optimal quantizer $(q_{i\cdot}^{(k+1)}(z_i))_i$, in terms of the most current quantizers with index $i' \neq i$, all reconstruction functions $\hat{x}_j^{(k)}(q_{\cdot j}, y_j)$, and the cost measure $c^{(k)}(q, x, \hat{x}, y, z)$.
5. Reconstruction update: For each j, obtain the next optimal reconstruction function $\hat{x}_j^{(k+1)}(q_{\cdot j}, y_j)$, given the most current version of the reconstruction functions with index $j' \neq j$, all quantizers $(q_{i\cdot}^{(k+1)}(z_i))_i$, and the cost measure $c^{(k)}(q, x, \hat{x}, y, z)$.
6. Increment k and go back to 2.

It can be proved that the sequence of costs $\mathcal{C}^{(k)}$ in the above algorithm is nonincreasing, and since it is nonnegative, it converges. In addition, any quantizer satisfying the optimality conditions (3.1) and (3.2), without ambiguity in any of the minimizations involved, is a fixed point of the algorithm. Even though these properties do not imply per se that the cost converges to a local or global minimum, the experimental results in the next section show good convergence properties, especially when the algorithm is combined with genetic search for initialization. Many variations on the algorithm are possible, such as constraining the reconstruction function, for instance, imposing linearity in Y for each Q when the distortion is the mean-squared error of the estimate $\hat{X}$, or any of the variations mentioned in [31, 37].

3.4 EXPERIMENTAL RESULTS

In the following section we report experimental results obtained by applying the extension of the Lloyd algorithm for distributed coding described in Section 3.3.2 to the WZ coding problem. We consider a simple, intuitive case, represented in Figure 3.5. Let $X \sim \mathcal{N}(0, 1)$ and $N_Y \sim \mathcal{N}(0, 1/\gamma_Y)$ be independent, and let $Y = X + N_Y$. We wish to design scalar WZ quantizers minimizing the Lagrangian cost $\mathcal{C} = \mathcal{D} + \lambda\,\mathcal{R}$, where the distortion is the mean-squared error (MSE) $\mathcal{D} = \mathrm{E}\,\|X - \hat{X}\|^2$, and the rate is that required by an ideal SW codec, $\mathcal{R} = \mathrm{H}(Q|\,Y)$. The information-theoretic distortion-rate function

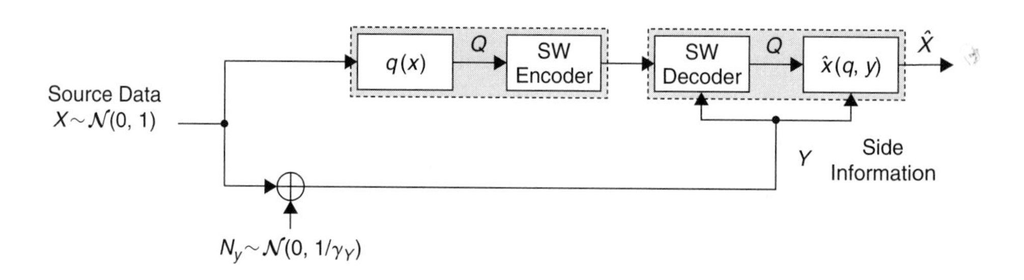

FIGURE 3.5

Experimental setup for clean WZ quantization.

for i.i.d. pairs drawn according to the statistics of X, Y is $\mathcal{D}^{\mathrm{WZ}}_{X|Y}(\mathcal{R}) = \sigma^2_{X|Y}\, 2^{-2\mathcal{R}}$ [42, §I.C](also [33,Theorem 2]).

For the case $\gamma_Y = 10$ and several values of λ, scalar WZ quantizers and reconstruction functions were designed using the extension of the Lloyd algorithm developed in this work. A simple genetic search method was combined with the Lloyd algorithm to select initial quantizers based on their cost after convergence. The algorithm was applied to a fine discretization of the joint PDF of X, Y, as a PMF with approximately $2.0 \cdot 10^5$ points (x, y) contained in a two-dimensional ellipsoid of probability $1 - 10^{-4}$ (according to the reference PDF), producing 919 different values for X and 919 different values for Y.

The corresponding distortion-rate points are shown in Figure 3.6, along with the information-theoretic distortion-rate function and the distortion bound $\mathcal{D}_0 = \sigma^2_{X|Y}$. Recall that in nondistributed quantization, at high rates, asymptotically optimal scalar quantizers introduce a distortion penalty of $10\log_{10}\frac{\pi e}{6} \simeq 1.53$ dB with respect to the information-theoretic bound [16]. Note that the experimental results obtained for this particular WZ quantization setup exhibit the same behavior.

Two scalar WZ quantizers for $\mathcal{R} \simeq 0.55$ and $\mathcal{R} \simeq 0.98$ bit are depicted in Figure 3.7. Conventional high-rate quantization theory shows that uniform quantizers are asymptotically optimal. Note that quantizer (b), corresponding to a higher rate than quantizer (a), is uniform, and perhaps surprisingly, no quantization index is reused. This experimental observation, together with the 1.53 dB distortion gap mentioned earlier,

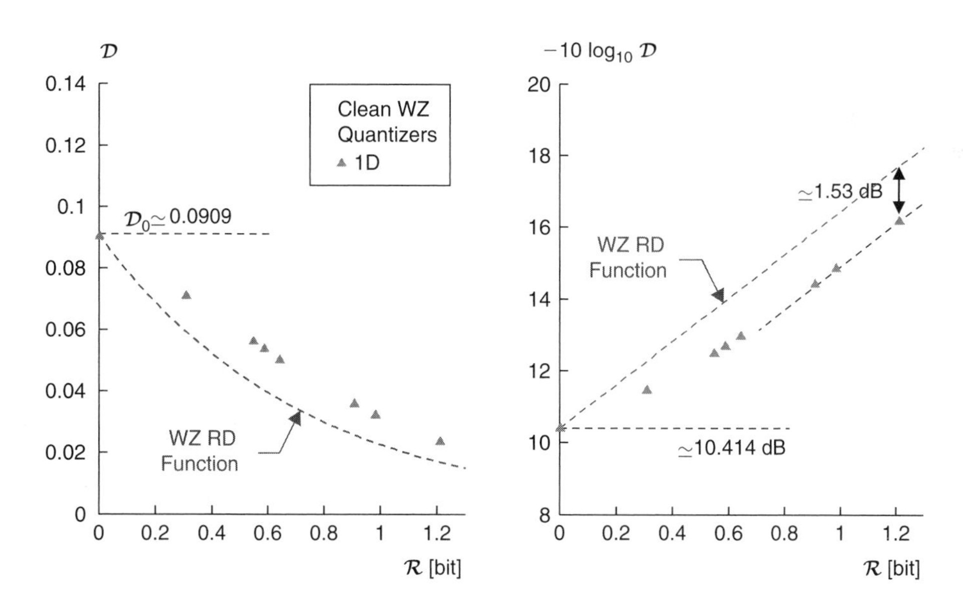

FIGURE 3.6

Distortion-rate performance of optimized scalar WZ quantizers with SW coding. $\mathcal{R} = \frac{1}{k}\mathrm{H}(Q\,|\,Y)$, $X \sim \mathcal{N}(0,1)$, $N_Y \sim \mathcal{N}(0,1/10)$ independent, $Y = X + N_Y$.

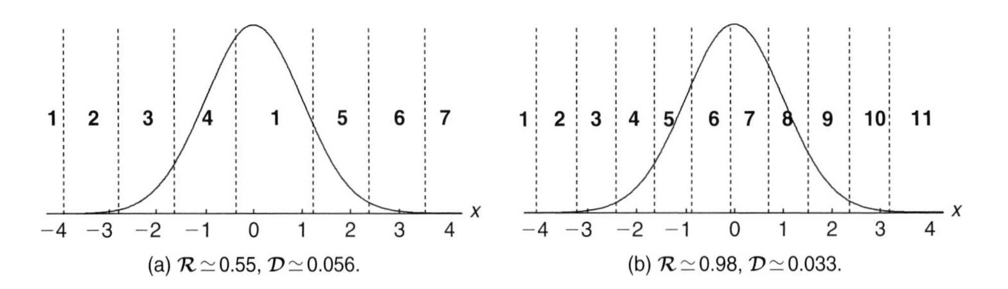

FIGURE 3.7

WZ quantizers with SW coding obtained by the Lloyd algorithm. $\mathcal{R}=\frac{1}{k}\mathrm{H}(Q|Y)$, $X\sim\mathcal{N}(0,1)$, $N_Y\sim\mathcal{N}(0,1/10)$ independent, $Y=X+N_Y$.

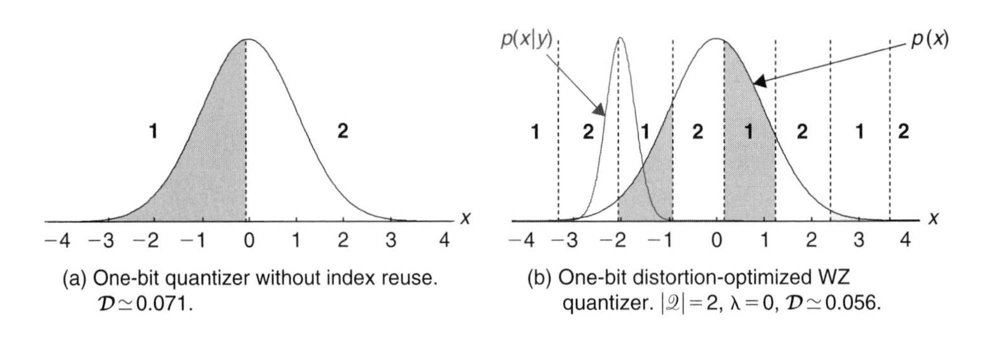

FIGURE 3.8

Example of distortion reduction in WZ quantization due to index reuse. $X\sim\mathcal{N}(0,1)$, $N_Y\sim\mathcal{N}(0,1/10)$ independent, $Y=X+N_Y$.

will be confirmed by the high-rate quantization theory presented in Section 3.5. These experiments are then revisited in Section 3.6 to explain these facts, show consistency with the theory developed, and confirm the usefulness of our extension of the Lloyd algorithm.

We observed that when SW coding is used, as long as the rate was sufficiently high, hardly any of the quantizers obtained in our experiments reused quantization indices. On the other hand, it may seem intuitive that disconnected quantization regions may improve the rate with little distortion impact due to the presence of side information at the decoder. In order to illustrate the benefit of index reuse in distributed quantization *without* SW coding, the extended Lloyd algorithm was run on several initial quantizers with $|\mathcal{Q}|=2$ quantization indices, setting $\lambda=0$. This models the situation of fixed-length coding, where the rate is fixed to one bit, and the distortion is to be minimized. As shown in Figure 3.8, the quantizer (b) obtained using the Lloyd algorithm leads to smaller distortion than a simple one-bit quantizer (a) with no index reuse. The conditional PDF of X given Y is superimposed to show that it is narrow enough for

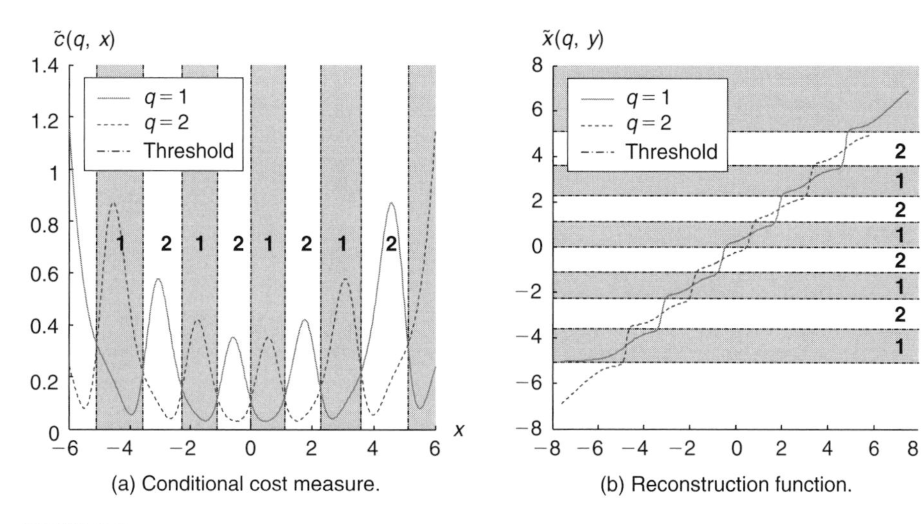

(a) Conditional cost measure.

(b) Reconstruction function.

FIGURE 3.9

Example of conditional cost and reconstruction function for WZ quantization with fixed-length coding. $X \sim \mathcal{N}(0,1)$, $N_Y \sim \mathcal{N}(0,1/10)$ independent, $Y = X + N_Y$. $|\mathcal{Q}| = 2$, $\lambda = 0$, $\mathcal{D} \simeq 0.056$.

the index repetition to have negligible impact on the distortion ($\sigma^2_{X|Y} = 1/(1+\gamma_Y) \simeq 0.091$).

Figure 3.9 shows an example of encoder conditional cost measures and reconstruction functions under the heavy index repetition scenario of fixed-length coding. Figure 3.9(a) clearly illustrates the quantizer optimality condition $q^*(x) = \arg\min_q \tilde{c}(q,x)$, and Figure 3.9(b), the centroid condition $\hat{x}(q,y) = \mathrm{E}[X|q,y]$. Observe that it is perfectly possible for the estimated reconstruction to fall on a region corresponding to a different quantization index, just as the best MSE estimate of a uniform binary r.v. is 0.5, despite being an impossible outcome.

3.5 HIGH-RATE DISTRIBUTED QUANTIZATION

In this section we investigate the properties of entropy-constrained optimal quantizers for distributed source coding at high rates, when MSE is used as a distortion measure. Increasingly complex settings are considered. We start with WZ quantization of directly observed data, continue by introducing noisy observations, and conclude with the general case of network distributed coding of noisy observations, which involves not only several encoders, but also several decoders with access to side information. Proofs and technical details are provided in [31, 32, 35, 36].

Our approach is inspired by traditional developments of high-rate quantization theory, specifically Bennett's distortion approximation [1, 28], Gish and Pierce's rate approximation [17], and Gersho's conjecture [15]. The reason why a less rigorous approach is followed is first and foremost simplicity and readability, but also the fact that rigorous nondistributed high-rate quantization theory is not without technical gaps. A comprehensive and detailed discussion of both heuristic and rigorous studies, along with the challenges still faced by the latter appears in [19] (§IV.H is particularly relevant). However, we believe that the ideas contained in this section might help to derive completely rigorous results on high-rate distributed quantization along the lines of [2, 7, 18, 24, 46], to cite a few.

3.5.1 High-rate WZ Quantization of Clean Sources

We study the properties of high-rate quantizers for the WZ coding setting in Figure 3.10. This is a special case of the network of Section 3.2.2 with $m=1$ encoders and $n=1$ decoders, when in addition the source data X is directly observed, that is, $X=Z$. Suppose that X takes values in the k-dimensional Euclidean space, $\mathbb{R}^k$. MSE is used as a distortion measure; thus the expected distortion per sample is $\mathcal{D}=\frac{1}{k}\ \mathrm{E}\ \|X-\hat{X}\|^2$. The expected rate per sample is defined as $\mathcal{R}=\frac{1}{k}\ \mathrm{H}(Q|\,Y)$ [37]. Recall from Section 3.2.3 that this rate models ideal SW coding of the index Q with side information Y, with negligible decoding error probability and rate redundancy.

We emphasize that the quantizer only has access to the source data, not to the side information. However, the joint statistics of X and Y are assumed to be known, and are exploited in the design of $q(x)$ and $\hat{x}(q,y)$. Consistently with Section 3.2.4, we consider the problem of characterizing the quantization and reconstruction functions that minimize the expected Lagrangian cost $\mathcal{C}=\mathcal{D}+\lambda\,\mathcal{R}$, with λ a nonnegative real number, for high rate $\mathcal{R}$.

We shall use the term *uniform tessellating quantizer* in reference to quantizers whose quantization regions are possibly rotated versions of a common convex polytope, with equal volume. Lattice quantizers are, strictly speaking, a particular case. In the following results, Gersho's conjecture for nondistributed quantizers, which allows rotations, will be shown to imply that optimal WZ quantizers are also tessellating quantizers, and the uniformity of the cell volume will be proved as well. M_k denotes the minimum normalized moment of inertia of the convex polytopes tessellating $\mathbb{R}^k$ (e.g., $M_1=1/12$) [16].

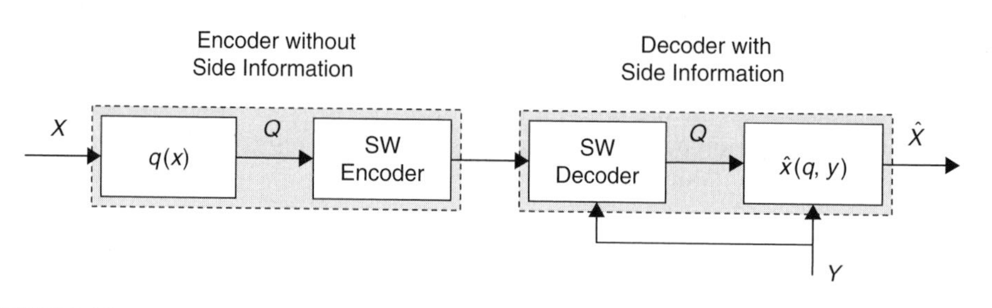

FIGURE 3.10

WZ quantization.

Suppose that *for each value* y in the alphabet of Y, the statistics of X given $Y = y$ are such that the conditional differential entropy $\mathrm{h}(X|y)$ exists and is finite. Suppose further that *for each* y, there exists an asymptotically optimal entropy-constrained uniform tessellating quantizer of x, $q(x|y)$, with rate $\mathcal{R}_{X|Y}(y)$ and distortion $\mathcal{D}_{X|Y}(y)$, with no two cells assigned to the same index and with cell volume $V(y) > 0$, which satisfies, for large $\mathcal{R}_{X|Y}(y)$,

$$\mathcal{D}_{X|Y}(y) \simeq M_k \, V(y)^{\frac{2}{k}}, \tag{3.3}$$

$$\mathcal{R}_{X|Y}(y) \simeq \tfrac{1}{k}\left(\mathrm{h}(X|y) - \log_2 V(y)\right), \tag{3.4}$$

$$\mathcal{D}_{X|Y}(y) \simeq M_k \, 2^{\frac{2}{k}\mathrm{h}(X|y)} \, 2^{-2\mathcal{R}_{X|Y}(y)}. \tag{3.5}$$

These hypotheses hold if the Bennett assumptions [1,28] apply to the conditional PDF $p_{X|Y}(x|y)$ *for each* value of the side information y, and if Gersho's conjecture [15] is true (known to be the case for $k = 1$), among other technical conditions, mentioned in [19].

We claim that under these assumptions, there exists an asymptotically optimal quantizer $q(x)$ for large $\mathcal{R}$, for the WZ coding setting considered such that:

i. $q(x)$ is a uniform tessellating quantizer with minimum moment of inertia M_k and cell volume V.

ii. No two cells of the partition defined by $q(x)$ need to be mapped into the same quantization index.

iii. The rate and distortion satisfy

$$\mathcal{D} \simeq M_k \, V^{\frac{2}{k}}, \tag{3.6}$$

$$\mathcal{R} \simeq \tfrac{1}{k}\left(\mathrm{h}(X|Y) - \log_2 V\right), \tag{3.7}$$

$$\mathcal{D} \simeq M_k \, 2^{\frac{2}{k}\mathrm{h}(X|Y)} \, 2^{-2\mathcal{R}}. \tag{3.8}$$

This fact can be proven using the quantization setting in Figure 3.11, which we shall refer to as a conditional quantizer, along with an argument of optimal rate allocation for $q(x|y)$, where $q(x|y)$ can be regarded as a quantizer on the values x and y taken by the source data and the side information, or a family of quantizers on x indexed by y. In this case, the side information Y is available to the sender, and the design of the quantizer $q(x|y)$ on x, for each value y, is a nondistributed entropy-constrained quantization problem. More precisely, for all y define

$$\mathcal{D}_{X|Y}(y) = \tfrac{1}{k}\,\mathrm{E}[\|X - \hat{X}\|^2 | y],$$

$$\mathcal{R}_{X|Y}(y) = \tfrac{1}{k}\,\mathrm{H}(Q|y),$$

$$\mathcal{C}_{X|Y}(y) = \mathcal{D}_{X|Y}(y) + \lambda\,\mathcal{R}_{X|Y}(y).$$

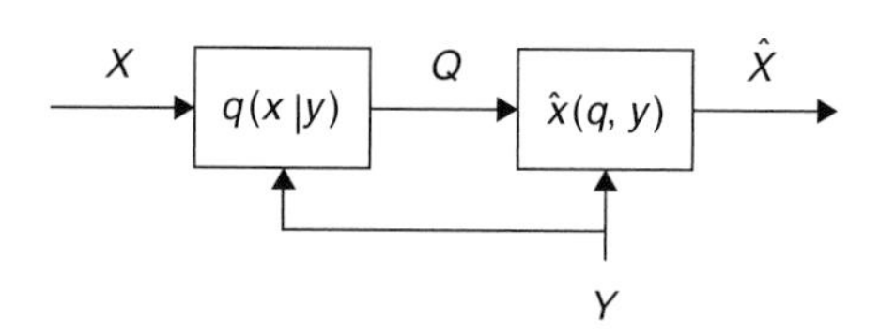

FIGURE 3.11

Conditional quantizer.

By iterated expectation, $\mathcal{D} = \mathrm{E}\,\mathcal{D}_{X|Y}(Y)$ and $\mathcal{R} = \mathrm{E}\,\mathcal{R}_{X|Y}(Y)$; thus the overall cost satisfies $\mathcal{C} = \mathrm{E}\,\mathcal{C}_{X|Y}(Y)$. As a consequence, a family of quantizers $q(x|y)$ minimizing $\mathcal{C}_{X|Y}(y)$ for each y also minimizes $\mathcal{C}$.

Since $\mathcal{C}_{X|Y}(y)$ is a convex function of $\mathcal{R}_{X|Y}(y)$ for all y, it has a global minimum where its derivative vanishes, or equivalently, at $\mathcal{R}_{X|Y}(y)$ such that $\lambda \simeq 2 \ln 2\, \mathcal{D}_{X|Y}(y)$. Suppose that λ is small enough for $\mathcal{R}_{X|Y}(y)$ to be large and for the approximations (3.3)–(3.5) to hold, for each y. Then, all quantizers $q(x|y)$ introduce the same distortion (proportional to λ) and consequently have a common cell volume $V(y) \simeq V$. This, together with the fact that $\mathrm{E}_Y[\mathrm{h}(X|y)]_{y=Y} = \mathrm{h}(X|Y)$, implies (3.6)–(3.8). Provided that a translation of the partition defined by $q(x|y)$ affects neither the distortion nor the rate, all uniform tessellating quantizers $q(x|y)$ may be set to be (approximately) the same, which we denote by $q(x)$. Since none of the quantizers $q(x|y)$ maps two cells into the same indices, neither does $q(x)$. Now, since $q(x)$ is asymptotically optimal for the conditional quantizer and does not depend on y, it is also optimal for the WZ quantizer in Figure 3.10. This proves our claim.

We would like to remark that Equation (3.8) means that, asymptotically, there is no loss in performance by not having access to the side information in the quantization. In addition, since index repetition is not required, the distortion (3.6) would be asymptotically the same if the reconstruction $\hat{x}(q, y)$ were of the form $\hat{x}(q) = \mathrm{E}[X|q]$. This means that under the hypotheses of this section, asymptotically, there is a quantizer that leads to no loss in performance by ignoring the side information in the reconstruction (but it is still used by the SW decoder).

Finally, let X and Y be jointly Gaussian random vectors. Recall that the conditional covariance $\Sigma_{X|Y}$ does not depend on y, that $\mathrm{h}(X|Y) = \frac{1}{2} \log_2 \left((2\pi e)^k \det \Sigma_{X|Y}\right)$, and that $M_k \to \frac{1}{2\pi e}$ [49]. Then, on account of the results of this section, for large $\mathcal{R}$,

$$\mathcal{D} \simeq M_k\, 2\pi e\, (\det \Sigma_{X|Y})^{\frac{1}{k}}\, 2^{-2\mathcal{R}} \underset{[k\to\infty]}{\longrightarrow} (\det \Sigma_{X|Y})^{\frac{1}{k}}\, 2^{-2\mathcal{R}}. \tag{3.9}$$

3.5.2 High-rate WZ Quantization of Noisy Sources

We will now examine the properties of high-rate quantizers of a noisy source with side information at the decoder, as illustrated in Figure 3.12, which we will refer to as WZ

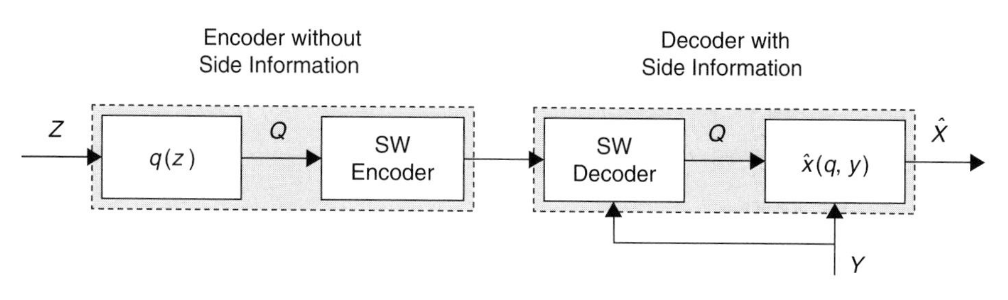

FIGURE 3.12

WZ quantization of a noisy source.

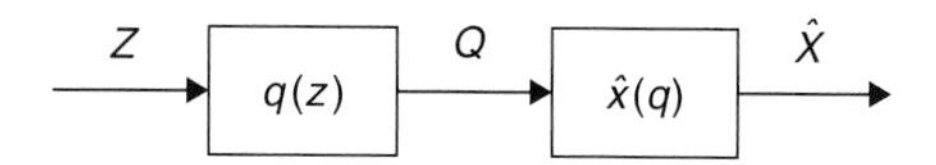

FIGURE 3.13

Quantization of a noisy source without side information.

quantizers of a noisy source. This is precisely the network in Section 3.2.2 with $m = 1$ encoders and $n = 1$ decoders. Unlike the problem in Section 3.5.1, this time a noisy observation Z is quantized in lieu of the unseen source data X. Clearly, this includes the problem in the previous section as the particular case $Z = X$.

The hypotheses of the previous section directly carry over to the more general problem studied here. Assume that X takes values in $\mathbb{R}^k$. The expected distortion *per sample of the unseen source* is defined as $\mathcal{D} = \frac{1}{k}\,\mathrm{E}\,\|X - \hat{X}\|^2$. Suppose further that the coding of the index Q is carried out by an ideal SW coder, at a rate per sample $\mathcal{R} = \frac{1}{k}\,\mathrm{H}(Q|\,Y)$. We emphasize that the quantizer only has access to the observation, not to the source data or the side information. However, the joint statistics of X, Y, and Z can be exploited in the design of $q(z)$ and $\hat{x}(q,y)$. Consistently with Section 3.2.4, we consider the problem of characterizing the quantizers and reconstruction functions that minimize the expected Lagrangian cost $\mathcal{C} = \mathcal{D} + \lambda\mathcal{R}$, with λ a nonnegative real number, for high-rate $\mathcal{R}$.

3.5.2.1 Nondistributed Case

We start by considering the simpler case of quantization of a noisy source *without* side information, depicted in Figure 3.13. The following extends the main result of [10, 41] to entropy-constrained quantization, valid for any rate $\mathcal{R} = \mathrm{H}(Q)$, not necessarily high.

Define $\bar{x}(z) = \mathrm{E}[X|\,z]$, the best MSE estimator of X given Z, and $\bar{X} = \bar{x}(Z)$. We show in [32, 35] that for any nonnegative λ and any Lagrangian-cost optimal quantizer of a noisy source *without* side information (Figure 3.13), there exists an implementation with the same cost in two steps:

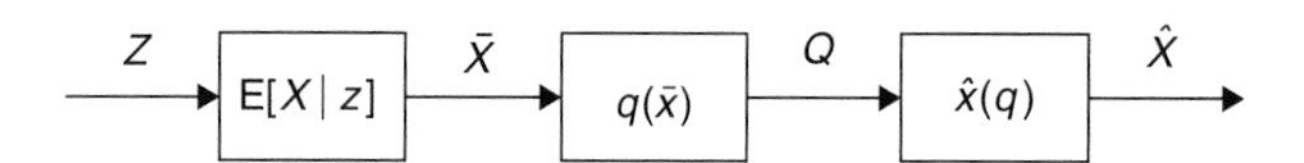

FIGURE 3.14

Optimal implementation of MSE quantization of a noisy source without side information.

1. Obtain the minimum MSE estimate $\bar{X}$.
2. Quantize the estimate $\bar{X}$ regarded as a clean source, using a quantizer $q(\bar{x})$ and a reconstruction function $\hat{x}(q)$, minimizing $\mathrm{E}\ \|\bar{X} - \hat{X}\|^2 + \lambda\ \mathrm{H}(Q)$.

This is illustrated in Figure 3.14. Furthermore, the total distortion per sample is

$$\mathcal{D} = \tfrac{1}{k}\left(\mathrm{E}\,\mathrm{tr}\,\mathrm{Cov}[X|Z] + \mathrm{E}\ \|\bar{X} - \hat{X}\|^2\right), \tag{3.10}$$

where the first term is the MSE of the estimation step.

The following observation is an immediate consequence of the previous, very intuitive result, and conventional theory of high-rate quantization of clean sources. Assume now that $\mathrm{h}(\bar{X}) < \infty$ and that there exists a uniform tessellating quantizer $q(\bar{x})$ of $\bar{X}$ with cell volume V that is asymptotically optimal in Lagrangian cost at high rates. Then, there exists an asymptotically optimal quantizer $q(z)$ of a noisy source in the setting of Figure 3.13 such that:

i. An asymptotically optimal implementation of $q(z)$ is that represented in Figure 3.14, with a uniform tessellating quantizer $q(\bar{x})$ having cell volume V.

ii. The rate and distortion per sample satisfy

$$\mathcal{D} \simeq \tfrac{1}{k}\,\mathrm{E}\,\mathrm{tr}\,\mathrm{Cov}[X|Z] + M_k\, V^{\frac{2}{k}},$$

$$\mathcal{R} \simeq \tfrac{1}{k}\left(\mathrm{h}(\bar{X}) - \log_2 V\right),$$

$$\mathcal{D} \simeq \tfrac{1}{k}\,\mathrm{E}\,\mathrm{tr}\,\mathrm{Cov}[X|Z] + M_k\, 2^{\frac{2}{k}\mathrm{h}(\bar{X})}\, 2^{-2\mathcal{R}}.$$

3.5.2.2 *Distributed Case*

We are now ready to consider WZ quantization of a noisy source, as shown in Figure 3.12. Define $\bar{x}(y,z) = \mathrm{E}[X|y,z]$, the best MSE estimator of X given Y and Z, $\bar{X} = \bar{x}(Y,Z)$, and $\mathcal{D}_\infty = \frac{1}{k}\,\mathrm{E}\,\mathrm{tr}\,\mathrm{Cov}[X|Y,Z]$. The following fact is an extension of the results on high-rate WZ quantization in Section 3.5, to noisy sources.

Suppose that the conditional expectation function $\bar{x}(y,z)$ is additively separable, that is, $\bar{x}(y,z) = \bar{x}_Y(y) + \bar{x}_Z(z)$, and define $\bar{X}_Z = \bar{x}_Z(Z)$. Suppose further that *for each value* y in the alphabet of Y, $\mathrm{h}(\bar{X}|y) < \infty$, and that there exists a uniform tessellating quantizer $q(\bar{x}, y)$ of $\bar{X}$, with no two cells assigned to the same index and cell volume $V(y) > 0$, with rate $\mathcal{R}_{\bar{X}|Y}(y)$ and distortion $\mathcal{D}_{\bar{X}|Y}(y)$, such that, at high rates, it is asymptotically optimal in Lagrangian cost and

$$\mathcal{D}_{\bar{X}|Y}(y) \simeq M_k \, V(y)^{\frac{2}{k}},$$
$$\mathcal{R}_{\bar{X}|Y}(y) \simeq \tfrac{1}{k}\left(\mathrm{h}(\bar{X}|y) - \log_2 V(y)\right),$$
$$\mathcal{D}_{\bar{X}|Y}(y) \simeq M_k \, 2^{\frac{2}{k}\mathrm{h}(\bar{X}|y)} \, 2^{-2\mathcal{R}_{\bar{X}|Y}(y)}.$$

It was proved in [35, 36] that under these assumptions there exists an asymptotically optimal quantizer $q(z)$ for large $\mathcal{R}$, for the WZ quantization setting represented in Figure 3.12, such that:

i. $q(z)$ can be implemented as an estimator $\bar{x}_Z(z)$ followed by a uniform tessellating quantizer $q(\bar{x}_Z)$ with cell volume V.

ii. No two cells of the partition defined by $q(\bar{x}_Z)$ need to be mapped into the same quantization index.

iii. The rate and distortion per sample satisfy

$$\mathcal{D} \simeq \mathcal{D}_\infty + M_k \, V^{\frac{2}{k}}, \tag{3.11}$$
$$\mathcal{R} \simeq \tfrac{1}{k}\left(\mathrm{h}(\bar{X}|\,Y) - \log_2 V\right), \tag{3.12}$$
$$\mathcal{D} \simeq \mathcal{D}_\infty + M_k \, 2^{\frac{2}{k}\mathrm{h}(\bar{X}|Y)} \, 2^{-2\mathcal{R}}. \tag{3.13}$$

iv. $\mathrm{h}(\bar{X}|\,Y) = \mathrm{h}(\bar{X}_Z|\,Y)$.

Clearly, this is a generalization of the clean case in Section 3.5.1, since $Z = X$ implies $\bar{x}(y, z) = \bar{x}_Z(z) = z$, trivially additively separable. Observe that, if the estimator $\bar{x}(y, z)$ is additively separable, there is no asymptotic loss in performance by not using the side information at the encoder.

The proof of this fact is very similar to that for clean sources in Section 3.5.1. More precisely, it analyzes the case when the side information is available at the encoder. The main difference is that it applies the separation result obtained for the noisy, nondistributed case, to the noisy, conditional setting, for each value of the side information.

The case in which X can be written as $X = f(Y) + g(Z) + N$, for any (measurable) functions f, g and any random variable N with $\mathrm{E}[N|\,y, z]$ constant with (y, z), gives an example of an additively separable estimator. This includes the case in which X, Y, and Z are jointly Gaussian. Furthermore, in the Gaussian case, since $\bar{x}_Z(z)$ is an affine function and $q(\bar{x}_Z)$ is a uniform tessellating quantizer, the overall quantizer $q(\bar{x}_Z(z))$ is also a uniform tessellating quantizer, and if Y and Z are uncorrelated, then $\bar{x}_Y(y) = \mathrm{E}[X|\,y]$ and $\bar{x}_Z(z) = \mathrm{E}[X|\,z]$, which does *not* hold in general.

Similarly to the observation made in the clean case, there is a WZ quantizer $q(\bar{x}_Z)$ that leads to no asymptotic loss in performance if the reconstruction function is

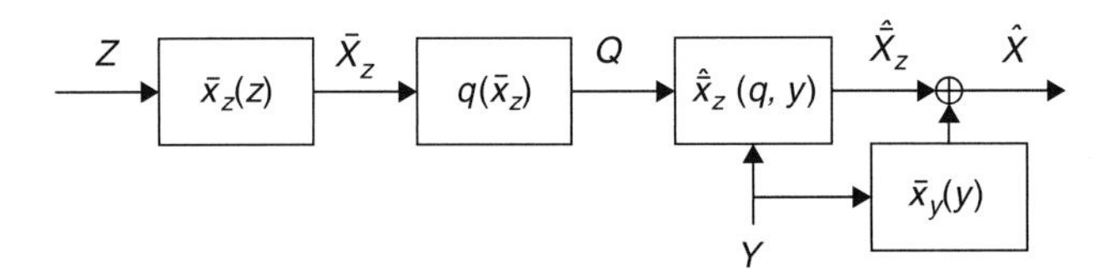

FIGURE 3.15

Asymptotically optimal implementation of MSE WZ quantization of a noisy source with additively separable $\bar{x}(y,z)$.

$\hat{x}(q,y) = \hat{\bar{x}}_Z(q) + \bar{x}_Y(y)$, where $\hat{\bar{x}}_Z(q)$ are the centroids of $q(\bar{x}_Z)$. Consequently, the WZ quantization setting of Figure 3.12 can be implemented as depicted in Figure 3.15, where $\hat{\bar{x}}_Z(q,y)$ can be made independent from y without asymptotic loss in performance, so that the pair $q(\bar{x}_Z)$, $\hat{\bar{x}}_Z(q)$ form a uniform tessellating quantizer and reconstructor for $\bar{X}_Z$.

3.5.3 High-rate Network Distributed Quantization

Thus far we have characterized optimal WZ quantization for clean and noisy sources at high rates, which addresses the special case of $m=1$ encoder and $n=1$ decoder for the network of Section 3.2.2. Next, we study two more general settings. First, we consider the clean and noisy cases with an arbitrary number of encoders but a single decoder, and second, the general case with arbitrary m and n represented in Figure 3.3.

3.5.3.1 *Clean, Single-decoder Case*

Consider the distributed quantization problem in Figure 3.16, where the data generated by m sources is observed directly, encoded separately, and decoded jointly with side information. For all $i=1,\ldots,m$, let X_i be a $\mathbb{R}^{k_i}$-valued r.v., representing source data in a k_i-dimensional Euclidean space, available only at encoder i. Define $k_\mathrm{T} = \sum_i k_i$, and $X = (X_i)_{i=1}^m$ (hence X is a r.v. in $\mathbb{R}^{k_\mathrm{T}}$). All quantization indices $Q = (Q_i)_{i=1}^m$ are used at the decoder to jointly estimate the source data. The expected distortion per sample is $\mathcal{D} = \frac{1}{k_\mathrm{T}} \mathrm{E}\,\|X - \hat{X}\|^2$. Coding of the quantization indices is carried out by an ideal SW codec, with expected rate per sample $\mathcal{R} = \frac{1}{k_\mathrm{T}}\,\mathrm{H}(Q|Y)$. Define M_{k_i} as Gersho's constant corresponding to the dimension k_i of X_i, and $\bar{M} = \prod_i M_{k_i}^{k_i}$.

The following result is proved in [31]. The conceptual summary of the assumptions is that for each value of the side information Y, a.s., the conditional statistics of $X|y$ are well behaved, in the sense that one could design a set of quantizers $(q_i(x_i|y))_i$ for each value of the side information, with the usual properties of conventional, nondistributed high-rate theory. The result in question asserts that in the problem of distributed quantization with side information, depicted in Figure 3.16, for any rate $\mathcal{R}$ sufficiently high:

i. Each $q_i(x_i)$ is approximately a uniform tessellating quantizer with cell volume V_i and normalized moment of inertia M_{k_i}.

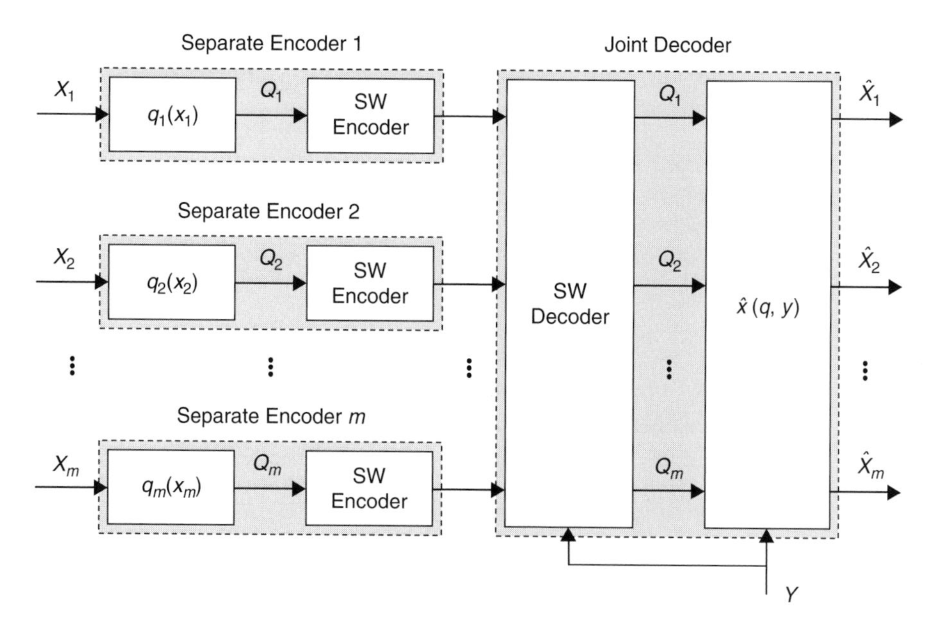

FIGURE 3.16

Distributed quantization of m clean sources with side information.

ii. No two cells of the partition defined by each $q_i(x_i)$ need to be mapped into the same quantization index.

iii. The distortion $\mathcal{D}_i$ introduced by each quantizer is approximately constant, that is, $\mathcal{D}_i \simeq \mathcal{D}$ for all i, and it satisfies $\mathcal{D}_i \simeq M_{k_i} V_i^{2/k_i}$.

iv. The overall distortion and the rate satisfy

$$\mathcal{D} \simeq \bar{M}\, V^{2/k_{\mathrm{T}}}, \qquad \mathcal{R} \simeq \frac{1}{k_{\mathrm{T}}}\left(\mathrm{h}(X|Y) - \log_2 V\right), \qquad \mathcal{D} \simeq \bar{M}\, 2^{\frac{2}{k_{\mathrm{T}}}\mathrm{h}(X|Y)}\, 2^{-2\mathcal{R}},$$

where $V = \prod_i V_i$ denotes the overall volume corresponding to the joint quantizer $q(x) = (q_i(x_i))_i$.

It is important to realize that this confirms that, asymptotically at high rates, there is no loss in performance by not having access to the side information in the quantization, and that the loss in performance due to separate encoding of the source data is completely determined by Gersho's constant. This implies that the rate-distortion loss due to distributed quantization with side information vanishes in the limit of high rates and high dimension.

3.5.3.2 Noisy, Single-decoder Case

The problem of distributed quantization of noisy sources with side information and a single decoder, depicted in Figure 3.17, turns out to be significantly more complex

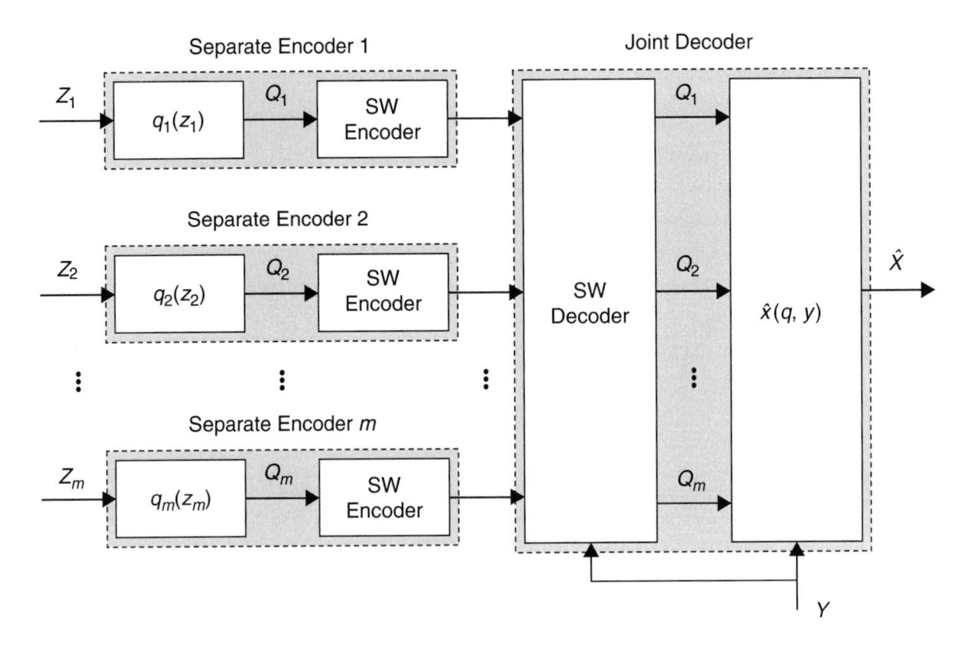

FIGURE 3.17

Distributed quantization of m noisy sources with side information.

when several encoders are involved. In fact, the analysis developed in [31] builds on the special case when the source data is the sum of the observations, in such a way that the clean case is no longer a special case of the noisy one, except for the WZ setting, that is, when $m=1$. The argument considers transformations of the observations that preserve optimal performance, conceptually not too distant from the separation result for noisy, nondistributed quantization, described in Section 3.5.2, albeit mathematically more sophisticated. Here, we present the main result.

As usual, X is a $\mathbb{R}^k$-valued r.v. We define $\mathcal{D}_\infty = \frac{1}{k} \operatorname{E} \operatorname{tr} \operatorname{Cov}[X | Y, Z]$. First, we make the following assumptions on the statistical dependence of X, Y, and Z:

i. There exist measurable functions $\bar{x}_Y : \mathcal{Y} \to \mathbb{R}^k$ and $\bar{x}_{Z_i} : \mathcal{Z}_i \to \mathbb{R}^k$, $i = 1, \ldots, m$, such that $\operatorname{E}[X | y, z] = \bar{x}_Y(y) + \sum_i \bar{x}_{Z_i}(z_i)$. Define $\bar{X}_Y = \bar{x}_Y(Y)$ and $\bar{X}_{Z_i} = \bar{x}_{Z_i}(Z_i)$, thus $\bar{X} = \bar{X}_Y + \sum_i \bar{X}_{Z_i}$.

ii. Either $m = 1$, or $\{Z_i\}_i$ are conditionally independent given Y, or the maps $\{x \mapsto \bar{x}_{Z_i}(x)\}_i$ are injective.

Furthermore, it is assumed that for almost every $y \in \mathcal{Y}$, the conditional statistics of $(\bar{X}_{Z_i})_i | y$ are well-behaved, in the sense of conventional, nondistributed high-rate quantization theory.

Under these assumptions, in the problem of noisy distributed quantization with side information, depicted in Figure 3.17, for any rate $\mathcal{R}$ sufficiently high:

i. There exists a collection of asymptotically optimal quantizers $(q_i(z_i))_i$, consisting of the transformations $(\bar{x}_{Z_i}(z_i))_i$, followed by approximately uniform tessellating quantizers with a common cell volume $V^{1/m}$ and normalized moment of inertia M_k, where V represents the volume of the corresponding joint tessellation.

ii. No two cells of the partition defined by each $q_i(z_i)$ need to be mapped into the same quantization index.

iii. The overall distortion and the rate satisfy

$$\mathcal{D} = \mathcal{D}_\infty + \bar{\mathcal{D}},$$

$$\frac{1}{m}\bar{\mathcal{D}} \simeq M_k\, V^{\frac{2}{km}},$$

$$\frac{1}{m}\mathcal{R} \simeq \frac{1}{km}\left(\mathrm{h}((\bar{X}_{Z_i})_{i=1}^m|Y) - \log_2 V\right),$$

$$\frac{1}{m}\bar{\mathcal{D}} \simeq M_k\, 2^{\frac{2}{km}\mathrm{h}((\bar{X}_{Z_i})_{i=1}^m|Y)}\, 2^{-2\frac{1}{m}\mathcal{R}}.$$

3.5.3.3 Network Case

Our final high-rate analysis is concerned with the general problem of network distributed quantization of noisy sources with side information, presented in Section 3.2.2 and depicted in Figure 3.3. We assume that it is possible to break down the network into $n \in \mathbb{Z}^+$ subnetworks such that:

i. The overall distortion $\mathcal{D}$ is a nonnegative linear combination of the distortions $\mathcal{D}_j$ of each subnetwork, and the overall rate $\mathcal{R}$ is a positive linear combination of the rates $\mathcal{R}_j$ of each subnetwork. Mathematically, $\mathcal{D} = \sum_j \delta_j \mathcal{D}_j$ and $\mathcal{R} = \sum_j \rho_j \mathcal{R}_j$, for some $\delta_j, \rho_j \in \mathbb{R}^+$.

ii. For each subnetwork j, the distortion and rate are (approximately) related by an exponential law of the form $\mathcal{D}_j = \mathcal{D}_{\infty j} + \bar{\mathcal{D}}_j$, where $\bar{\mathcal{D}}_j = \alpha_j\, 2^{-\mathcal{R}_j/\beta_j}$, for some $\mathcal{D}_{\infty j} \in \mathbb{R}$ and $\alpha_j, \beta_j \in \mathbb{R}^+$.

The first assumption holds, for example, when the overall distortion for the network distributed codec is a nonnegative linear combination of distortions corresponding to reconstructions at each decoder, and ideal SW codecs are used for lossless coding of the quantization indices. The second assumption holds approximately for each of the high-rate quantization problems studied in this section, including the complicated exponential law for the rate-distortion performance in the single-decoder noisy network, which did not obey the 6 dB/bit rule. Simply speaking, these assumptions mean that we can basically break down a network distributed coding problem into subproblems that satisfy the hypotheses of the cases considered thus far.

The next result characterizes the overall rate-distortion performance of the network distributed codec by means of an optimal rate allocation. Specifically, we consider the problem of minimizing $\mathcal{D}$ subject to a constraint on $\mathcal{R}$. We would like to remark that although it is possible to incorporate nonnegativity constraints on each of the rates $\mathcal{R}_j$ into the problem, and to solve it using the Karush–Kuhn–Tucker (KKT) conditions, our second assumption makes the result practical only at high rates, which implicitly introduce nonnegativity.

Define

$$\mathcal{D}_\infty = \sum_j \delta_j \mathcal{D}_{\infty j}, \qquad \bar{\mathcal{D}} = \sum_j \delta_j \bar{\mathcal{D}}_j,$$

$$\beta = \sum_j \beta_j \rho_j > 0, \qquad \text{and} \qquad \alpha = \beta \prod_j \left(\frac{\alpha_j \delta_j}{\beta_j \rho_j} \right)^{\frac{\beta_j \rho_j}{\beta}} > 0.$$

It is clear that $\mathcal{D} = \mathcal{D}_\infty + \bar{\mathcal{D}}$. It can be shown [31] that the minimum $\mathcal{D}$ subject to a constraint on $\mathcal{R}$ is given by

$$\bar{\mathcal{D}} = \alpha\, 2^{-\mathcal{R}/\beta}. \tag{3.14}$$

Furthermore, the minimum is achieved if, and only if,

$$\bar{\mathcal{D}}_j = \frac{\beta_j \rho_j}{\beta \delta_j} \bar{\mathcal{D}}, \tag{3.15}$$

or equivalently,

$$\frac{\mathcal{R}_j}{\beta_j} = \frac{\mathcal{R}}{\beta} + \log_2 \frac{\alpha_j \beta \delta_j}{\alpha \beta_j \rho_j}. \tag{3.16}$$

Observe that the overall rate-distortion performance of the network distributed codec follows the same exponential law of each of the subnetworks, which permits a hierarchical application, considering networks of networks when convenient. This does not come as a surprise since a common principle lies behind all of our high-rate quantization proofs: a rate allocation problem where rate and distortion are related exponentially.

3.6 EXPERIMENTAL RESULTS REVISITED

In this section we proceed to revisit the experimental results of Section 3.4, where the Lloyd algorithm described in Section 3.3.2 was applied to a scalar, quadratic-Gaussian WZ coding problem. Precisely, $X \sim \mathcal{N}(0, 1)$ and $N_Y \sim \mathcal{N}(0, 1/10)$ were independent, and $Y = X + N_Y$.

We saw in Section 3.5.1 that the high-rate approximation to the operational distortion-rate function is given by (3.9). Figure 3.6, which showed the rate-distortion

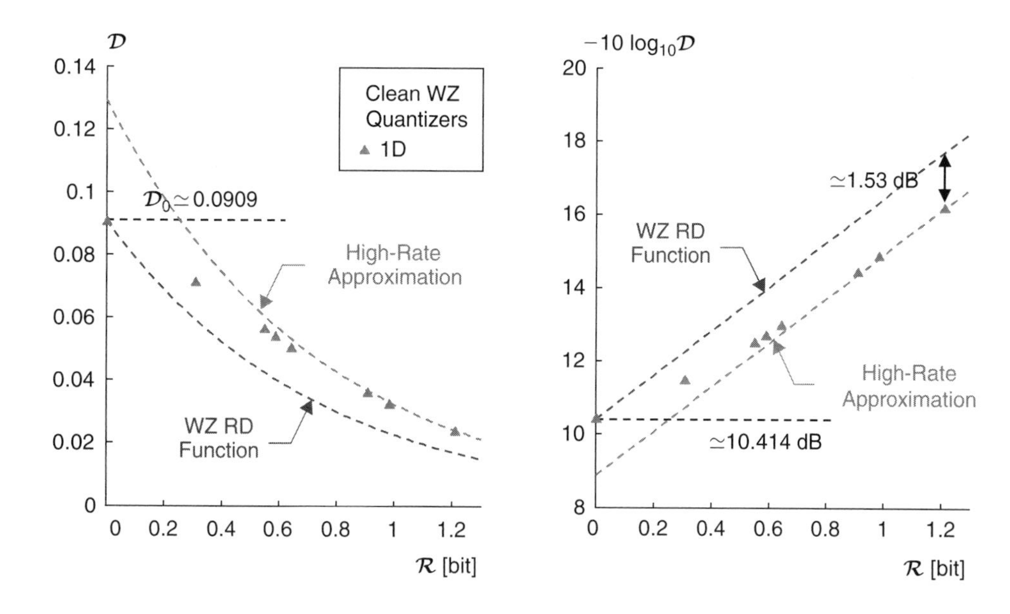

FIGURE 3.18

Distortion-rate performance of optimized scalar WZ quantizers with SW coding. $\mathcal{R} = \frac{1}{k}\mathrm{H}(Q|Y)$, $X \sim \mathcal{N}(0,1)$, $N_Y \sim \mathcal{N}(0,1/10)$ independent, $Y = X + N_Y$.

performance of several quantizers obtained with the Lloyd algorithm, can now be completed by superimposing the high-rate approximation to the optimal performance. The new plot is shown in Figure 3.18. The close match in distortion at high rates confirms the theory developed in this chapter and the usefulness of our extension of the Lloyd algorithm, for the statistics of the example. The distortion gap observed in the plot is now explained by the theory. Just as in the nondistributed case, 1.53 dB corresponds to the fact that the scalar moment of inertia $M_1 = 1/12$ is greater than its limit as the quantizer dimension tends to infinity (3.9).

Section 3.5.1 on clean distributed quantization with side information and ideal SW coding concluded that uniform quantizers with no index reuse are asymptotically optimal at high rates. This enables us to explain the uniformity and lack of index repetition observed in quantizer (b) in Figure 3.7. According to the distortion-rate performance in Figure 3.18, quantizer (a) in Figure 3.7 may be intuitively considered a low-rate quantizer, and quantizer (b), a high-rate quantizer.

3.7 CONCLUSIONS

We have established necessary optimality conditions for network distributed quantization of noisy sources and extended the Lloyd algorithm for its design. The concept of cost measure enables us to model the lossless coding method used for the quantization indices, particularly ideal SW coding. In addition, cost measures are the key to a very general formulation that unifies a number of related problems, such as broadcast with side information, distributed classification, or quantization of side information.

If ideal SW coders are used, uniform tessellating quantizers without index repetition are asymptotically optimal at high rates. It is known [47] that the rate loss in the WZ problem for smooth continuous sources and quadratic distortion vanishes as $\mathcal{D} \to 0$. Our work shows that this is also true for the operational rate loss and for each finite dimension k. These approximations have been applied to develop a theory of transforms for distributed source coding, which assumes high rates [32, 35], but not arbitrarily high dimension or Gaussian statistics [13, 14].

Suppose that the conditional expectation of the unseen source data X given the side information Y and the noisy observation Z is additively separable. Then, at high rates, optimal WZ quantizers of Z can be decomposed into estimators and uniform tessellating quantizers for clean sources, achieving the same rate-distortion performance as if the side information were available at the encoder. The additive separability condition for high-rate WZ quantization of noisy sources, albeit less restrictive, is similar to the condition required for zero rate loss in the quadratic Gaussian noisy WZ problem [33], which applies exactly for any rate but requires arbitrarily large dimension.

Experimental results confirm the high-rate approximation theory for distributed quantization. In addition, they suggest that the convergence properties of the extended Lloyd algorithm are similar to those of the classical one and can benefit from a genetic search algorithm for initialization.

REFERENCES

[1] W. R. Bennett, "Spectra of quantized signals," *Bell Syst., Tech. J.* 27, July 1948.

[2] J. A. Bucklew and G. L. Wise, "Multidimensional asymptotic quantization theory with rth power distortion measures," *IEEE Trans. Inform. Theory*, vol. IT-28, pp. 239–247, Mar. 1982.

[3] J. Cardinal, "Compression of side information," in *Proc. IEEE Int. Conf. Multimedia, Expo (ICME)*, vol. 2, Baltimore, MD, July 2003, pp. 569–572.

[4] ———, "Quantization of side information," unpublished.

[5] J. Cardinal and G. V. Asche, "Joint entropy-constrained multiterminal quantization," in *Proc. IEEE Int. Symp. Inform. Theory (ISIT)*, Lausanne, Switzerland, June 2002, p. 63.

[6] T. M. Cover, "A proof of the data compression theorem of Slepian and Wolf for ergodic sources," *IEEE Trans. Inform. Theory*, vol. 21, no. 2, pp. 226–228, Mar. 1975, (Corresp.).

[7] I. Csiszár, "Generalized entropy and quantization problems," in *Proc. Prague Conf. Inform. Theory, Stat. Decision Functions, Random Processes*, Prague, Czech Republic, 1973, pp. 159–174.

[8] R. L. Dobrushin and B. S. Tsybakov, "Information transmission with additional noise," *IRE Trans. Inform. Theory*, vol. IT-8, pp. S293–S304, 1962.

[9] M. Effros and D. Muresan, "Codecell contiguity in optimal fixed-rate and entropy-constrained network scalar quantizers," in *Proc. IEEE Data Compression Conf. (DCC)*, Snowbird, UT, Apr. 2002, pp. 312–321.

[10] Y. Ephraim and R. M. Gray, "A unified approach for encoding clean and noisy sources by means of waveform and autoregressive vector quantization," *IEEE Trans. Inform. Theory*, vol. IT-34, pp. 826–834, July 1988.

[11] M. Fleming and M. Effros, "Network vector quantization," in *Proc. IEEE Data Compression Conf. (DCC)*, Snowbird, UT, Mar. 2001, pp. 13–22.

[12] M. Fleming, Q. Zhao, and M. Effros, "Network vector quantization," *IEEE Trans. Inform. Theory*, vol. 50, no. 8, pp. 1584–1604, Aug. 2004.

[13] M. Gastpar, P. L. Dragotti, and M. Vetterli, "The distributed Karhunen-Loève transform," *IEEE Trans. Inform. Theory*, 2004, submitted for publication.

[14] ———, "On compression using the distributed Karhunen–Loève transform," in *Proc. IEEE Int. Conf. Acoust., Speech, Signal Processing (ICASSP)*, vol. 3, Montréal, Canada, May 2004, pp. 901–904.

[15] A. Gersho, "Asymptotically optimal block quantization," *IEEE Trans. Inform. Theory*, vol. IT-25, pp. 373–380, July 1979.

[16] A. Gersho and R. M. Gray, *Vector Quantization and Signal Compression*. Boston, MA: Kluwer Academic Publishers, 1992.

[17] H. Gish and J. N. Pierce, "Asymptotically efficient quantizing," *IEEE Trans. Inform. Theory*, vol. IT-14, pp. 676–683, Sept. 1968.

[18] R. M. Gray, T. Linder, and J. Li, "A Lagrangian formulation of Zador's entropy-constrained quantization theorem," *IEEE Trans. Inform. Theory*, vol. 48, no. 3, pp. 695–707, March 2002.

[19] R. M. Gray and D. L. Neuhoff, "Quantization," *IEEE Trans. Inform. Theory*, vol. 44, pp. 2325–2383, Oct. 1998.

[20] J. A. Gubner, "Distributed estimation and quantization," *IEEE Trans. Inform. Theory*, vol. 39, no. 4, pp. 1456–1459, July 1993.

[21] J. Kusuma, L. Doherty, and K. Ramchandran, "Distributed compression for sensor networks," in *Proc. IEEE Int. Conf. Image Processing (ICIP)*, vol. 1, Thessaloniki, Greece, Oct. 2001, pp. 82–85.

[22] W. M. Lam and A. R. Reibman, "Quantizer design for decentralized estimation systems with communication constraints," in *Proc. Annual Conf. Inform. Sci. Syst. (CISS)*, Baltimore, MD, Mar. 1989.

[23] ———, "Design of quantizers for decentralized estimation systems," *IEEE Trans. Inform. Theory*, vol. 41, no. 11, pp. 1602–1605, Nov. 1993.

[24] T. Linder and K. Zeger, "Asymptotic entropy-constrained performance of tessellating and universal randomized lattice quantization," *IEEE Trans. Inform. Theory*, vol. 40, pp. 575–579, Mar. 1993.

[25] Z. Liu, S. Cheng, A. D. Liveris, and Z. Xiong, "Slepian-Wolf coded nested quantization (SWC-NQ) for Wyner-Ziv coding: Performance analysis and code design," in *Proc. IEEE Data Compression Conf. (DCC)*, Snowbird, UT, Mar. 2004, pp. 322–331.

[26] S. P. Lloyd, "Least squares quantization in PCM," *IEEE Trans. Inform. Theory*, vol. IT-28, pp. 129–137, Mar. 1982.

[27] D. Muresan and M. Effros, "Quantization as histogram segmentation: Globally optimal scalar quantizer design in network systems," in *Proc. IEEE Data Compression Conf. (DCC)*, Snowbird, UT, Apr. 2002, pp. 302–311.

[28] S. Na and D. L. Neuhoff, "Bennett's integral for vector quantizers," *IEEE Trans. Inform. Theory*, vol. 41, pp. 886–900, July 1995.

[29] S. S. Pradhan, J. Kusuma, and K. Ramchandran, "Distributed compression in a dense microsensor network," *IEEE Signal Processing Mag.*, vol. 19, pp. 51–60, Mar. 2002.

[30] S. S. Pradhan and K. Ramchandran, "Distributed source coding using syndromes (DISCUS): Design and construction," in *Proc. IEEE Data Compression Conf. (DCC)*, Snowbird, UT, Mar. 1999, pp. 158–167.

[31] D. Rebollo-Monedero, "Quantization and transforms for distributed source coding," Ph.D. dissertation, Stanford Univ., 2007.

[32] D. Rebollo-Monedero, A. Aaron, and B. Girod, "Transforms for high-rate distributed source coding," in *Proc. Asilomar Conf. Signals, Syst., Comput.*, vol. 1, Pacific Grove, CA, Nov. 2003, pp. 850–854, invited paper.

[33] D. Rebollo-Monedero and B. Girod, "A generalization of the rate-distortion function for Wyner–Ziv coding of noisy sources in the quadratic-Gaussian case," in *Proc. IEEE Data Compression Conf. (DCC)*, Snowbird, UT, Mar. 2005, pp. 23–32.

[34] ———, "Network distributed quantization," in *Proc. IEEE Inform. Theory Workshop (ITW)*, Lake Tahoe, CA, Sept. 2007, invited paper.

[35] D. Rebollo-Monedero, S. Rane, A. Aaron, and B. Girod, "High-rate quantization and transform coding with side information at the decoder," *EURASIP J. Signal Processing, Special Issue Distrib. Source Coding*, vol. 86, no. 11, pp. 3160–3179, Nov. 2006, invited paper.

[36] D. Rebollo-Monedero, S. Rane, and B. Girod, "Wyner–Ziv quantization and transform coding of noisy sources at high rates," in *Proc. Asilomar Conf. Signals, Syst., Comput.*, vol. 2, Pacific Grove, CA, Nov. 2004, pp. 2084–2088.

[37] D. Rebollo-Monedero, R. Zhang, and B. Girod, "Design of optimal quantizers for distributed source coding," in *Proc. IEEE Data Compression Conf. (DCC)*, Snowbird, UT, Mar. 2003, pp. 13–22.

[38] S. D. Servetto, "Lattice quantization with side information," in *Proc. IEEE Data Compression Conf. (DCC)*, Snowbird, UT, Mar. 2000, pp. 510–519.

[39] J. D. Slepian and J. K. Wolf, "Noiseless coding of correlated information sources," *IEEE Trans. Inform. Theory*, vol. IT-19, pp. 471–480, July 1973.

[40] H. Viswanathan, "Entropy coded tesselating quantization of correlated sources is asymptotically optimal," 1996, unpublished.

[41] J. K. Wolf and J. Ziv, "Transmission of noisy information to a noisy receiver with minimum distortion," *IEEE Trans. Inform. Theory*, vol. IT-16, no. 4, pp. 406–411, July 1970.

[42] A. D. Wyner and J. Ziv, "The rate-distortion function for source coding with side information at the decoder," *IEEE Trans. Inform. Theory*, vol. IT-22, no. 1, pp. 1–10, Jan. 1976.

[43] Z. Xiong, A. Liveris, S. Cheng, and Z. Liu, "Nested quantization and Slepian–Wolf coding: A Wyner–Ziv coding paradigm for i.i.d. sources," in *Proc. IEEE Workshop Stat. Signal Processing (SSP)*, St. Louis, MO, Sept. 2003, pp. 399–402.

[44] Z. Xiong, A. D. Liveris, and S. Cheng, "Distributed source coding for sensor networks," *IEEE Signal Processing Mag.*, vol. 21, no. 5, pp. 80–94, Sept. 2004.

[45] Y. Yang, S. Cheng, Z. Xiong, and W. Zhao, "Wyner–Ziv coding based on TCQ and LDPC codes," in *Proc. Asilomar Conf. Signals, Syst., Comput.*, vol. 1, Pacific Grove, CA, Nov. 2003, pp. 825–829.

[46] P. L. Zador, "Asymptotic quantization error of continuous signals and the quantization dimension," *IEEE Trans. Inform. Theory*, vol. IT-28, no. 2, pp. 139–149, Mar. 1982.

[47] R. Zamir, "The rate loss in the Wyner–Ziv problem," *IEEE Trans. Inform. Theory*, vol. 42, no. 6, pp. 2073–2084, Nov. 1996.

[48] R. Zamir and T. Berger, "Multiterminal source coding with high resolution," *IEEE Trans. Inform. Theory*, vol. 45, no. 1, pp. 106–117, Jan. 1999.

[49] R. Zamir and M. Feder, "On lattice quantization noise," *IEEE Trans. Inform. Theory*, vol. 42, no. 4, pp. 1152–1159, July 1996.

[50] R. Zamir and S. Shamai, "Nested linear/lattice codes for Wyner–Ziv encoding," in *Proc. IEEE Inform. Theory Workshop (ITW)*, Killarney, Ireland, June 1998, pp. 92–93.

[51] R. Zamir, S. Shamai, and U. Erez, "Nested linear/lattice codes for structured multiterminal binning," *IEEE Trans. Inform. Theory*, vol. 48, no. 6, pp. 1250–1276, June 2002.

CHAPTER

Zero-error Distributed Source Coding

4

Ertem Tuncel

Department of Electrical Engineering, University of California, Riverside, CA

Jayanth Nayak

Mayachitra, Inc., Santa Barbara, CA

Prashant Koulgi and Kenneth Rose

Department of Electrical and Computer Engineering, University of California, Santa Barbara, CA

CHAPTER CONTENTS

4.1 INTRODUCTION

This chapter tackles the classical distributed source coding (DSC) problem from a different perspective. As usual, let X^n and Y^n be a pair of correlated sequences observed

Distributed Source Coding: Theory, Algorithms, and Applications

by two sensors $\mathcal{S}_X$ and $\mathcal{S}_Y$. Also, let the samples $(X_t, Y_t) \in \mathcal{X} \times \mathcal{Y}$ be i.i.d.$\sim P_{XY}(x, y)$, where without loss of generality, it can be assumed that $P_X(x) > 0$ and $P_Y(y) > 0$ for all $(x, y) \in \mathcal{X} \times \mathcal{Y}$. The sensors are required to convey these sequences to $\mathcal{S}_Z$, a third sensor or a central processing unit, without communicating with each other. What is the region of achievable rates R_X and R_Y if the reconstruction $(\hat{X}^n, \hat{Y}^n)$ at $\mathcal{S}_Z$ is to be lossless? The answer heavily depends on what is meant by "lossless," as well as on whether or not we are allowed to use variable-length codes. More specifically, the sensors could be operating under three different requirements, which are listed here in increasing order of strength:

a. *Coding with vanishingly small error:* This is the original DSC regime as introduced by Slepian and Wolf [14], and requires only that $\Pr[(\hat{X}^n, \hat{Y}^n) \neq (X^n, Y^n)] \to 0$ as $n \to \infty$. Fixed- or variable-length codes do not differ in the achieved rate performance.

b. *Variable-length zero-error (VLZE) coding:* This regime has the more stringent requirement that $\Pr[(\hat{X}^n, \hat{Y}^n) \neq (X^n, Y^n)] = 0$ for all $n \geqslant 1$ [2]. Thus, every possible value of (X^n, Y^n) has to be accounted for no matter how small the probability. As a result, the achieved rate region is smaller [7,8]. This is in contrast to point-to-point communication, where coding with vanishingly small error and zero-error variable-length coding are known to achieve the same minimum rate, namely, $H(P_X)$, the entropy of the source X.

c. *Fixed-length zero-error (FLZE) coding:* The zero-error codes are further required to be fixed-length [16]. Unlike the case in point-to-point coding, where the achieved rate is simply the logarithm of the size of the source alphabet $\mathcal{X}$, the problem continues to be nontrivial because, if sufficiently strong, the correlation structure between the sequences can still be exploited for compression.

The primary motivation for zero-error coding is that in delay- and/or complexity-sensitive applications (e.g., most sensor network scenarios) where the blocklength n must be very small, the error incurred by the vanishingly small error regime may be too large to tolerate. Despite the small n motivation, even asymptotic results are of interest as they can be used as benchmarks for practical zero-error schemes. Apart from its practical significance, zero-error DSC has also been studied owing to its connection to graph theory, which undoubtedly provides the most suitable tools for the problem.

Here, we discuss VLZE coding in detail. We focus on the case commonly referred to as *asymmetric coding*, where Y^n is sent in a point-to-point fashion and then treated as side information available only to $\mathcal{S}_Z$. For this case, the parameters of the problem are completely captured by a properly defined *characteristic graph* G and the marginal probability mass function P_X. The problem then reduces to designing codes such that no vertex can be assigned a codeword that prefixes those of its neighbors.

Unlike the case in the regime of vanishingly small error, there is no known single-letter asymptotic characterization in general. One possible infinite-letter characterization for the minimum achievable rate is given in terms of *chromatic entropy* of

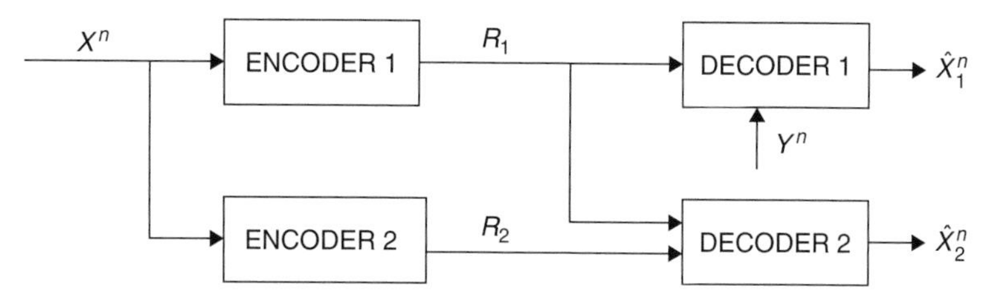

FIGURE 4.1

The block diagram for Extension 1.

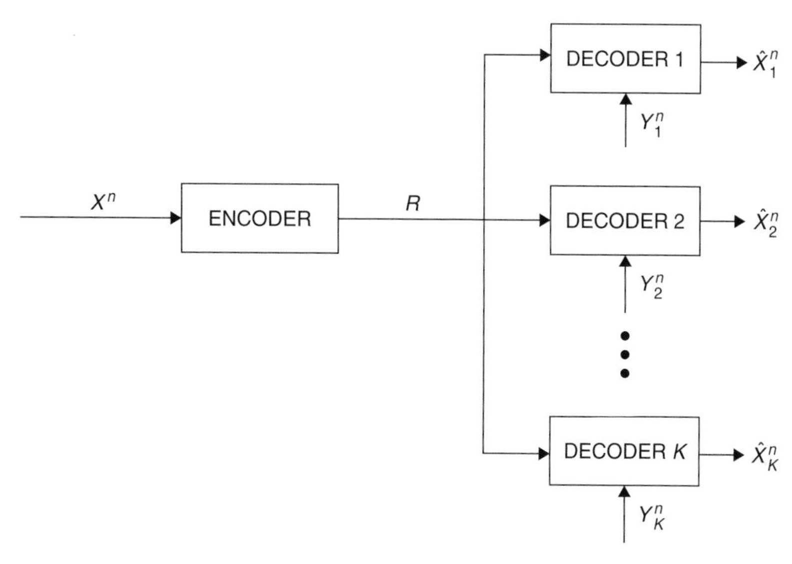

FIGURE 4.2

The block diagram for Extension 2.

G under the distribution P_X [2]. Koulgi et al. [7] showed that another such characterization is given by $\overline{H}(G, P_X)$, the *complementary graph entropy* of G under P_X. Even though neither characterization is computable in general, the latter one provides some additional insight into the problem. In particular, known single-letter lower and upper bounds for complementary graph entropy immediately translate to bounds for the minimum rate. Also, connections with zero-error channel coding is strengthened via a formula due to Marton [12]. These connections are useful in tackling two "network" extensions shown in Figures 4.1 and 4.2 and described as follows.

1. *Side information may be absent:* In the first stage, expending rate R_1, the encoder describes X^n in sufficient detail to ensure a reconstruction $\hat{X}_1^n$ with

$\Pr[\hat{X}_1^n \neq X^n] = 0$, in the presence of the side information Y^n. The complementary second-stage description of rate R_2 allows the decoder to reconstruct $\hat{X}_2^n$ so that $\Pr[\hat{X}_2^n \neq X^n] = 0$, even in the absence of Y^n.

2. *Compound side information:* It is desired to broadcast X^n to K receivers, each equipped with a different side information vector Y_k^n, so that at each receiver k the reconstruction $\hat{X}_k^n$ satisfies $\Pr[\hat{X}_k^n \neq X^n] = 0$. An equivalent scenario is that there is only one receiver that may have access to any one of the side information vectors Y_k^n.

We show that one can further benefit from the connection between VLZE coding and complementary graph entropy to derive the minimum rates for these extensions. Although the minimum rates we derive here are still in terms of the complementary graph entropy, and hence not computable in general, our results imply that there is no *rate loss* incurred on the coders in these scenarios. That is, in Extension 1, both R_1 and $R_1 + R_2$ can simultaneously be kept at their minimum, that is, $\overline{H}(G, P_X)$ and $H(P_X)$, respectively. Similarly in Extension 2, $R = \max_k \overline{H}(G_k, P_X)$ can be achieved, where G_k is the characteristic graph for the joint distribution P_{XY_k}. It should be noted that both of these results become trivial if we only require vanishingly small error probability, for simple lower bounds can be shown to be achievable using the techniques introduced by Slepian and Wolf [14]. However, VLZE coding is considerably more involved.

We then turn to finite n and discuss difficulties of optimal code design. In contrast to point-to-point coding where the optimal design algorithm due to Huffman is polynomial time ($O(|\mathcal{X}| \log |\mathcal{X}|)$ to be more specific), the optimal VLZE code design in DSC is NP-hard; that is, an NP-complete problem can be reduced to it in polynomial time [7]. Thus, if the widely held conjecture that P $\neq$ NP is true, no polynomial-time optimal code design algorithm exists. In fact, even the simpler problem of determining whether a code exists with given codeword lengths for each vertex is NP-complete, as shown by Yan and Berger [17]. This is again in contrast to point-to-point coding where the same question can be answered by simply checking whether Kraft's inequality is satisfied (which is of complexity $O(|\mathcal{X}|)$).

The rest of the chapter is organized as follows. We begin by discussing the connections between VLZE coding and graph theory in the next section. In Section 4.3, we show that complementary graph entropy is indeed an asymptotic characterization of the minimum achievable rate. The network extensions in Figures 4.1 and 4.2 are then discussed in Section 4.4. Section 4.5 tackles optimal code design. The chapter concludes in Section 4.6.

4.2 GRAPH THEORETIC CONNECTIONS

4.2.1 VLZE Coding and Graphs

Although our focus is on the VLZE coding problem, the connections to graph theory can be most easily seen by considering FLZE coding first and defining a *confusability* relation between the source symbols [16]. Distinct $x, x' \in \mathcal{X}$ are said to be confusable if there is a $y \in \mathcal{Y}$ such that $P_{XY}(x, y) > 0$ and $P_{XY}(x', y) > 0$. The key observation is that two confusable letters cannot be assigned the same codeword in any valid FLZE code with blocklength $n = 1$. This motivated Witsenhausen to define a *characteristic graph* $G = (\mathcal{X}, \mathcal{E})$ such that $\{x, x'\} \in \mathcal{E}$ if $x, x' \in \mathcal{X}$ are confusable, thereby reducing FLZE coding for given P_{XY} to *coloring* G so that two adjacent vertices are not assigned the same color. Given y, the decoding is then performed by uniquely identifying the vertex x, among all vertices with the encoded color, that satisfies $P_{XY}(x, y) > 0$. Thus, the minimum achievable FLZE coding rate is given by $\lceil \log_2 \chi(G) \rceil$, where $\chi(G)$ is the minimum number of colors needed for a valid coloring of G and is referred to as the *chromatic number* of G.

One can generalize this concept to any blocklength $n > 1$ by defining the characteristic graph $G_n = (\mathcal{X}^n, \mathcal{E}_n)$ for the underlying probability mass function

$$P_{XY}^n(x^n, y^n) = \prod_{t=1}^{n} P_{XY}(x, y).$$

It can be observed that distinct x^n and x'^n are adjacent in G_n if and only if either $x_t = x'_t$ or $\{x_t, x'_t\} \in \mathcal{E}$, for all $1 \leq t \leq n$. This implies that G_n is in fact the same as what is commonly referred to in graph theory as the nth AND-power of G, denoted G^n.

As an example, consider $\mathcal{X} = \mathcal{Y} = \{0, 1, 2, 3, 4\}$ with $P_X(x) = \frac{1}{5}$ for all $x \in \mathcal{X}$ and

$$P_{Y|X}(y|x) = \begin{cases} \frac{1}{2} & y = x \text{ or } y = x + 1 \bmod 5 \\ 0 & \text{otherwise} \end{cases}$$

This induces a "pentagon" graph G shown in Figure 4.3. Figure 4.3(a) shows an optimal coloring scheme that uses $\chi(G) = 3$ colors, thereby achieving rate $\lceil \log_2 \chi(G) \rceil = 2$ bits. Now, consider $n = 2$. We can visualize the characteristic graph G^2 as shown in Figure 4.4, where the five individual pentagons represent the confusability relation for x_2, each for a fixed x_1, and the *meta*-pentagon represents the confusability relation for x_1 alone. There are a total of 100 edges in G^2, and for clarity, most are omitted in

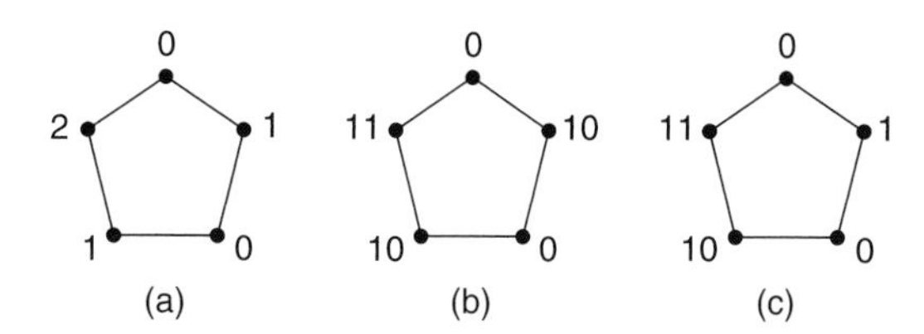

FIGURE 4.3

The characteristic graph for the example P_{XY} in the text. (a) shows the optimal coloring scheme, and (b) shows the resultant VLZE code obtained by encoding the colors in (a). By instantaneous coding, a better code can be obtained as shown in (c).

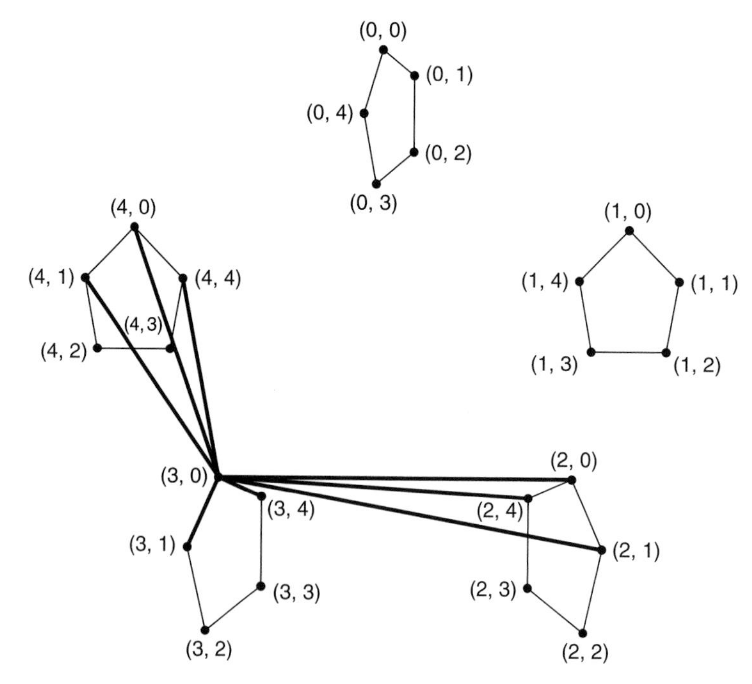

FIGURE 4.4

Partial visualization of the graph G^2.

Figure 4.4. Those indicated in bold correspond to the edges that connect (3, 0) to its adjacent vertices. It can be shown that

$$c(x_1, x_2) = 2x_1 + x_2 \mod 5$$

is a valid coloring scheme. This immediately shows the benefit of block-coding even in the zero-error context since the bit rate per symbol drops to $\frac{1}{2}\lceil \log_2 \chi(G^2) \rceil = 1.5$ bits.

For VLZE coding, one possible strategy is to color G^n and design an optimal prefix-free code in the color space. For the pentagon graph with $n = 1$, this results in a code as shown in Figure 4.3(b). However, a strategy that is potentially more general is *instantaneous* coding: If x^n is adjacent to x'^n in G^n, then

$$\phi_n(x^n) \neq_P \phi_n(x'^n)$$

where

$$\phi_n : \mathcal{X}^n \rightarrow \{0, 1\}^*$$

is the variable-length encoder, with $\{0, 1\}^*$ and $\neq_P$ indicating all binary sequences of finite length (including length 0) and prefix-free relation, respectively [2]. That this is necessary and sufficient follows from the observation that all vertices in the *fan-out* set of any given y^n, that is, the set of all $x^n \in \mathcal{X}^n$ with $P^n_{XY}(x^n, y^n) > 0$, must be connected. It follows by construction that coloring followed by prefix-free coding of colors is itself an instantaneous coding scheme. However, as Figure 4.3(c) shows with an example, the latter is indeed a more general class; that is, there are instantaneous codes that cannot be implemented by prefix-free coding of colors. The bit rate expended by any VLZE code is determined by

$$R(\phi_n) = \frac{1}{n} \sum_{x \in \mathcal{X}^n} P^n_X(x^n)|\phi_n(x^n)|$$

and therefore, once G is built and P_X is known, there is no further dependence of the minimum rate on $P_{Y|X}$. For this reason, the minimum achievable rate with blocklength n will henceforth be denoted as $R^n_{\min}(G, P_X)$. Similarly, we denote by $R_{\min}(G, P_X)$ the asymptotic minimum rate.

4.2.2 Basic Definitions and Notation

Before proceeding further, we need to make several definitions. Let $\mathcal{P}_2(\mathcal{A})$ denote the set of all distinct pairs of vertices in $\mathcal{A}$. We say that $G(\mathcal{A})$ is a subgraph of G *induced* by $\mathcal{A} \subset \mathcal{X}$ if its vertex set is $\mathcal{A}$ and its edge set is $\mathcal{E} \cap \mathcal{P}_2(\mathcal{A})$. For $G_1 = (\mathcal{X}, \mathcal{E}_1)$ and $G_2 = (\mathcal{X}, \mathcal{E}_2)$, we use the notation $G_1 \subset G_2$ to indicate that $\mathcal{E}_1 \subset \mathcal{E}_2$. Similarly, $G_1 \cup G_2$ is the graph with vertex set $\mathcal{X}$ and edge set $\mathcal{E}_1 \cup \mathcal{E}_2$. The complement of $G = (\mathcal{X}, \mathcal{E})$ is $\overline{G} = (\mathcal{X}, \mathcal{E}^c)$, where $\mathcal{E}^c = \mathcal{P}_2(\mathcal{X}) \setminus \mathcal{E}$. A graph is called *empty* (denoted $E_{\mathcal{X}}$) or *complete* (denoted $K_{\mathcal{X}}$) if $\mathcal{E} = \emptyset$ or $\mathcal{E} = \mathcal{P}_2(\mathcal{X})$, respectively. It is obvious that $\overline{E}_{\mathcal{X}} = K_{\mathcal{X}}$ and $G \cup \overline{G} = K_{\mathcal{X}}$.

A set $\mathcal{X}'$ is called *independent* in G if $G(\mathcal{X}') = E_{\mathcal{X}'}$. The set of all independent sets in G is denoted as $\Gamma(G)$. The *stability number*, $\alpha(G)$, is the cardinality of the largest independent set in G. Similarly, a set $\mathcal{X}'$ is called a *clique* in G if $G(\mathcal{X}') = K_{\mathcal{X}'}$ and the set of all cliques is denoted as $\Omega(G)$. Clearly, $\Omega(G) = \Gamma(\overline{G})$.

Let us also formally define the AND- and OR-products and powers. The AND-product of $G_1 = (\mathcal{X}_1, \mathcal{E}_1)$ and $G_2 = (\mathcal{X}_2, \mathcal{E}_2)$, denoted by $G_1 \times G_2$, has the vertex set $\mathcal{X}_1 \times \mathcal{X}_2$, and distinct (x_1, x_2) and (x_1', x_2') are connected if both $x_1 \overset{G_1}{\sim} x_1'$ and $x_2 \overset{G_2}{\sim} x_2'$, where $a \overset{G}{\sim} b$ stands for "either $a = b$ or a is adjacent to b in G." The OR-product of G_1 and G_2, denoted by $G_1 \cdot G_2$, has the same vertex set $\mathcal{X}_1 \times \mathcal{X}_2$, but distinct (x_1, x_2) and (x_1', x_2') are connected if either $\{x_1, x_1'\} \in \mathcal{E}_1$ or $\{x_2, x_2'\} \in \mathcal{E}_2$. We denote by $G^n = (\mathcal{X}^n, \mathcal{E}_n)$ and $G^{(n)} = (\mathcal{X}^n, \mathcal{E}_{(n)})$ the n-fold AND- and OR-products of G with itself, or in other words, nth AND- and OR-powers, respectively. Note that $\overline{G_1 \cdot G_2} = \overline{G}_1 \times \overline{G}_2$, and more specifically, $\overline{G^{(n)}} = \overline{G}^n$. Also, $G^n \subset G^{(n)}$.

Finally, the strongly typical set of sequences $\mathcal{T}^n_{P_X,\varepsilon}$ is defined in [4] as

$$\mathcal{T}^n_{P_X,\varepsilon} = \left\{ x^n \in \mathcal{X}^n : \left| \frac{1}{n} N(a|x^n) - P_X(a) \right| \leq \varepsilon \right\}$$

where $N(a|x^n)$ denotes the number of occurrences of a in x^n. Among the many useful properties of strong typicality, we particularly use the fact that $\mathcal{T}^n_{P_X,\varepsilon}$ captures most of the probability [4, p. 34]:

$$P^n_X(\mathcal{T}^n_{P_X,\varepsilon}) \geqslant 1 - \frac{|\mathcal{X}|}{4n\varepsilon^2}. \tag{4.1}$$

Other useful properties will be referred to as needed in the sequel.

4.2.3 Graph Entropies

Definition 4.1. The graph entropy of G under P_X is defined as

$$H(G, P_X) = \lim_{n\to\infty} \frac{1}{n} \log_2 \left[\min_{\mathcal{A}: P^n_X(\mathcal{A}) > 1-\varepsilon} \chi(G^{(n)}(\mathcal{A})) \right]. \tag{4.2}$$

In words, $H(G, P_X)$ is the normalized logarithm of the minimum number of colors needed to color any high probability subset of $G^{(n)}$. Exactly how high the probability, which is determined by the value of $0 < \varepsilon < 1$, is immaterial [9]. Körner also derived a single-letter characterization of $H(G, P_X)$, given by

$$H(G, P_X) = \min_{X \in Z \in \Gamma(G)} I(X; Z) \tag{4.3}$$

where $X \in Z \in \Gamma(G)$ is a shorthand notation for $z \in \Gamma(G)$ and $P_{Z|X}(z|x) = 0$ if $x \notin z$.

Definition 4.2. The complementary graph entropy of G under P_X is defined as

$$\overline{H}(G, P_X) = \lim_{\varepsilon\to 0} \overline{H}_\varepsilon(G, P_X)$$

where

$$\overline{H}_\varepsilon(G, P_X) = \limsup_{n\to\infty} \frac{1}{n} \log_2 \left[\chi(G^n(\mathcal{T}^n_{P_X,\varepsilon})) \right]. \tag{4.4}$$

This definition was made in [10]. It is very similar to that of graph entropy, except that AND-powers instead of OR-powers are colored. This difference prohibited a single-letter formula for $\overline{H}(G, P_X)$. It even remains unknown whether $\lim_{\varepsilon\to 0}$ is necessary, or whether lim sup can be replaced by a regular limit.

Both the graph entropy and the complementary graph entropy were defined in different contexts than VLZE distributed source coding. The first link between graph entropies and DSC was built by Alon and Orlitsky [2] through their *chromatic* entropy.

Definition 4.3. The chromatic entropy of G under P_X is given by

$$H_\chi(G, P_X) = \min\{H(c(X)) : c(\cdot) \text{ is a valid coloring of } G\}. \tag{4.5}$$

Alon and Orlitsky [2] proved the following theorem. We include a sketch of the proof here, since the proof of Theorem 4.4 in this chapter also uses the same techniques.

Theorem 4.1.

$$R_{\min}(G, P_X) = \lim_{n\to\infty} \frac{1}{n} H_\chi(G^n, P_X^n). \tag{4.6}$$

Remark. That $R_{\min}(G, P_X)$ is upper bounded by the right-hand side of (4.6) follows using the aforementioned simple strategy of first coloring G^n and designing prefix-free codes for colors. It is somewhat surprising that this simple strategy is asymptotically optimal.

PROOF. The limit in (4.6) exists because $H_\chi(G^n, P_X^n)$ is subadditive in n. That is,

$$H_\chi(G^{m+n}, P_X^{m+n}) \leqslant H_\chi(G^m, P_X^m) + H_\chi(G^n, P_X^n).$$

That, in turn, follows from

$$H_\chi(G_1 \times G_2, P_{X_1} P_{X_2}) \leqslant H_\chi(G_1, P_{X_1}) + H_\chi(G_2, P_{X_2})$$

as $c_1 \times c_2$ is a valid coloring for $G_1 \times G_2$ if c_1 and c_2 are valid for G_1 and G_2, respectively.

The preceding Remark implies that we only need to prove a converse. The key observation is that an arbitrary code ϕ_n can be decomposed as $\phi_n = \theta_n \circ c_n$ where θ_n is a one-to-one code; that is, it assigns distinct (not necessarily prefix-free) codewords to colors. Utilizing a lower bound for one-to-one codes derived in [1], one can show

$$nR(\phi_n) \geqslant H_\chi(G^n, P_X^n) - \log_2 \left(H_\chi(G^n, P_X^n) + 1 \right) - \log_2 e,$$

which can be further bounded using the subadditivity as

$$nR(\phi_n) \geqslant H_\chi(G^n, P_X^n) - \log_2 \left(H_\chi(G, P_X) + 1 \right) - \log_2 n - \log_2 e.$$

This implies for an arbitrary sequence ϕ_n of variable-length codes that

$$\liminf_{n\to\infty} R(\phi_n) \geqslant \lim_{n\to\infty} \frac{1}{n} H_\chi(G^n, P_X^n)$$

completing the proof. ■

Alon and Orlitsky [2] further showed implicitly that the following single-letter bound for $R_{\min}(G, P_X)$ holds:

$$R_{\min}(G, P_X) \leqslant H(G, P_X).$$

They also showed

$$R_{\min}^n(G, P_X) \geqslant \frac{1}{n} H(G^n, P_X^n).$$

Finally, they defined the *clique* entropy $H_\omega(G, P_X)$, which is intimately related to the graph entropy:

$$\begin{aligned} H_\omega(G, P_X) &= \max_{X \in Z \in \Omega(G)} H(X|Z) \\ &= H(P_X) - H(\overline{G}, P_X). \end{aligned} \tag{4.7}$$

The relation between these four graph entropies is well-studied (see, e.g., [2, 5, 10, 11]. It is in particular known that

$$H_\omega(G, P_X) \leqslant \overline{H}(G, P_X) \leqslant H(G, P_X) \leqslant H_\chi(G, P_X) \tag{4.8}$$

for all G and P_X. Furthermore, for the important class of *perfect* graphs (cf. Berge [3], Chapter 16), Equation (4.8) can be strengthened as

$$H_\omega(G, P_X) = \overline{H}(G, P_X) = H(G, P_X) \tag{4.9}$$

for all P_X.

4.2.4 Graph Capacity

A closely related concept in zero-error information theory is the graph capacity problem. Roughly, if we replace coloring, which can be looked at as *covering* using independent sets (because vertices that are assigned the same color must form an independent set), with *packing* of independent sets such that they do not overlap, we obtain the graph capacity problem. Similar to the correspondence of coloring to zero-error DSC, this corresponds to a *channel* coding problem where the probability of error is set to zero at all blocklengths n. Although graph capacity by itself deserves a detailed discussion, for the purposes of this chapter, we only need to define

$$C_\varepsilon(G, P_X) = \limsup_{n \to \infty} \frac{1}{n} \log_2 \left[\alpha(G^n(T^n_{P_X,\varepsilon})) \right] \tag{4.10}$$

and

$$C(G, P_X) = \lim_{\varepsilon \to 0} C_\varepsilon(G, P_X).$$

The intuitive covering/packing duality between the zero-error DSC and channel coding is then concretized by the formula

$$C(G, P_X) + \overline{H}(G, P_X) = H(P_X) \tag{4.11}$$

derived by Marton [12]. The relation (4.11) in particular shows that any single-letter formula for either $C(G, P_X)$ or $\overline{H}(G, P_X)$ would immediately yield a formula for the other. Unfortunately, both are long-standing open problems.

4.3 COMPLEMENTARY GRAPH ENTROPY AND VLZE CODING

In the next theorem from Koulgi et al., [7], $\overline{H}(G, P_X)$ is shown to be an alternative infinite-letter characterization for $R_{\min}(G, P_X)$. We include the proof for completeness.

Theorem 4.2.

$$R_{\min}(G, P_X) = \overline{H}(G, P_X) \tag{4.12}$$

PROOF. We first show that

$$\overline{H}(G, P_X) \geqslant \lim_{n\to\infty} \frac{1}{n} H_\chi(G^n, P_X^n). \tag{4.13}$$

Toward that end, fix $\varepsilon > 0$ and observe from (4.4) that for any $n > n_0(\varepsilon)$ there is a coloring c of G^n satisfying:

$$|c(\mathcal{T}^n_{P_X,\varepsilon})| \leqslant 2^{n[\overline{H}_\varepsilon(G,P_X)+\varepsilon]}. \tag{4.14}$$

Define the indicator function $\Phi : \mathcal{X}^n \to \{0, 1\}$ as

$$\Phi(x^n) = \begin{cases} 1 & \text{if } x^n \in \mathcal{T}^n_{P_X,\varepsilon} \\ 0 & \text{else.} \end{cases}$$

We then have

$$\begin{aligned} H_\chi(G^n, P^n) &\leqslant H(c(X^n)) \\ &\leqslant H(\Phi) + H(c(X^n)|\Phi) \\ &\leqslant H(\Phi) + H(c(X^n)|X^n \in \mathcal{T}^n_{P_X,\varepsilon}) + \varepsilon H(c(X^n)|X^n \notin \mathcal{T}^n_{P_X,\varepsilon}) \\ &\leqslant 1 + n\left[\overline{H}_\varepsilon(G, P_X) + \varepsilon(1 + \log|\mathcal{X}|)\right], \end{aligned}$$

where we used (4.14) in the last step. Normalizing by n and taking limits, (4.13) follows.

Now consider the reversed inequality in (4.13). Fix $\varepsilon > 0$ and let the coloring function c on G^n achieve $H_\chi(G^n, P_X^n)$, so that

$$H_\chi(G^n, P_X^n) = H(c(X^n)).$$

To lower bound $H(c(X^n))$, we use the following elementary lower bound for the entropy function: if Q is a probability distribution over the set $\mathcal{Q}$, and $S \subseteq \mathcal{Q}$, then

$$H(Q) \geqslant -\left[\sum_{j\in S} Q(j)\right] \log \max_{j\in S} Q(j).$$

Thus we have the following estimate for $H_\chi(G^n, P_X^n)$:

$$H(c(X^n)) \geqslant -P_X^n(\mathcal{T}^n_{P_X,\varepsilon}) \log\left[\max_{x^n\in\mathcal{T}^n_{P_X,\varepsilon}} P_X^n(c(x^n))\right]. \tag{4.15}$$

The probability $P_X^n(\mathcal{T}^n_{P_X,\varepsilon})$ can be lower bounded as in (4.1). In any coloring of G^n, the maximum cardinality of a single-colored subset of $\mathcal{T}^n_{P_X,\varepsilon}$ cannot exceed $\alpha(G^n(\mathcal{T}^n_{P_X,\varepsilon})$, the size of the largest independent set induced by $\mathcal{T}^n_{P_X,\varepsilon}$ in G^n. Thus,

$$\max_{x^n\in\mathcal{T}^n_{P_X,\varepsilon}} P_X^n(c(x^n)) \leqslant \alpha(G^n(\mathcal{T}^n_{P_X,\varepsilon})) \max_{x^n\in\mathcal{T}^n_{P_X,\varepsilon}} P_X^n(x^n). \tag{4.16}$$

But using the definition of $\mathcal{T}^n_{P_X,\varepsilon}$ and well-known properties of types and typical sequences, and from uniform continuity of entropy ([4, pp. 32–33]), we have

$$\begin{aligned} -\frac{1}{n}\log \max_{x^n\in\mathcal{T}^n_{P_X,\varepsilon}} P_X^n(x^n) &= \min_{Q:|Q(x)-P_X(x)|\leqslant\varepsilon\ \forall x\in\mathcal{X}} \{H(Q) + D(Q||P_X)\} \\ &\geqslant \min_{Q:|Q(x)-P_X(x)|\leqslant\varepsilon\ \forall x\in\mathcal{X}} H(Q) \\ &\geqslant \left[H(P_X) + \varepsilon|\mathcal{X}|\log\varepsilon\right]. \end{aligned} \tag{4.17}$$

Substituting (4.1), (4.16), and (4.17) in (4.15),

$$\frac{1}{n}H_{\chi}(G^n, P_X^n) \geqslant \left(1 - \frac{|\mathcal{X}|}{4n\varepsilon^2}\right)\left\{H(P_X) - \frac{1}{n}\log \alpha(G^n(\mathcal{T}^n_{P_X,\varepsilon})) + \varepsilon|\mathcal{X}|\log \varepsilon\right\}$$

for any n and $\varepsilon > 0$. Taking the lim inf of both sides,

$$\lim_{n\to\infty}\frac{1}{n}H_{\chi}(G^n, P_X^n) \geqslant H(P_X) - C_{\varepsilon}(G, P_X) + \varepsilon|\mathcal{X}|\log \varepsilon.$$

Since this is true for every ε, the result follows by letting $\varepsilon \to 0$ and using (4.11). ■

Even though neither (4.6) nor (4.12) provides a single-letter characterization for $R_{\min}(G, P_X)$ in general, (4.12) gives additional insight into the problem:

1. Known single-letter lower and upper bounds for complementary graph entropy given in (4.8) can immediately be translated to bounds for the minimum rate. Also, (4.9) provides a single-letter expression for $R_{\min}(G, P_X)$ when G is perfect.
2. $R_{\min}(G, P_X) = \overline{H}(G, P_X)$ reveals another asymptotically optimal variable-length coding scheme: Encode all the vertices in $\mathcal{T}^n_{P_X,\varepsilon}$ using roughly $n\overline{H}(G, P_X)$ bits and the rest of the vertices with roughly $n\log|\mathcal{X}|$ bits. This insight will be very useful when we analyze Extension 1 in the next section.

4.4 NETWORK EXTENSIONS

4.4.1 Extension 1: VLZE Coding When Side Information May Be Absent

We now consider the scenario depicted in Figure 4.1. Let $\phi_n^{(1)} : \mathcal{X}^n \to \{0,1\}^*$ and $\phi_n^{(2)} : \mathcal{X}^n \to \{0,1\}^*$ be the first- and the second-stage encoders, respectively. Also denote by $\varphi_n^{(1)} : \{0,1\}^* \times \mathcal{Y}^n \to \mathcal{X}^n$ and $\varphi_n^{(2)} : \{0,1\}^* \times \{0,1\}^* \to \mathcal{X}^n$ the decoders at the two stages. The corresponding rates are

$$R_i(\phi_n^{(i)}) = \frac{1}{n}\sum_{x^n\in\mathcal{X}^n} P_X^n(x^n)|\phi_n^{(i)}(x^n)|$$

for $i = 1, 2$. Our goal is to characterize the region of rates achieved by instantaneous codes that ensure both

$$\Pr[X^n \neq \varphi_n^{(1)}(\phi_n^{(1)}(X^n), Y^n)] = 0$$

and

$$\Pr[X^n \neq \varphi_n^{(2)}(\phi_n^{(1)}(X^n), \phi_n^{(2)}(X^n))] = 0.$$

We will focus on a restricted class of instantaneous coding schemes where the first-stage encoded bitstream can be uniquely parsed even without the help of the side information, to which the second-stage decoder does not have access. Note that in contrast to the single-stage problem, this prohibits possible use of codewords that prefix each other. Instead, we adopt coloring of G^n followed by prefix-free coding of colors, that is, $\phi_n^{(1)} = \psi_n \circ c_n$, where ψ_n is a prefix-free code. Both $\varphi_n^{(1)}$ and $\varphi_n^{(2)}$ can then instantaneously decode the color and must utilize it together with Y^n and $\phi_n^{(2)}(X^n)$, respectively, to successfully decode X^n. This in particular implies that $\phi_n^{(2)}(X^n)$ must be instantaneously decodable when the color is known.

Although this restriction could potentially result in increased rates, it does not, as the following theorem implies.

Theorem 4.3. There exists an instantaneous two-step coding strategy $(\phi_n^{(1)}, \phi_n^{(2)})$ that achieves

$$\lim_{n\to\infty} R_1(\phi_n^{(1)}) = \overline{H}(G, P_X) \tag{4.18}$$

$$\lim_{n\to\infty} \left[R_1(\phi_n^{(1)}) + R_2(\phi_n^{(2)})\right] = H(P_X) \tag{4.19}$$

where G is the confusability graph for P_{XY}.

PROOF. The definitions (4.4) and (4.10) imply, for any $\varepsilon > 0$ and $n > n_0(\varepsilon)$, the existence of a coloring $c_{\varepsilon,n}(\cdot)$ of G^n satisfying

$$|c_{\varepsilon,n}(\mathcal{T}^n_{P_X,\varepsilon})| \leqslant 2^{n[\overline{H}_\varepsilon(G,P_X)+\varepsilon]}$$

and

$$|c^{-1}_{\varepsilon,n}(i)| \leqslant 2^{n[C_\varepsilon(G,P_X)+\varepsilon]}$$

for all $i \in c_{\varepsilon,n}(\mathcal{T}^n_{P_X,\varepsilon})$. The latter follows from the fact that each color class is, by definition, an independent set. We can then use the following coding strategy: Denote by $\beta(\cdot|\mathcal{A}) : \mathcal{A} \to \{0,1\}^{\lceil \log_2 |\mathcal{A}| \rceil}$ the fixed-length canonical representation of elements in $\mathcal{A}$. Set the first-stage encoder to

$$\phi_n^{(1)}(x^n) = \begin{cases} 0 \cdot \beta(x^n|\mathcal{X}^n) & x^n \notin \mathcal{T}^n_{P_X,\varepsilon} \\ 1 \cdot \beta(c_{\varepsilon,n}(x^n)|c_{\varepsilon,n}(\mathcal{T}^n_{P_X,\varepsilon})) & x^n \in \mathcal{T}^n_{P_X,\varepsilon} \end{cases}$$

where $\cdot$ denotes concatenation. The decoder $\varphi_n^{(1)}$ can instantaneously recover all $x^n \notin \mathcal{T}^n_{P_X,\varepsilon}$, and $i = c_{\varepsilon,n}(x^n)$ whenever $x^n \in \mathcal{T}^n_{P_X,\varepsilon}$. In the latter case, x^n is also easily recovered since it is the unique vertex that is simultaneously in both $c^{-1}_{\varepsilon,n}(i)$ and the clique in G^n induced by y^n. The resultant first-stage rate is bounded by

$$\begin{aligned} R_1(\phi_n^{(1)}) &\leqslant \frac{1}{n} + P_X^n(\mathcal{T}^n_{P_X,\varepsilon})\left[\overline{H}_\varepsilon(G,P_X) + \varepsilon\right] + \left[1 - P_X^n(\mathcal{T}^n_{P_X,\varepsilon})\right]\log_2|\mathcal{X}| \\ &\leqslant \overline{H}_\varepsilon(G,P_X) + \frac{1}{n} + \varepsilon + \frac{|\mathcal{X}|}{4n\varepsilon^2}\log_2|\mathcal{X}| \end{aligned}$$

for sufficiently large n. Setting the second-stage encoder to

$$\phi_n^{(2)}(x^n) = \begin{cases} \lambda & x^n \notin \mathcal{T}^n_{P_X,\varepsilon} \\ \beta(x^n|c^{-1}_{\varepsilon,n}(i)) & x^n \in \mathcal{T}^n_{P_X,\varepsilon} \text{ and } c_{\varepsilon,n}(x^n) = i \end{cases}$$

where λ is the null codeword, we ensure that x^n is instantaneously recovered in the absence of the side information. Therefore, the second-stage rate is bounded by

$$R_2(\phi_n^{(2)}) \leq P_X^n(\mathcal{T}_{P_X,\varepsilon}^n)\left[C_\varepsilon(G,P_X)+\varepsilon\right]$$
$$\leq C_\varepsilon(G,P_X)+\varepsilon.$$

Letting $n \to \infty$ and then $\varepsilon \to 0$ yields an asymptotic rate pair not larger than $(\overline{H}(G,P_X), C(G,P_X))$. The proof is complete observing (4.11) ■

4.4.2 Extension 2: VLZE Coding with Compound Side Information

We will utilize $\mathcal{G}=\{G_1,G_2,\ldots,G_K\}$, the family of confusability graphs for pairs (X,Y_k), $k=1,\ldots,K$. For fixed-length coding, it has been observed in [13] that valid block codes have a one-to-one correspondence to colorings of

$$\mathcal{G}^n \triangleq \bigcup_{k=1}^{K} G_k^n.$$

That is because the encoder cannot send the same message for two different outcomes of X^n if they correspond to adjacent vertices in *any* G_k^n. On the other hand, a coloring of $\cup_k G_k^n$ is automatically a coloring for all G_k^n, and hence, each receiver can uniquely decode X^n upon knowing its color in $\cup_k G_k^n$ and the side information Y_k^n.

Turning to variable-length coding, an instantaneous code $\phi_n : \mathcal{X}^n \to \{0,1\}^*$ must satisfy

$$\phi_n(x^n) \neq_P \phi_n(x'^n)$$

for all $\{x^n, x'^n\}$ adjacent in $\mathcal{G}^n$. Therefore, similar to the original DSC scenario, the minimum achievable bit rate is purely characterized by $(\mathcal{G},P_X)$, and there is no further dependence on $P_{X,Y_1,\ldots,Y_K}$. Denote by $R_{\min}(\mathcal{G},P_X)$ the minimum asymptotically achievable rate. We now characterize this rate using chromatic entropy, thereby extending (4.6) to more than one receiver.

Theorem 4.4.

$$R_{\min}(\mathcal{G},P_X) = \lim_{n\to\infty} \frac{1}{n} H_\chi(\mathcal{G}^n, P_X^n). \tag{4.20}$$

PROOF. The proof follows the exact same lines as in the proof of Theorem 4.1. The only detail that needs to be filled in is the proof of subadditivity of the chromatic entropy with respect to AND-power for families of graphs, that is,

$$H_\chi(\mathcal{G}^{m+n}, P_X^{m+n}) \leq H_\chi(\mathcal{G}^m, P_X^m) + H_\chi(\mathcal{G}^n, P_X^n). \tag{4.21}$$

It is clear that the Cartesian product of colorings $c_m(\cdot)$ and $c_n(\cdot)$ for $\mathcal{G}^m$ and $\mathcal{G}^n$, respectively, constitutes a coloring for $\mathcal{G}^m \times \mathcal{G}^n$. Also observe that

$$\bigcup_k (G_k \times H_k) \subset \bigcup_k G_k \times \bigcup_k H_k$$

which implies

$$\bigcup_k G_k^{m+n} \subset \bigcup_k G_k^m \times \bigcup_k G_k^n.$$

Hence, any coloring for $\mathcal{G}^m \times \mathcal{G}^n$ is also a coloring for $\mathcal{G}^{m+n}$. In particular, if c_m and c_n above achieve $H_\chi(\mathcal{G}^m, P_X^m)$ and $H_\chi(\mathcal{G}^n, P_X^n)$, respectively, then $c_m \times c_n$ is a (potentially suboptimal) coloring for $\mathcal{G}^{m+n}$, proving (4.21). ■

Simonyi [13] generalized complementary graph entropy and graph capacity definitions for a family of graphs as

$$\overline{H}_\varepsilon(\mathcal{G}, P_X) = \limsup_{n\to\infty} \frac{1}{n} \log_2 \chi(\mathcal{G}^n(\mathcal{T}_{P_X,\varepsilon}^n))$$

$$C_\varepsilon(\mathcal{G}, P_X) = \limsup_{n\to\infty} \frac{1}{n} \log_2 \alpha(\mathcal{G}^n(\mathcal{T}_{P_X,\varepsilon}^n))$$

followed by

$$\overline{H}(\mathcal{G}, P_X) = \lim_{\varepsilon\to 0} \overline{H}_\varepsilon(\mathcal{G}, P_X)$$

$$C(\mathcal{G}, P_X) = \lim_{\varepsilon\to 0} C_\varepsilon(\mathcal{G}, P_X).$$

It was also proved in [13] that

$$C(\mathcal{G}, P_X) + \overline{H}(\mathcal{G}, P_X) = H(P_X) \tag{4.22}$$

which is a generalization of Marton's identity (4.11).

We now state the following theorem, which generalizes Theorem 4.2. We omit the proof because it is but a repetition of the proof of Theorem 4.2 except that G is to be replaced with $\mathcal{G}$ everywhere. That, in turn, is possible owing to (4.20) and (4.22).

Theorem 4.5.

$$R_{\min}(\mathcal{G}, P_X) = \overline{H}(\mathcal{G}, P_X).$$

The following theorem is our main result.

Theorem 4.6.

$$R_{\min}(\mathcal{G}, P_X) = \max_k R_{\min}(G_k, P_X)$$

where $R_{\min}(G_k, P_X)$ is the minimum achievable VLZE coding rate in the original DSC problem for the pair (X, Y_k).

PROOF. The proof is trivially similar to that of Theorem 1 in [13]. However, we provide it here for completeness. We need the key result of [6], stating that for any family of graphs $\mathcal{G}$ and probability distribution P,

$$C(\mathcal{G}, P) = \min_k C(G_k, P). \tag{4.23}$$

Then, using Theorems 4.2 and 4.5, (4.11), (4.22), and (4.23), we obtain

$$\begin{aligned} R_{\min}(\mathcal{G}, P_X) &= \overline{H}(\mathcal{G}, P_X) \\ &= H(P_X) - C(\mathcal{G}, P_X) \\ &= H(P_X) - \min_k C(G_k, P_X) \\ &= \max_k [H(P_X) - C(G_k, P_X)] \\ &= \max_k \overline{H}(G_k, P_X) \\ &= \max_k R_{\min}(G_k, P_X). \end{aligned}$$

■

Thus, asymptotically, no rate loss is incurred because of the need to transmit information to more than one receiver. In other words, conveying X^n to multiple receivers is no harder than conveying it to the most needy receiver.

4.5 VLZE CODE DESIGN

4.5.1 Hardness of Optimal Code Design

In this section we show that optimal VLZE code design in DSC is NP-hard. This essentially means that all NP-complete problems can be transformed to it in polynomial time. An NP-hard problem is "at least as hard" as all the NP-complete problems, since it cannot be solved in polynomial time unless P = NP, which is widely believed to be untrue.

To prove the NP-hardness of optimal VLZE code design, it suffices to prove that the following version of it is NP-complete:

PROBLEM: VLZE L-codability
INSTANCE: Graph $G = (\mathcal{X}, \mathcal{E})$, distribution P_X on $\mathcal{X}$, and a positive real number L.
QUESTION: Is there a VLZE code ϕ for (G, P_X) of rate $R(\phi) \leq L$?

That, in turn, requires a reduction from an NP-complete problem to this problem. Toward that end, we pick the problem of Graph N-colorability (GNC) for $N \geq 3$:

PROBLEM: GNC
INSTANCE: Graph $G = (\mathcal{X}, \mathcal{E})$.
QUESTION: Is G colorable with N colors?

This problem is well known to be NP-complete. Throughout this section, we will assume, without loss of generality, that the graphs under consideration do not have isolated vertices.

Theorem 4.7. [7] The VLZE L-codability problem is NP-complete.

PROOF. Let $G=(\mathcal{X},\mathcal{E})$ be an instance of GNC, with $N=4$. Defining auxiliary vertex sets $\mathcal{I}=\{i_1,i_2,i_3,i_4\}$, $\mathcal{J}=\{j_1,j_2,j_3,j_4\}$, and $\mathcal{K}=\{k_1,k_2,k_3,k_4\}$, construct a new graph $G'=(\mathcal{X}',\mathcal{E}')$ as follows:

$$\mathcal{X}'=\mathcal{X}\cup\mathcal{I}\cup\mathcal{J}\cup\mathcal{K}$$

$$\mathcal{E}'=\mathcal{E}\cup\mathcal{P}_2(\mathcal{I})\cup\mathcal{P}_2(\mathcal{J})\cup\mathcal{P}_2(\mathcal{K})\cup\Big\{\{i_a,j_b\},\{i_a,k_b\}:1\leq a,b\leq 4,a\neq b\Big\}$$
$$\cup\Big\{\{j_a,k_b\}:1\leq a,b\leq 4\Big\}\cup\Big\{\{j_a,x\}:1\leq a\leq 4,x\in\mathcal{X}\Big\}.$$

It is in fact easier to describe $\overline{G'}$, which has the edge set

$$\mathcal{E}'^c=\mathcal{E}^c\cup\Big\{\{i_a,j_a\},\{i_a,k_a\},1\leq a\leq 4\Big\}\cup\Big\{\{i_a,x\},\{k_a,x\}:1\leq a\leq 4,x\in\mathcal{X}\Big\}.$$

Figure 4.5 visualizes $\overline{G'}$. It can be verified that the edge structure of G' enforces the following constraints on the codewords $\phi(x)$:

1. $\phi(i_a)\neq_P\phi(i_b),\phi(j_a)\neq_P\phi(j_b),\phi(k_a)\neq_P\phi(k_b)$, for $1\leq a,b\leq 4,a\neq b$.
2. $\phi(j_a)\neq_P\phi(k_b)$ for $1\leq a,b\leq 4$.
3. $\phi(i_a)\neq_P\phi(j_b),\phi(i_a)\neq_P\phi(k_b)$ for $1\leq a,b\leq 4,a\neq b$.
4. $\phi(j_a)\neq_P\phi(x)$ for $1\leq a\leq 4$ and $x\in\mathcal{X}$.

Also assign the probability distribution P on $\mathcal{X}'$ such that, for $x\in\mathcal{X}'$,

$$P_X(x)=\begin{cases}\frac{\varepsilon}{|\mathcal{X}|} & \text{if } x\in\mathcal{X}\\ \frac{1-\varepsilon}{12} & \text{else.}\end{cases}$$

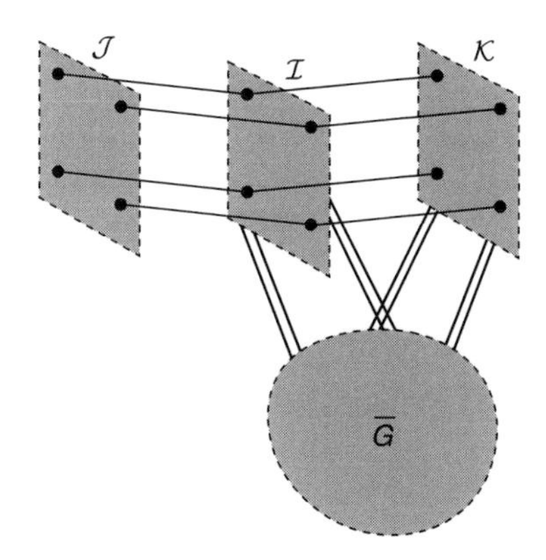

FIGURE 4.5

The graph $\overline{G'}$. Pairs of double lines between two subgraphs indicate that every vertex in one subgraph is connected to every vertex in the other.

We claim that the optimal VLZE code ϕ^* for G' also minimizes

$$R_{\text{aux}}(\phi) = \sum_{x\in\mathcal{X}'-\mathcal{X}} |\phi(x)|$$

provided $\varepsilon < \frac{1}{12|\mathcal{X}|+25}$. That is because for such a choice of ε, if

$$\sum_{x\in\mathcal{X}'-\mathcal{X}} |\phi^*(x)| > \sum_{x\in\mathcal{X}'-\mathcal{X}} |\phi(x)|$$

for some other code ϕ, then

$$\begin{aligned}
\sum_{x\in\mathcal{X}'} P_X(x)\Big(|\phi^*(x)|-|\phi(x)|\Big) &= \frac{1-\varepsilon}{12}\sum_{x\in\mathcal{X}'-\mathcal{X}}\Big(|\phi^*(x)|-|\phi(x)|\Big) \\
&\quad -\frac{\varepsilon}{|\mathcal{X}|}\sum_{x\in\mathcal{X}}\Big(|\phi(x)|-|\phi^*(x)|\Big) \\
&\geqslant \frac{1-\varepsilon}{12}-\frac{\varepsilon}{|\mathcal{X}|}|\mathcal{X}|\Big((|\mathcal{X}|+3)-1\Big) \qquad (4.24) \\
&> 0
\end{aligned}$$

resulting in a contradiction, where (4.24) follows from the (impossible) worst case scenario where (i) for $x\in\mathcal{X}'-\mathcal{X}, \phi^*(x)$ and $\phi(x)$ are identical except for one place where they differ by 1 bit, and (ii) for $x\in\mathcal{X}, \phi^*(x)=1$ and $\phi(x)=|\mathcal{X}|+3$. $|\mathcal{X}|+3$ bits are indeed sufficient for $x\in\mathcal{X}$ as one can always choose a code such that

$$\phi(i_1)=00,\ \phi(i_2)=01,\ \phi(i_3)=10,\ \phi(i_4)=11, \qquad (4.25)$$

$$\phi(j_1)=000,\ \phi(j_2)=010,\ \phi(j_3)=100,\ \phi(j_4)=110, \quad \text{and} \qquad (4.26)$$

$$\phi(k_1)=001,\ \phi(k_2)=011,\ \phi(k_3)=101 \text{ and } \phi(k_4)=111 \qquad (4.27)$$

together with $\phi(x)=001\cdot 1_x$ for $x\in\mathcal{X}$, where 1_x is a length-$|\mathcal{X}|$ binary indicator vector assuming the value 1 only at a location corresponding to x.

In fact, it is easy to show that (4.25)-(4.27) minimize $R_{\text{aux}}(\phi)$. Using the above argument, one can therefore assume, without loss of generality, $\phi^*(x)=\phi(x)$ as given in (4.25)-(4.27) for all $x\in\mathcal{X}'-\mathcal{X}$. The corresponding minimum rate is then given by

$$\begin{aligned}
R(\phi^*) &= \frac{\varepsilon}{|\mathcal{X}|}\sum_{x\in\mathcal{X}}|\phi^*(x)| + \frac{1-\varepsilon}{12}\{4\cdot 2+8\cdot 3\} \\
&= \frac{\varepsilon}{|\mathcal{X}|}\sum_{x\in\mathcal{X}}|\phi^*(x)| + \frac{8(1-\varepsilon)}{3}.
\end{aligned}$$

Now, suppose that G is 4-colorable. Then the four color classes may be assigned the four 3-bit codewords that do not appear in $\mathcal{J}$, namely, 001, 011, 101, and 111, so that

$$R(\phi^*) = \frac{\varepsilon}{|\mathcal{X}|}\cdot 3|\mathcal{X}| + \frac{8(1-\varepsilon)}{3} = \frac{8+\varepsilon}{3}.$$

Conversely, suppose G is not 4-colorable. Then $|\phi^*(x)|>3$ for at least one node $x\in\mathcal{X}$, so that

$$R(\phi^*) > \frac{8+\varepsilon}{3}.$$

Thus G is 4-colorable if and only if there exists a VLZE code for (G', P_X) of rate $\frac{8+\varepsilon}{3}$ bits. Setting $L = \frac{8+\varepsilon}{3}$ in the instance of the VLZE L-codability problem completes the reduction. ■

4.5.2 Hardness of Coding with Length Constraints

Yan and Berger [17] considered an even simpler problem in which the instance is as follows.

PROBLEM: VLZE Coding with Length Constraints
INSTANCE: Graph $G = (\mathcal{X}, \mathcal{E})$ and nonnegative integers l_x for each $x \in \mathcal{X}$.
QUESTION: Is there a VLZE code ϕ for (G, P_X) so that $|\phi(x)| = l_x$ for all $x \in \mathcal{X}$?

The corresponding problem in point-to-point coding is well known to have a simple solution. That is, it suffices to check whether the Kraft inequality holds, that is,

$$\sum_{x \in \mathcal{X}} 2^{-l_x} \leqslant 1.$$

In contrast, Yan and Berger [17] proved the following theorem with a somewhat involved proof. We provide a very simple alternative proof.

Theorem 4.8. The problem of VLZE Coding with Length Constraints is NP-complete.

PROOF. Consider the GNC problem with $N = 4$. Given $G = (\mathcal{X}, \mathcal{E})$, we can set $l_x = 2$ for all $x \in \mathcal{X}$ and create an instance of the current problem. The straightforward observation that G is 4-colorable if and only if there exists a VLZE code ϕ with $|\phi(x)| = 2$ for all $x \in \mathcal{X}$ finishes the proof. ■

In [15], VLZE coding with given $|\phi(x)|$ was analyzed from a clique-based Kraft-sum point of view. The following theorem, whose proof we do not include due to space constraints, was proven.

Theorem 4.9. If for all graphs G in some family $\mathcal{G}$

$$\sum_{x \in z} 2^{-l_x} \leqslant \alpha \quad \forall z \in \Omega(G) \tag{4.28}$$

is sufficient for the existence of a VLZE code ϕ with given codeword lengths $|\phi(x)| = l_x$, then for any $G \in \mathcal{G}$, one can construct a code ϕ satisfying

$$R(\phi) - R_{\min}(G, P_X) \leqslant 1 - \log \alpha.$$

Tuncel [15] also showed that Equation (4.28) with $\alpha = \frac{1}{2}$ and $\alpha = \frac{1}{4}$ are indeed sufficient conditions for codes to exist for the classes of graphs with three and four cliques, respectively. Previously, Yan and Berger [17] had shown the sufficiency of $\alpha = 1$ for the class of graphs with only two cliques.

4.5.3 An Exponential-time Optimal VLZE Code Design Algorithm

In this section, we provide the algorithm proposed by Koulgi et al. [7] for optimal VLZE code design. The algorithm is inevitably exponential-time. Before describing the algorithm, we extend the notion of an induced subgraph by also inducing probabilities of vertices. To that end, for a graph $G=(\mathcal{X},\mathcal{E})$ and P_X, define for any subset $\mathcal{X}'$

$$P(\mathcal{X}')=\sum_{x\in\mathcal{X}'}P_X(x)$$

and

$$P_{X|\mathcal{X}'}(x)=\frac{P_X(x)}{P(\mathcal{X}')}.$$

Also define the weighted codeword length of the subgraph $G(\mathcal{X}')$ as

$$L(\mathcal{X}')=P(\mathcal{X}')R^1_{\min}(G(\mathcal{X}'),P_{X|\mathcal{X}'}).$$

Then, for any code ϕ and codeword i, we have

$$L(\phi^{-1}(i*))=L(\phi^{-1}(i0*))+L(\phi^{-1}(i1*))+P(\phi^{-1}(i*))-P(\phi^{-1}(i)) \tag{4.29}$$

where $i*$ denotes the set of all codewords that are prefixed by i, provided that none of the sets above are empty. If, on the other hand, ϕ is the optimal code achieving $R^1_{\min}(G,P_X)$, Equation (4.29) can be recast as

$$\begin{aligned}L(\phi^{-1}(i*))=&\min_{\mathcal{D}\subseteq\phi^{-1}(i*)-\phi^{-1}(i)}\{L(\mathcal{D})+L(\phi^{-1}(i*)-\phi^{-1}(i)-\mathcal{D})\}\\&+P(\phi^{-1}(i*))-P(\phi^{-1}(i)).\end{aligned} \tag{4.30}$$

Note that the vertices in the set $\phi^{-1}(i)$ must be isolated in $G(\phi^{-1}(i))$, because otherwise they would violate the prefix condition of VLZE codes. Using this observation together with (4.30) suggests an iterative algorithm whereby given any $\mathcal{X}'\subseteq\mathcal{X}$, isolated vertices $\mathcal{I}(\mathcal{X}')$ in $G(\mathcal{X}')$ are identified and

$$L(\mathcal{X}')=\min_{\mathcal{D}\subseteq\mathcal{X}'-\mathcal{I}(\mathcal{X}')}\left\{L(\mathcal{D})+L(\mathcal{X}'-\mathcal{I}(\mathcal{X}')-\mathcal{D})\right\}+P(\mathcal{X}')-P(\mathcal{I}(\mathcal{X}'))$$

is computed, with the terminating condition that $L(\mathcal{X}')=0$ if $\mathcal{I}(\mathcal{X}')=\mathcal{X}'$.

In fact, it is not necessary to go over all subsets $\mathcal{D}\subseteq\mathcal{X}'-\mathcal{I}(\mathcal{X}')$. As was argued in [7], it suffices to consider only those $\mathcal{D}$ that induce a *dominating 2-partition*, that is, where every vertex in $\mathcal{D}$ is connected to some vertex in $\mathcal{X}'-\mathcal{I}(\mathcal{X}')-\mathcal{D}$ and vice versa. This follows by the observation that (i) if a vertex in $\phi^{-1}(0*)-\phi^{-1}(0)$ is not connected to any vertex in $\phi^{-1}(1*)$, then it can instead be assigned the codeword 1, and (ii) if a vertex in $\phi^{-1}(0)$ is not connected to any vertex in $\phi^{-1}(1*)$, then it can instead be assigned the null codeword since it must be isolated. Since both actions reduce the rate, ϕ cannot be optimal.

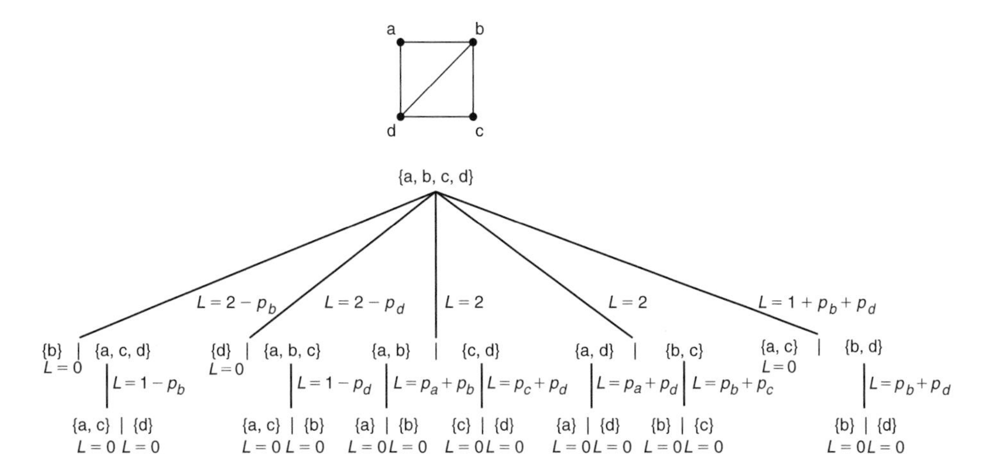

FIGURE 4.6

The recursive algorithm, which is terminated within two levels for this example. Values indicated on branches of the tree represent values of $L(\mathcal{X}')$ returned for each dominating 2-partition.

In Figure 4.6, this recursive algorithm is demonstrated using a very small perfect graph with vertices $\mathcal{X} = \{a, b, c, d\}$ and associated probabilities $\{p_a, p_b, p_c, p_d\}$. Considering the simplest case, if $p_a = p_b = p_c = p_d = 0.25$, one obtains

$$L(\mathcal{X}) = R^1_{\min}(G, P_X) = 1.5$$

achieved by $\phi(a) = \phi(c) = 0$, $\phi(b) = 10$, $\phi(d) = 11$. In this case, the rate can be shown to coincide with $R_{\min}(G, P_X) = \overline{H}(G, P_X) = H(G, P_X)$. On the other hand, if $p_a = \frac{1}{6}, p_b = \frac{1}{3}, p_c = \frac{1}{6}, p_d = \frac{1}{3}$, the optimal code $\phi(a) = 10$, $\phi(b) = 0$, $\phi(c) = 10$, $\phi(d) = 11$ achieves $R^1_{\min}(G, P_X) = \frac{5}{3} \approx 1.667$, whereas

$$R_{\min}(G, P_X) = \log_2 3 \approx 1.585.$$

4.6 CONCLUSIONS

As we discussed in this chapter, the behavior of distributed source coding with zero error is significantly different from the better-known vanishingly small error regime. Unfortunately, it is also considerably less understood. Single-letter characterization of achievable rates seems impossible because of the connection of zero-error coding to the notoriously difficult problem of graph capacity. The task of designing optimal VLZE codes is also difficult in that unless P = NP, optimal VLZE code design takes exponential time.

What is discovered so far is not insignificant, however. In particular, as we argued, the fact that complementary graph entropy characterizes the asymptotically achievable minimum rate for VLZE coding is instrumental in proving that there is no rate loss in certain network extensions of the main DSC problem. Also, single-letter bounds on complementary graph entropy, which are known to be tight for perfect graphs, are directly inherited for the minimum rate. On the code design front, the exponential time algorithm outlined in Section 4.5.3 is still useful if the size of the graph is small. Koulgi et al. [7] also proposed approximate polynomial time algorithms. Also, as Theorem 4.9 states, one can upper bound the redundancy of optimal codes for a class of graphs if the Kraft sum on each clique can be shown to be sufficient for the class.

REFERENCES

[1] Alon, N., and Orlitsky, A. A lower bound on the expected length of one-to-one codes. *IEEE Transactions on Information Theory*, 40(5):1670–1672, 1994.

[2] Alon, N., and Orlitsky, A. Source coding and graph entropies. *IEEE Transactions on Information Theory*, 42(5):1329–1339, 1996.

[3] Berge, C. *Graphs and Hypergraphs*. North-Holland, Amsterdam, 1973.

[4] Csiszár, I., and Körner, J. *Information Theory, Coding Theorems for Discrete Memoryless Systems*, Academic Press, New York, 1982.

[5] Csiszár, I., Körner, J., Lovász, L., Marton, K., and Simonyi, G. Entropy splitting for antiblocking corners and perfect graphs. *Combinatorica*, 10(1), 1990.

[6] Gargano, L., Körner, J., and Vaccaro, U. Capacities: From information theory to extremal set theory. *Journal of Combinatorial Theory Series A*, 68(2):296–316, 1994.

[7] Koulgi, P., Tuncel, E., Regunathan, S., and Rose, K. On zero-error source coding with decoder side information. *IEEE Transactions on Information Theory*, 49(1):99–111, 2003.

[8] Koulgi, P., Tuncel, E., Regunathan, S., and Rose, K. On zero-error coding of correlated sources. *IEEE Transactions on Information Theory*, 49(11):2856–2873, 2003.

[9] Körner, J. Coding of an information source having ambiguous alphabet and the entropy of graphs. *Trans. of the 6th Prague Conference on Information Theory, etc.*, Academia, Prague, 411–425, 1973.

[10] Körner, J., and Longo, G. Two-step encoding of finite memoryless sources. *IEEE Transactions on Information Theory*, 19(6):778–782, 1973.

[11] Körner, J., and Marton, K. Graphs that split entropies. *SIAM J. Discrete Math.*, 1, 71–79, 1988.

[12] Marton, K. On the Shannon capacity of probabilistic graphs. *Journal of Combinatorial Theory, Series B*, 57(2):183–195, 1993.

[13] Simonyi, G. On Witsenhausen's zero-error rate for multiple sources. *IEEE Transactions on Information Theory*, 49(12):3258–3261, 2003.

[14] Slepian, D., and Wolf, J. K. Noiseless coding of correlated information sources. *IEEE Transactions on Information Theory*, 19(4):471–480, 1973.

[15] Tuncel, E. Kraft inequality and zero-error source coding with decoder side information. *IEEE Transactions on Information Theory*, 53(12):4810–4816, 2007.

[16] Witsenhausen, H. S. The zero-error side information problem and chromatic numbers. *IEEE Transactions on Information Theory*, 22(5):592–593, 1976.

[17] Yan, Y., and Berger, T. Zero-error instantaneous coding of correlated sources with length constraints is NP-complete. *IEEE Transactions on Information Theory*, 52(4):1705–1708, 2006.

CHAPTER

Distributed Coding of Sparse Signals

Vivek K Goyal
Department of Electrical Engineering and Computer Science, Massachusetts Institute of Technology, Cambridge, MA

Alyson K. Fletcher
Department of Electrical Engineering and Computer Sciences, University of California, Berkeley, CA

Sundeep Rangan
Qualcomm Flarion Technologies, Bridgewater, NJ

CHAPTER CONTENTS

5.1 INTRODUCTION

Recent results in compressive sampling [1] have shown that sparse signals can be recovered from a small number of random measurements. The concept of random measurement, which we review in this section, is such that the generation of measurements may be considered to be distributed. Thus one is led to the question of whether distributed coding of random measurements can provide an efficient representation of sparse signals in an information-theoretic sense. Through both theoretical and experimental results, we show that encoding a sparse signal through distributed *regular* quantization of random measurements incurs a significant penalty relative to centralized encoding of the sparse signal. Information theory provides alternative quantization and encoding strategies, but they come at the cost of much greater estimation complexity.

5.1.1 Sparse Signals

Since the 1990s, modeling signals through sparsity has emerged as an important and widely applicable technique in signal processing. Its most well-known success is in image processing, where great advances in compression and estimation have come from modeling images as sparse in a wavelet domain [2].

In this chapter we use a simple, abstract model for sparse signals. Consider an N-dimensional vector x that can be represented as $x = Vu$, where V is some orthogonal N-by-N matrix and $u \in \mathbb{R}^N$ has only K nonzero entries. In this case, we say that u is *K-sparse* and that x is *K-sparse with respect to V*. The set of positions of nonzero coefficients in u is called the *sparsity pattern*, and we call $\alpha = K/N$ the *sparsity ratio*.

Knowing that x is K-sparse with respect to a given basis V can be extremely valuable for signal processing. For example, in compression, x can be represented by the K positions and values of the nonzero elements in u, as opposed to the N elements of x. When the sparsity ratio α is small, the compression gain can be significant. Similarly, in estimating x in the presence of noise, one only has to estimate K (as opposed to N) real parameters.

Another important property of sparse signals has recently been uncovered: they can be recovered in a computationally tractable manner from a relatively small number of random samples. The method, known as *compressive sampling* (sometimes called *compressed sensing* or *compressive sensing*), was developed in [3–5] and is detailed nicely in several articles in the March 2008 issue of *IEEE Signal Processing Magazine*. (This chapter is adapted from [6] with permission of the IEEE.)

A basic model for compressive sampling is shown in Figure 5.1. The N-dimensional signal x is assumed to be K-sparse with respect to some orthogonal matrix V. The "sampling" of x is represented as a linear transformation by a matrix Φ yielding a sample vector $y = \Phi x$. Let the size of Φ be M-by-N, so y has M elements; we call each element of y a measurement of x. A decoder must recover the signal x from y knowing V and Φ, but not necessarily the sparsity pattern of the unknown signal u.

Since u is K-sparse, x must belong to one of $\binom{N}{K}$ subspaces in $\mathbb{R}^N$. Similarly, y must belong to one of $\binom{N}{K}$ subspaces in $\mathbb{R}^M$. For almost all Φs with $M \geqslant K+1$, an exhaustive search through the subspaces can determine which subspace x belongs to and thereby recover the signal's sparsity pattern and values. Therefore, in principle, a K-sparse signal can be recovered from as few as $M = K+1$ random samples.

Unfortunately, the exhaustive search we have described above is not tractable for interesting sizes of problems since the number of subspaces to search, $\binom{N}{K}$, can be enormous; if α is held constant as N is increased, the number of subspaces grows

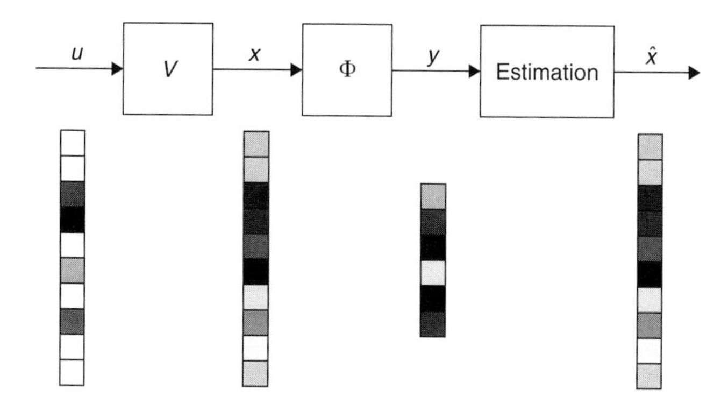

FIGURE 5.1

Block diagram representation of compressive sampling. The signal x is sparse with respect to V, meaning that $u = V^{-1}x$ has only a few nonzero entries. $y = \Phi x$ is "compressed" in that it is shorter than x. (White boxes represent zero elements.)

exponentially with N. The remarkable main result of compressive sampling is to exhibit recovery methods that are computationally feasible, numerically stable, and robust against noise while requiring a number of measurements not much larger than K.

5.1.2 Signal Recovery with Compressive Sampling

Compressive sampling is based on recovering x via convex optimization. When we observe $y = \Phi x$ and x is sparse with respect to V, we are seeking x consistent with y and such that $V^{-1}x$ has few nonzero entries. To try to minimize the number of nonzero entries directly yields an intractable problem [7]. Instead, solving the optimization problem

$$\text{(LP reconstruction)} \qquad \hat{x}_{\text{LP}} = \underset{x\,:\,y=\Phi x}{\operatorname{argmin}} \|V^{-1}x\|_1$$

often gives exactly the desired signal recovery, and there are simple conditions that guarantee exact recovery. Following pioneering work by Logan in the 1960s, Donoho and Stark [8] obtained results that apply, for example, when V is the N-by-N identity matrix and the rows of Φ are taken from the matrix representation of the length-N discrete Fourier transform (DFT). Subsequent works considered randomly selected rows from the DFT matrix [3] and then certain other random matrix ensembles [4, 5]. In this chapter, we will concentrate on the case when Φ has independent Gaussian entries.

A central question is: How many measurements M are needed for LP reconstruction to be successful? Since Φ is random, there is always a chance that reconstruction will fail. We are interested in how M should scale with signal dimension N and sparsity K so that the probability of success approaches 1. A result of Donoho and Tanner [9] indicates that $M \sim 2K\log(N/K)$ is a sharp threshold for successful recovery. Compared to the intractable exhaustive search through all possible subspaces, LP recovery requires only a factor $2\log(N/K)$ more measurements.

If measurements are subject to additive Gaussian noise so that $\hat{y} = \Phi x + \eta$ is observed, with $\eta \sim \mathcal{N}(0, \sigma^2)$, then the LP reconstruction should be adjusted to allow slack in the constraint $y = \Phi x$. A typical method for reconstruction is the following convex optimization:

$$\text{(Lasso reconstruction)} \qquad \hat{x}_{\text{Lasso}} = \underset{x}{\operatorname{argmin}} \left(\|\hat{y} - \Phi x\|_2^2 + \lambda \|V^{-1}x\|_1 \right),$$

where the parameter $\lambda > 0$ trades off data fidelity and reconstruction sparsity. The best choice for λ depends on the variance of the noise and problem size parameters. Wainwright [10] has shown that, when the signal-to-noise ratio (SNR) scales as SNR $= O(N)$, the scaling $M \sim 2K \log(N - K) + K$ is a sharp threshold for $V^{-1}\hat{x}_{\text{Lasso}}$ to have the correct sparsity pattern with high probability. While this M may be much smaller than N, it is significantly more measurements than required in the noiseless case. The same scaling with respect to N and K has been shown to be necessary for the success of any algorithm and sufficient for the success of a very simple algorithm [11]; various reconstruction algorithms exhibit different dependence only on SNR and relative sizes of the nonzero components of x.

5.2 COMPRESSIVE SAMPLING AS DISTRIBUTED SOURCE CODING

In the remainder of this chapter, we will be concerned with the trade-off between quality of approximation and the number of bits of storage for a signal x that is K-sparse with respect to orthonormal basis V. In contrast to the emphasis on number of measurements M in Section 5.1, the currency in which we denominate the cost of a representation is bits rather than real coefficients.

In any compression involving scalar quantization, the choice of coordinates is key. Traditionally, signals to be compressed are modeled as jointly Gaussian vectors. These vectors can be visualized as lying in an ellipsoid, since this is the shape of the level curves of their probability density (see left panel of Figure 5.2). Source coding theory for jointly Gaussian vectors suggests choosing orthogonal coordinates aligned with the principal axes of the ellipsoid (the Karhunen–Loève basis) and then allocating bits to the dimensions based on their variances. This gives a *coding gain* relative to arbitrary coordinates [12]. For high-quality (low-distortion) coding, the coding gain is a constant number of bits per dimension that depends on the eccentricity of the ellipse. Geometrically, the results are similar if changes of coordinates can only be applied separately to disjoint subsets of the coordinates [13].

Sparse signal models are geometrically quite different from jointly Gaussian vector models. Instead of being visualized as ellipses, they yield unions of subspaces (see right panel of Figure 5.2). A natural encoding method for a signal x that is K-sparse with respect to V is to identify the subspace containing x and then quantize within the subspace, spending a number of bits proportional to K. Note that doing this requires that the encoder know V and that there is a cost to communicating the subspace

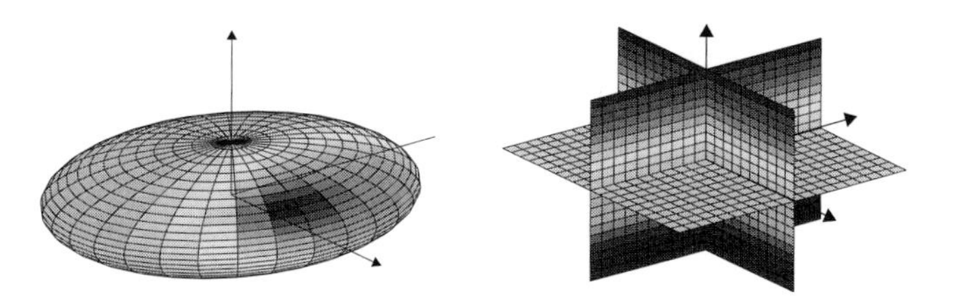

FIGURE 5.2

Left: Depiction of Gaussian random vectors as an ellipsoid. Classical rate-distortion theory and transform coding results are for this sort of source, which serves as a good model for discrete cosine transform (DCT) coefficients of an image or modified discrete cosine transform (MDCT) coefficients of audio. Right: Depiction of 2-sparse signals in $\mathbb{R}^3$, which form a union of three subspaces. This serves as a good conceptual model for wavelet coefficients of images.

Table 5.1 Performance Summary: Distortions for Several Scenarios When N Is Large with $\alpha = K/N$ Held Constant. Rate R and Distortion D Are Both Normalized by K. J Represents the Sparsity Pattern of $u = V^{T}x$. The Boxed Entry Is a Heuristic Analysis of the Compressive Sampling Case. $H(\cdot)$ Represents the Binary Entropy Function, and the Rotational Loss R^* Satisfies $R^* = O(\log R)$

		Encoder	
		Centralized (code J and nonzeros of u)	**Distributed (scalar coding of Φx)**
	Knows J a priori	$c2^{-2R}$	$c2^{-2(R-R^*)}$
Decoder	Is told J	$c2^{-2(R-H(\alpha)/\alpha)}$	$c2^{-2(R-H(\alpha)/\alpha-R^*)}$
	Infers J	N/A	$\boxed{c\delta(\log N)2^{-2\delta R}}$

index, denoted J, which will be detailed later. With all the proper accounting, when $K \ll N$, the savings is more dramatic than just a constant number of bits.

Following the compressive sampling framework, one obtains a rather different way to compress x: quantize the measurements $y = \Phi x$, with Φ and V known to the decoder. Since Φ spreads the energy of the signal uniformly across the measurements, each measurement should be allocated the same number of bits. The decoder should estimate x as well as it can; we will not limit the computational capability of the decoder.

How well will compressive sampling work? It depends both on how much it matters to use the best basis (V) rather than a set of random vectors (Φ) and on how much the quantization of y affects the decoder's ability to infer the correct subspace. We separate these issues, and our results are previewed and summarized in Table 5.1. We will derive the first three entries and then the boxed result, which requires much more explanation. But first we will establish the setting more concretely.

5.2.1 Modeling Assumptions

To reflect the concept that the orthonormal basis V is not used in the sensor/encoder, we model V as random and available only at the estimator/decoder. It is chosen uniformly at random from the set of orthogonal matrices. The source vector x is also random; to model it as K-sparse with respect to V, we let $x = Vu$ where $u \in \mathbb{R}^N$ has K nonzero entries in positions chosen uniformly at random. As depicted in Figure 5.3, we denote the nonzero entries of u by $u_K \in \mathbb{R}^K$ and let the discrete random variable J represent the sparsity pattern. Note that both V and Φ can be considered side information available at the decoder but not at the encoder.

Let the components of u_K be independent and Gaussian $\mathcal{N}(0, 1)$. Observe that $\mathbf{E}\left[\|u\|^2\right] = K$, and since V is orthogonal we also have $\mathbf{E}\left[\|x\|^2\right] = K$. For the measurement matrix Φ, let the entries be independent $\mathcal{N}(0, 1/K)$ and independent of V and u. This normalization makes each entry of y have unit variance.

Let us now establish some notation to describe scalar quantization. When scalar y_i is quantized to yield $\hat{y}_i$, it is convenient to define the relative quantization error $\beta = \mathbf{E}\left[|y_i - \hat{y}_i|^2\right] / \mathbf{E}\left[|y_i|^2\right]$ and then further define $\rho = 1 - \beta$ and $v_i = \hat{y}_i - \rho y_i$. These definitions yield a gain-plus-noise notation $\hat{y}_i = \rho y_i + v_i$, where

$$\sigma_v^2 = \mathbf{E}\left[|v_i|^2\right] = \beta(1-\beta)\mathbf{E}\left[|y_i|^2\right], \tag{5.1}$$

to describe the effect of quantization. Quantizers with optimal (centroid) decoders result in v being uncorrelated with y [14, Lemma 5.1]; other precise justifications are also possible [15].

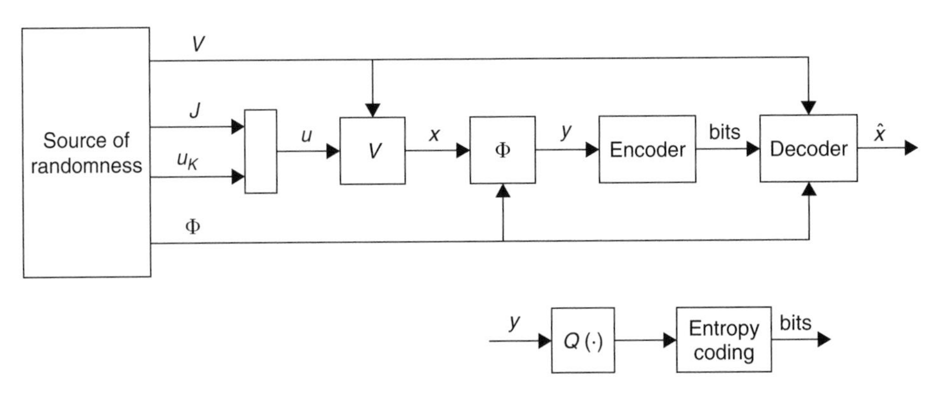

FIGURE 5.3

Block diagram representation of the compressive sampling scenario analyzed information-theoretically. V is a random orthogonal matrix; u is a K-sparse vector with $\mathcal{N}(0,1)$ nonzero entries, and Φ is a Gaussian measurement matrix. More specifically, the sparsity pattern of u is represented by J, and the nonzero entries are denoted u_K. In the initial analysis, the encoding of $y = \Phi x$ is by scalar quantization and scalar entropy coding.

In subsequent analyses, we will want to relate β to the rate (number of bits) of the quantizer. The exact value of β depends not only on the rate R, but also on the distribution of y_i and the particular quantization method. However, the scaling of β with R is as 2^{-2R} under many different scenarios (see the Appendix to this chapter). We will write

$$\beta = c2^{-2R} \tag{5.2}$$

without repeatedly specifying the constant $c \geqslant 1$.

With the established notation, the overall quantizer output vector can be written as

$$\hat{y} = \rho\Phi Vu + v = Au + v, \tag{5.3}$$

where $A = \rho\Phi V$. The overall source coding and decoding process, with the gain-plus-noise representation for quantization, is depicted in Figure 5.4. Our use of (5.3) is to enable easy analysis of linear estimation of x from $\hat{y}$.

5.2.2 Analyses

Since the sparsity level K is the inherent number of degrees of freedom in the signal, we will let there be KR bits available for the encoding of x and also normalize the distortion by K: $D = \frac{1}{K}\mathbf{E}\left[\|x - \hat{x}\|^2\right]$. Where applicable, the number of measurements M is a design parameter that can be optimized to give the best distortion-rate trade-off. In particular, increasing M gives better conditioning of certain matrices, but it reduces the number of quantization bits per measurement.

Before analyzing the compressive sampling scenario (Figure 5.3), we consider some simpler alternatives, yielding the first three entries in Table 5.1.

5.2.2.1 *Signal in a Known Subspace*

If the sparsifying basis V and subspace J are fixed and known to both (centralized) encoder and decoder, the communication of x can be accomplished by sending quantized versions of the nonzero entries of $V^{-1}x$. Each of the K nonzero entries has unit variance and is allotted R bits, so $D(R) = c2^{-2R}$ performance is obtained, as given by the first entry in Table 5.1.

5.2.2.2 *Adaptive Encoding with Communication of J*

Now suppose that V is known to both encoder and decoder, but the subspace index J is random, uniformly selected from the $\binom{N}{K}$ possibilities. A natural *adaptive* (and

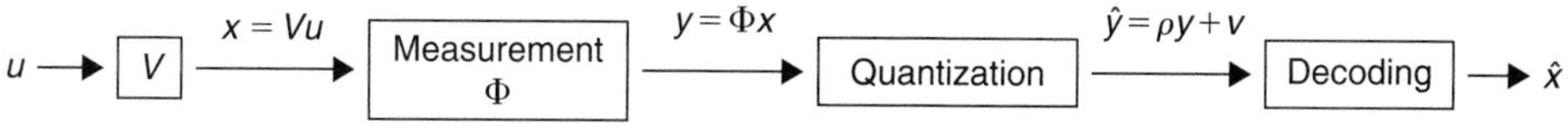

u → V → $x = Vu$ → Measurement Φ → $y = \Phi x$ → Quantization → $\hat{y} = \rho y + v$ → Decoding → $\hat{x}$

FIGURE 5.4

Source coding of x with additive noise representation for quantization.

centralized) approach is to spend $\log_2 \binom{N}{K}$ bits to communicate J and the remaining available bits to quantize the nonzero entries of $V^{-1}x$. Defining $R_0 = \frac{1}{K}\log_2 \binom{N}{K}$, the encoder has $KR - KR_0$ bits for the K nonzero entries of $V^{-1}x$ and thus attains performance

$$D_{\text{adaptive}}(R) = c2^{-2(R-R_0)}, \qquad R \geqslant R_0. \tag{5.4}$$

When K and N are large with the ratio $\alpha = K/N$ held constant, $\log_2 \binom{N}{K} \approx NH(\alpha)$ where $H(p) = -p\log_2 p - (1-p)\log_2(1-p)$ is the *binary entropy function* [16, p. 530]. Thus $R_0 \approx H(\alpha)/\alpha$, giving a second entry in Table 5.1.

If R does not exceed R_0, then the derivation above does not make sense, and even if R exceeds R_0 by a small amount, it may not pay to communicate J. A *direct* approach is to simply quantize each component of x with KR/N bits. Since the components of x have variance K/N, performance of $\mathbf{E}\left[(x_i - \hat{x}_i)^2\right] \leqslant c(K/N)2^{-2KR/N}$ can be obtained, yielding overall performance

$$D_{\text{direct}}(R) = c2^{-2KR/N}. \tag{5.5}$$

By choosing the better between (5.4) and (5.5) for a given rate, one obtains a simple baseline for the performance using V at the encoder. A convexification by time sharing could also be applied, and more sophisticated techniques are presented in [17].

5.2.2.3 *Loss from Random Measurements*

Now let us try to understand in isolation the effect of observing x only through Φx. The encoder sends a quantized version of $y = \Phi x$, and the decoder knows V and the sparsity pattern J.

From Equation (5.3), the decoder has $\hat{y} = \rho\Phi Vu + v$ and knows which K elements of u are nonzero. The performance of a linear estimate of the form $\hat{x} = F(J)\hat{y}$ will depend on the singular values of the M-by-K matrix formed by the K relevant columns of ΦV.[1] Using elementary results from random matrix theory, one can find how the distortion varies with M and R. (The distortion does not depend on N because the zero components of u are known.) The analysis given in [21] shows that for moderate to high R, the distortion is minimized when $K/M \approx 1 - ((2\ln 2)R)^{-1}$. Choosing the number of measurements accordingly gives performance

$$D_J(R) \approx 2(\ln 2)eR \cdot c2^{-2R} = c2^{-2(R-R^*)} \tag{5.6}$$

where $R^* = \frac{1}{2}\log_2(2\,(\ln 2)eR)$, giving the third entry in Table 5.1. Comparing to $c2^{-2R}$, we see that having access only to separately quantized random measurements induces a significant performance loss.

One interpretation of this analysis is that the coding rate has effectively been reduced by R^* bits per degree of freedom. Since R^* grows sublinearly with R, the

---

[1] One should expect a small improvement—roughly a multiplication of the distortion by K/M—from the use of a nonlinear estimate that exploits the boundedness of quantization noise [18, 19]. The dependence on ΦV is roughly unchanged [20].

situation is not too bad—at least the performance does not degrade with increasing K or N. The analysis when J is not known at the decoder—that is, it must be inferred from $\hat{y}$—reveals a much worse situation.

5.2.2.4 Loss from Sparsity Recovery

As we have mentioned before, compressive sampling is motivated by the idea that the sparsity pattern J can be detected, through a computationally tractable convex optimization, with a "small" number of measurements M. However, the number of measurements required depends on the noise level. We saw $M \sim 2K \log(N-K)+K$ scaling is required by lasso reconstruction; if the noise is from quantization and we are trying to code with KR total bits, this scaling leads to a vanishing number of bits per measurement.

Unfortunately, the problem is more fundamental than the suboptimality of lasso decoding. We will show that trying to code with KR total bits makes reliable recovery of the sparsity pattern impossible as the signal dimension N increases. In this analysis, we assume that the sparsity ratio $\alpha = K/N$ is held constant as the problems scale, and we see that no number of measurements M can give good performance.

To see why the sparsity pattern cannot be recovered, consider the problem of estimating the sparsity pattern of u from the noisy measurement y in (5.3). Let $E_{\text{signal}} = \mathbf{E}\left[\|Au\|^2\right]$ and $E_{\text{noise}} = \mathbf{E}\left[\|v\|^2\right]$ be the signal and noise energies, respectively, and define the SNR as $\text{SNR} = E_{\text{signal}}/E_{\text{noise}}$. The number of measurements M required to recover the sparsity pattern of u from y can be bounded below with the following theorem.

Theorem 5.1. Consider any estimator for recovering the sparsity pattern of a K-sparse vector u from measurements y of the form (5.3), where v is a white Gaussian vector uncorrelated with y. Let P_{error} be the probability of misdetecting the sparsity pattern, averaged over the realizations of the random matrix A and noise v. Suppose M, K, and $N-K$ approach infinity with

$$M < \frac{2K}{\text{SNR}}\left[(1-\varepsilon)\log(N-K)+1\right] \tag{5.7}$$

for some $\varepsilon > 0$. Then $P_{\text{error}} \to 1$; that is, the estimator will asymptotically always fail.

PROOF. This result follows from a stronger result [11, Theorem 1]. Define the minimum-to-average ratio as

$$\text{MAR} = \frac{\min_{j:u_j \neq 0} |u_j|^2}{\|u\|^2/K}.$$

Asymptotically in the required sense, according to [11, Theorem 1], a necessary condition for any algorithm to succeed in recovering the sparsity pattern of u is

$$M \geqslant \frac{2}{\text{MAR}\cdot\text{SNR}} K \log(N-K) + K - 1.$$

Since $\text{MAR} \leqslant 1$, the theorem follows. ∎

Under certain assumptions, the quantization error v in our problem will be asymptotically Gaussian, so we can apply the bound (see the Appendix). The theorem shows that to attain any nonvanishing probability of success, we need the scaling

$$M \geqslant \frac{2K}{\mathsf{SNR}}\left[(1-\varepsilon)\log(N-K)+1\right]. \tag{5.8}$$

Now, using the normalization assumptions described above, the expression $\rho = 1-\beta$, and σ_v^2 given in (5.1), it can be shown that the signal and noise energies are given by $E_{\text{signal}} = M(1-\beta)^2$ and $E_{\text{noise}} = M\beta(1-\beta)$. Therefore, the SNR is

$$\mathsf{SNR} = (1-\beta)/\beta. \tag{5.9}$$

Now, let $\delta = K/M$ be the "measurement ratio," that is, the ratio of degrees of freedom in the unknown signal to number of measurements. From (5.2), $\beta \geqslant 2^{-2\delta R}$ for any quantizer, and therefore, from (5.9), $\mathsf{SNR} \leqslant 2^{2\delta R} - 1$. Substituting this bound for the SNR into (5.8), we see that for the probability of error to vanish (or even become a value less than 1) will require

$$\frac{2^{2\delta R} - 1 - 2\delta}{2\delta(1-\varepsilon)} > \log(N-K). \tag{5.10}$$

Notice that, for any fixed R, the left-hand side of (5.10) is bounded above uniformly over all $\delta \in (0, 1]$. However, if the sparsity ratio $\alpha = K/N$ is fixed and $N \to \infty$, then $\log(N-K) \to \infty$. Consequently, the bound (5.10) is impossible to satisfy. We conclude that *for a fixed rate R and sparsity ratio α, as $N \to \infty$, there is no number of measurements M that can guarantee reliable sparsity recovery. In fact, the probability of detecting the sparsity pattern correctly approaches zero.* This conclusion applies not just to compressive sampling with basis pursuit, lasso, or matching pursuit detection, but even to exhaustive search methods.

How bad is this result for distributed coding of the random measurements? We have shown that exact sparsity recovery is fundamentally impossible when the total number of bit scales linearly with the degrees of freedom of the signal and the quantization is regular. However, exact sparsity recovery may not be necessary for good performance. What if the decoder can detect, say, 90 percent of the elements in the sparsity pattern correctly? One might think that the resulting distortion might still be small.

Unfortunately, when we translate the best known error bounds for reconstruction from nonadaptively encoded undersampled data,[2] we do not even obtain distortion that approaches zero as the rate is increased with K, M, and N fixed. For example, Candès and Tao [22] prove that an estimator similar to the lasso estimator attains a distortion

$$\frac{1}{K}\|x-\hat{x}\|^2 \leqslant c_1 \frac{K}{M}(\log N)\sigma^2, \tag{5.11}$$

[2] Remember that without undersampling, one can at least obtain the performance (5.5).

with large probability, from M measurements with noise variance σ^2, provided that the number of measurements is adequate. There is a constant $\delta \in (0, 1)$ such that $M = K/\delta$ is sufficient for (5.11) to hold with probability approaching 1 as N is increased with K/N held constant; but for any finite N, there is a nonzero probability of failure. Spreading RK bits among the measurements and relating the number of bits to the quantization noise variance gives

$$D = \frac{1}{K}\mathbf{E}\left[\|x - \hat{x}\|^2\right] \leq c_2\delta(\log N)2^{-2\delta R} + D_{\text{err}}, \tag{5.12}$$

where D_{err} is the distortion due to the failure event.[3] Thus if D_{err} is negligible, the distortion will decrease exponentially in the rate, but with an exponent reduced by a factor δ. However, as N increases to infinity, the distortion bound increases and is not useful.

5.2.3 Numerical Simulation

To illustrate possible performance, we performed numerical experiments with low enough dimensions to allow us to use a near-optimal decoder. We fixed the signal dimension to $N = 16$ with sparsity $K = 4$, and we quantized y by rounding each component to the nearest multiple of Δ to obtain $\hat{y}$. Rates were computed using approximations of $K^{-1}\sum_{n=1}^{N} H(\hat{y}_n)$ via numerical integration to correspond to optimal entropy coding of $\hat{y}_1, \hat{y}_2, \ldots, \hat{y}_M$. Distortions were computed by Monte Carlo estimation of $K^{-1}\mathbf{E}\left[\|x - \hat{x}\|^2\right]$, where the computation of $\hat{x}$ is described below. We varied M and Δ widely, and results for favorable (M, Δ) pairs are shown in Figure 5.5. Each data point represents averaging of distortion over 10,000 independent trials, and curves are labeled by the value of M.

Optimal estimation of x from $\hat{y}$ is conceptually simple but computationally infeasible. The value of $\hat{y}$ specifies an M-dimensional hypercube that contains y. This hypercube intersects one or more of the K-dimensional subspaces that contain y; denote the number of subspaces intersected by P. Each subspace corresponds to a unique sparsity pattern for x, and the intersection between the subspace and the hypercube gives a convex set containing the nonzero components of x. The set of values $S_{\hat{y}} \subset \mathbb{R}^N$ consistent with $\hat{y}$ is thus the union of the P convex sets and is nonconvex whenever $P > 1$. The optimal estimate is the conditional expectation of x given $\hat{y}$, which is the centroid of $S_{\hat{y}}$ under the Gaussian weighting of the nonzero components of x.[4] Finding the subspace nearest to $\hat{y}$ is NP-hard [7], and finding the P required subspaces is even harder.

The results reported here are for reconstructions computed by first projecting $\hat{y}$ to the nearest of the K-dimensional subspaces described above and then computing the

---

[3] Haupt and Nowak [23] consider optimal estimators and obtain a bound similar to (5.12) in that it has a term that is constant with respect to the noise variance. See also [24] for related results.

[4] Note that this discussion shows $\mathbf{E}\left[x \mid \hat{y}\right]$ generally has more nonzero entries than x.

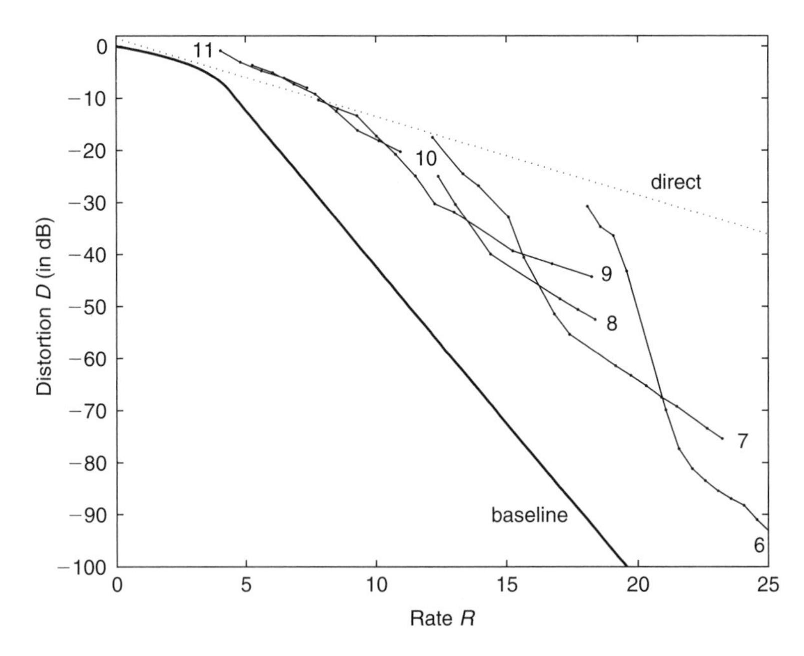

FIGURE 5.5

Rate-distortion performance of compressive sampling using entropy-constrained uniform scalar quantization of random measurements and reconstruction via maximum likelihood estimation of the sparsity pattern. In all simulations, $N=16$ and $K=4$. Quantization step size and number of measurements M are varied. Also plotted are the theoretical distortion curves for direct and baseline quantization.

least-squares approximation $\hat{x}$ assuming the sparsity pattern has been determined correctly. (The first step would be maximum likelihood detection of the sparsity pattern if $y-\hat{y}$ were white Gaussian, but this is of course not the case.) This reconstruction procedure generally produces a better estimate than a convex optimization or greedy algorithm. For our purposes of illustration, it also has the advantage of not requiring additional parameters (such as the λ in lasso). We reported results for lasso and orthogonal matching pursuit in [6].

Optimizing M trades off errors in the sparsity pattern against errors in the estimated values for the components. The simulation results show that the optimal value of M decreases as the rate increases. Our interpretation is that, except at a very low rate, instances in which the sparsity pattern recovery fails dominate the distortion. Thus, when there is more quantization noise, M must be larger to keep the probability of sparsity pattern detection failure low. If ignoring the possibility of sparsity detection failure was justified, then the distortion would decay as $\sim 2^{-2(K/M)R}$. At high rates, the simulations show steeper slopes for smaller M, but note that it is not reasonable to ignore sparsity pattern detection failure; recall the discussion leading to (5.12) and also see [25] for more extensive simulations consistent with these.

The "baseline" curve in Figure 5.5 shows the performance obtained by choosing judiciously—depending on the rate—between direct coding of x and adaptively coding the sparsity pattern and nonzero values. Specifically, it is the convexification of (5.4) with the point ($R = 0, D = 1$). The value of c used in (5.4) corresponds to scalar quantization of the nonzero entries of x. Scalar quantization of random measurements is not competitive with this baseline, but it provides improvement over simply coding each element of x independently (see (5.5)).

Let us now more closely compare the baseline method against compressive sampling. From (5.4), the distortion with adaptive quantization decreases exponentially with the rate R through the multiplicative factor 2^{-2R}. This appears in Figure 5.5 as a decrease in distortion of approximately 6 dB per bit. The best one could hope for with compressive sampling is that at a very high rate, $M = K + 1 = 5$ becomes the optimal choice, and the distortion decays as $\sim 2^{-2(K/M)R} = 2^{-2(4/5)R}$, or 4.8 dB per bit. This is a multiplicative rate penalty that is large by source coding standards, and it applies only at very high rates; the gap is larger at lower rates.

We see that, in the very high-rate regime (greater than about 15 bits per dimension), compressive sampling with near-optimal decoding achieves an MSE reduction of approximately 4 dB per bit. While better than simple direct quantization, this performance is significantly worse than the 6 dB per bit achieved with adaptive quantization. Moreover, this simulation in some sense represents the "best case" for compressive sampling since we are using an exhaustive-search decoder. Any practical decoder such as orthogonal matching pursuit or lasso will do much worse. Also, based on the analysis above, which suggests that compressive sampling incurs a $\log N$ penalty, the gap between adaptive quantization and compressive sampling will grow as the dimensions increase.

5.3 INFORMATION THEORY TO THE RESCUE?

We have thus far used information theory to provide context and analysis tools. It has shown us that compressing sparse signals by distributed quantization of random measurements incurs a significant penalty. Can information theory also suggest alternatives to compressive sampling? In fact, it does provide techniques that would give much better performance for source coding, but the complexity of decoding algorithms becomes even higher.

Let us return to Figure 5.3 and interpret it as a communication problem where x is to be reproduced approximately and the number of bits that can be used is limited. We would like to extract *source coding with side information* and *distributed source coding* problems from this setup. This will lead to results much more positive than those developed above.

In developing the baseline quantization method, we discussed how an encoder that knows V can recover J and u_K from x and thus send J exactly and u_K approximately. Compressive sampling is to apply when the encoder does not know (or want to use) the sparsifying basis V. In this case, an information theorist would say that we have a problem of lossy source coding of x with side information V available at the decoder—an instance of the *Wyner-Ziv problem* [26]. In contrast to the analogous lossless coding problem [27], the unavailability of the side information at the encoder does in general hurt the best possible performance. Specifically, let $L(D)$ denote the rate loss (increased rate because V is unavailable) to achieve distortion D. Then there are upper bounds to $L(D)$ that depend only on the source alphabet, the way distortion is measured, and the value of the distortion—not on the distribution of the source or side information [28]. For the scenario of interest to us (continuous-valued source and MSE distortion), $L(D) \leq 0.5$ bits for all D. The techniques to achieve this are complicated, but note that the *constant additive rate penalty* is in dramatic contrast to Figure 5.5.

Compressive sampling not only allows side information V to be available only at the decoder, but it also allows the components of the measurement vector y to be encoded separately. The way to interpret this information theoretically is to consider $y_1, y_2, \ldots, y_M$ as distributed sources whose joint distribution depends on side information (V, Φ) available at the decoder. Imposing a constraint of distributed encoding of y (while allowing joint decoding) generally creates a degradation of the best possible performance. Let us sketch a particular strategy that is not necessarily optimal but exhibits only a small additive rate penalty. This is inspired by [28, 29].

Suppose that each of M distributed encoders performs scalar quantization of its own y_i to yield $q(y_i)$. Earlier, this seemed to immediately get us in trouble (recall our interpretation of Theorem 5.1), but now we will do further encoding. The quantized values give us a lossless distributed compression problem with side information (V, Φ) available at the decoder. Using Slepian-Wolf coding, we can achieve a total rate arbitrarily close to $H(q(y))$. The remaining question is how the rate and distortion relate.

For sake of analysis, let us assume that the encoder and decoder share some randomness Z so that the scalar quantization above can be subtractively dithered (see, e.g., [30]). Then following the analysis in [29, 31], encoding the quantized samples $q(y)$ at rate $H(q(y) \mid V, Z)$ is within 0.755 bit of the conditional rate-distortion bound for source x given V. Thus the combination of universal dithered quantization with Slepian-Wolf coding gives a method of distributed coding with only a constant additive rate penalty. These methods inspired by information theory depend on coding across independent signal acquisition instances, and they generally incur large decoding complexity.

Let us finally interpret the "quantization plus Slepian-Wolf" approach described above when limited to a single instance. Suppose the y_is are separately quantized as described above. The main negative result of this article indicates that ideal separate entropy coding of each $q(y_i)$ is not nearly enough to get to good performance. The rate must be reduced by replacing an ordinary entropy code with one that collapses some distinct quantized values to the same index. The hope has to be that in the joint decoding of $q(y)$, the dependence between components will save the day. This is equivalent to saying that the quantizers in use are not regular [30], much like multiple description quantizers [32]. This approach is developed and simulated in [25].

5.4 CONCLUSIONS—WHITHER COMPRESSIVE SAMPLING?

To an information theorist, "compression" is the efficient representation of data with bits. In this chapter, we have looked at compressive sampling from this perspective, in order to determine if random measurements of sparse signals provide an efficient method of representing sparse signals.

The source coding performance depends sharply on how the random measurements are encoded into bits. Using familiar forms of quantization (*regular quantizers*; see [30]), even very weak forms of universality are precluded. One would want to spend a number of bits proportional to the number of degrees of freedom of the sparse signal, but this does not lead to good performance. In this case, we can conclude analytically that recovery of the sparsity pattern is asymptotically impossible. Furthermore, simulations show that the MSE performance is far from optimal.

Information theory provides alternatives based on universal versions of distributed lossless coding (Slepian-Wolf coding) and entropy-coded dithered quantization. These information-theoretic constructions indicate that it is reasonable to ask for good performance with merely linear scaling of the number of bits with the sparsity of the signal. However, practical implementation of such schemes remains an open problem. Other chapters of this volume present practical schemes that exploit different sorts of relationships between the distributed sources.

It is important to keep our mainly negative results in proper context. We have shown that compressive sampling combined with ordinary quantization is a bad compression technique, but our results say nothing about whether compressive sampling is an effective initial step in data acquisition. A good analogy within the realm of signal acquisition is oversampling in analog-to-digital conversion. Since MSE distortion in oversampled ADC drops only polynomially (not exponentially) with the oversampling factor, high oversampling alone—without other processing—leads to poor rate-distortion performance. Nevertheless, oversampling is ubiquitous. Similarly, compressive sampling is useful in contexts where sampling itself is very expensive, but the subsequent storage and communication of quantized samples is less constricted.

APPENDIX

5.5 QUANTIZER PERFORMANCE AND QUANTIZATION ERROR

A quantity that takes on uncountably many values—like a real number—cannot have an exact digital representation. Thus digital processing always involves quantized values. The relationships between the number of bits in a representation (rate R), the accuracy of a representation (distortion D), and properties of quantization error are central to this chapter and are developed in this Appendix.

The simplest form of quantization—*uniform scalar quantization*—is to round $x \in \mathbb{R}$ to the nearest multiple of some fixed resolution parameter Δ to obtain quantized version $\hat{x}$. For this type of quantizer, rate and distortion can be easily related through the step size Δ. Suppose x has a smooth distribution over an interval of length C. Then the quantizer produces about C/Δ intervals, which can be indexed with $R \approx \log_2(C/\Delta)$ bits. The error $x - \hat{x}$ is approximately uniformly distributed over $[-\Delta/2, \Delta/2]$, so the mean-squared error is $D = \mathbf{E}\left[(x-\hat{x})^2\right] \approx \frac{1}{12}\Delta^2$. Eliminating Δ, we obtain $D \approx \frac{1}{12}C^2 2^{-2R}$.

The 2^{-2R} dependence on rate is fundamental for compression with respect to MSE distortion. For any distribution of x, the best possible distortion as a function of rate (obtained with high-dimensional vector quantization [30]) satisfies

$$(2\pi e)^{-1} 2^{2h} 2^{-2R} \leqslant D(R) \leqslant \sigma^2 2^{-2R},$$

where h and σ^2 are the differential entropy and variance of x. Also, under high-resolution assumptions and with entropy coding, $D(R) \approx \frac{1}{12} 2^{2h} 2^{-2R}$ performance is obtained with uniform scalar quantization, which for a Gaussian random variable is $D(R) \approx \frac{1}{6}\pi e \sigma^2 2^{-2R}$. Covering all of these variations together, we write the performance as $D(R) = c\sigma^2 2^{-2R}$ without specifying the constant c.

More subtle is to understand the quantization error $e = x - \hat{x}$. With uniform scalar quantization, e is in the interval $[-\Delta/2, \Delta/2]$, and it is convenient to think of it as a uniform random variable over this interval, independent of x. This is merely a convenient fiction, since $\hat{x}$ is a deterministic function of x. In fact, as long as quantizers are regular and estimation procedures use linear combinations of many quantized values, second-order statistics (which are well understood [15]) are sufficient for understanding estimation performance. When x is Gaussian, a rather counterintuitive model where e is Gaussian and independent of x can be justified precisely: optimal quantization of a large block of samples is described by the *optimal test channel*, which is additive Gaussian [33].

ACKNOWLEDGMENTS

Support provided by a University of California President's Postdoctoral Fellowship, the Texas Instruments Leadership Universities Program, NSF CAREER Grant CCF-643836, and the Centre Bernoulli of École Polytechnique Fédérale de Lausanne. The authors thank Sourav Dey, Lav Varshney, Claudio Weidmann, Joseph Yeh, and an anonymous reviewer for thoughtful comments on early drafts of [6] and the editors of this volume for careful reading and constructive suggestions.

REFERENCES

[1] E. J. Candès and M. B. Wakin. An introduction to compressive sampling. *IEEE Sig. Process. Mag.*, 25(2):21–30, March 2008.

[2] D. L. Donoho, M. Vetterli, R. A. DeVore, and I. Daubechies. Data compression and harmonic analysis. *IEEE Trans. Inform. Theory*, 44(6):2435–2476, October 1998.

[3] E. J. Candès, J. Romberg, and T. Tao. Robust uncertainty principles: Exact signal reconstruction from highly incomplete frequency information. *IEEE Trans. Inform. Theory*, 52(2):489–509, February 2006.

[4] E. J. Candès and T. Tao. Near-optimal signal recovery from random projections: Universal encoding strategies? *IEEE Trans. Inform. Theory*, 52(12):5406–5425, December 2006.

[5] D. L. Donoho. Compressed sensing. *IEEE Trans. Inform. Theory*, 52(4):1289–1306, April 2006.

[6] V. K. Goyal, A. K. Fletcher, and S. Rangan. Compressive sampling and lossy compression. *IEEE Sig. Process. Mag.*, 25(2):48–56, March 2008.

[7] B. K. Natarajan. Sparse approximate solutions to linear systems. *SIAM J. Computing*, 24(2): 227–234, April 1995.

[8] D. L. Donoho and P. B. Stark. Uncertainty principles and signal recovery. *SIAM J. Appl. Math.*, 49(3):906–931, June 1989.

[9] D. L. Donoho and J. Tanner. Counting faces of randomly-projected polytopes when the projection radically lowers dimension. *J. Amer. Math. Soc.*, 22(1):1–53, January 2009.

[10] M. J. Wainwright. Sharp thresholds for high-dimensional and noisy recovery of sparsity. Technical report, Univ. of California, Berkeley, Dept. of Statistics, May 2006. arXiv:math.ST/0605740 v1 30 May 2006.

[11] A. K. Fletcher, S. Rangan, and V. K. Goyal. Necessary and sufficient conditions on sparsity pattern recovery. arXiv:0804.1839v1 [cs.IT]. April 2008.

[12] V. K. Goyal. Theoretical foundations of transform coding. *IEEE Sig. Process. Mag.*, 18(5):9–21, September 2001.

[13] M. Gastpar, P. L. Dragotti, and M. Vetterli. The distributed Karhunen–Loève transform. *IEEE Trans. Inform. Theory*, 52(12):5177–5196, December 2006.

[14] A. K. Fletcher. *A Jump Linear Framework for Estimation and Robust Communication with Markovian Source and Channel Dynamics*. PhD thesis, Univ. California, Berkeley, November 2005.

[15] H. Viswanathan and R. Zamir. On the whiteness of high-resolution quantization errors. *IEEE Trans. Inform. Theory*, 47(5):2029–2038, July 2001.

[16] R. G. Gallager. *Information Theory and Reliable Communication*. John Wiley & Sons, New York, 1968.

[17] C. Weidmann and M. Vetterli. Rate-distortion analysis of spike processes. In *Proc. IEEE Data Compression Conf.*, pp. 82–91, Snowbird, Utah, March 1999.

[18] V. K. Goyal, M. Vetterli, and N. T. Thao. Quantized overcomplete expansions in $\mathbb{R}^N$: Analysis, synthesis, and algorithms. *IEEE Trans. Inform. Theory*, 44(1):16–31, January 1998.

[19] S. Rangan and V. K. Goyal. Recursive consistent estimation with bounded noise. *IEEE Trans. Inform. Theory*, 47(1):457–464, January 2001.

[20] V. K. Goyal, J. Kovačević, and J. A. Kelner. Quantized frame expansions with erasures. *Appl. Comput. Harm. Anal.*, 10(3):203–233, May 2001.

[21] A. K. Fletcher, S. Rangan, and V. K. Goyal. On the rate-distortion performance of compressed sensing. In *Proc. IEEE Int. Conf. Acoust., Speech, and Signal Process.*, volume III, pp. 885–888, Honolulu, HI, April 2007.

[22] E. J. Candès and T. Tao. The Dantzig selector: Statistical estimation when p is much larger than n. *Ann. Stat.*, 35(6):2313–2351, December 2007.

[23] J. Haupt and R. Nowak. Signal reconstruction from noisy random projections. *IEEE Trans. Inform. Theory*, 52(9):4036–4048, September 2006.

[24] N. Meinshausen and B. Yu. Lasso-type recovery of sparse representations for high-dimensional data. *Ann. Stat.* To appear.

[25] R. J. Pai. Nonadaptive lossy encoding of sparse signals. Master's thesis, Massachusetts Inst. of Tech., Cambridge, MA, August 2006.

[26] A. D. Wyner and J. Ziv. The rate-distortion function for source coding with side information at the decoder. *IEEE Trans. Inform. Theory*, IT-22(1):1–10, January 1976.

[27] D. Slepian and J. K. Wolf. Noiseless coding of correlated information sources. *IEEE Trans. Inform. Theory*, IT-19(4):471–480, July 1973.

[28] R. Zamir. The rate loss in the Wyner–Ziv problem. *IEEE Trans. Inform. Theory*, 42(6):2073–2084, November 1996.

[29] J. Ziv. On universal quantization. *IEEE Trans. Inform. Theory*, IT-31(3):344–347, May 1985.

[30] R. M. Gray and D. L. Neuhoff. Quantization. *IEEE Trans. Inform. Theory*, 44(6):2325–2383, October 1998.

[31] R. Zamir and M. Feder. On universal quantization by randomized uniform/lattice quantization. *IEEE Trans. Inform. Theory*, 38(2):428–436, March 1992.

[32] V. K. Goyal. Multiple description coding: Compression meets the network. *IEEE Sig. Process. Mag.*, 18(5):74–93, September 2001.

[33] T. Berger. *Rate Distortion Theory*. Prentice-Hall, Englewood Cliffs, NJ, 1971.

PART

Algorithms and Applications

2

CHAPTER

Toward Constructive Slepian–Wolf Coding Schemes

Christine Guillemot and Aline Roumy
INRIA Rennes-Bretagne Atlantique, Campus Universitaire de Beaulieu, Rennes Cédex, France

CHAPTER CONTENTS

6.1 INTRODUCTION

This chapter deals with practical solutions for the Slepian–Wolf (SW) coding problem, which refers to the problem of lossless compression of correlated sources with

Distributed Source Coding: Theory, Algorithms, and Applications

coders that do not communicate. Here, we will consider the case of two binary correlated sources X and Y, characterized by their joint distribution. If the two coders communicate, it is well known from Shannon's theory that the minimum lossless rate for X and Y is given by the joint entropy $H(X, Y)$. In 1973 [29] Slepian and Wolf established that this lossless compression rate bound can be approached with a vanishing error probability for infinitely long sequences, even if the two sources are coded separately, provided that they are decoded jointly and that their correlation is known to both the encoder and the decoder. Hence, the challenge is to construct a set of encoders that do not communicate and a joint decoder that can achieve the theoretical limit.

This chapter gives an overview of constructive solutions for both the asymmetric and nonasymmetric SW coding problems. Asymmetric SW coding refers to the case where one source, for example Y, is transmitted at its entropy rate and used as side information to decode the second source X. Nonasymmetric SW coding refers to the case where both sources are compressed at a rate lower than their respective entropy rates. Sections 6.2 and 6.3 recall the principles and then describe practical schemes for asymmetric and symmetric coding, respectively. Practical solutions for which the compression rate is a priori fixed according to the correlation between the two sources are first described. In this case, the correlation between the two sources needs to be known—or estimated—at the transmitter. Rate-adaptive schemes in which the SW code is incremental are then presented. This chapter ends with Section 6.4 covering various advanced SW coding topics such as the design of schemes based on source codes and the generalization to the case of nonbinary sources, and to the case of M sources.

6.2 ASYMMETRIC SW CODING

The SW region for two discrete sources is an unbounded polygon with two corner points (see points A and B in Figure 6.1(b)). At these points, one source (say Y for point A) is compressed at its entropy rate and can therefore be reconstructed at the decoder independently of the information received from the other source X. The source Y is called the *side information* (SI) (available at the decoder only). X is compressed at a smaller rate than its entropy. More precisely, X is compressed at the conditional entropy $H(X|Y)$ and can therefore be reconstructed only if Y is available at the decoder. The sources X and Y play different roles in this scheme, and therefore the scheme is usually referred to as *asymmetric SW coding*.

6.2.1 Principle of Asymmetric SW Coding

6.2.1.1 The Syndrome Approach

Because of the random code generation in the SW theorem, the SW proof is nonconstructive. However, some details in the proof do give insights into how to construct SW bound achieving codes. More precisely, the proof relies on binning—that is, a partition

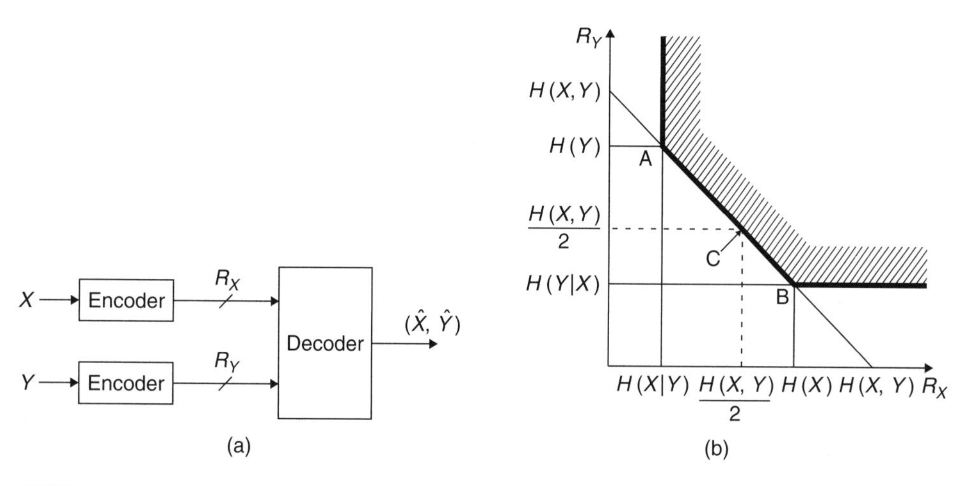

FIGURE 6.1

Distributed source coding of statistically dependent i.i.d. discrete random sequences X and Y. Setup (left); achievable rate region (right).

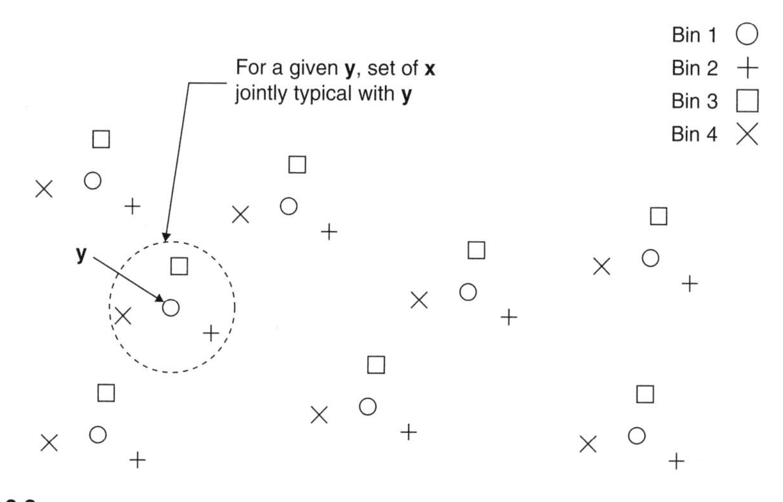

FIGURE 6.2

The partition of the sequences **x** into four subsets (called bins) such that the sequences in each bin are as far apart as possible. The set of sequences **x** jointly typical with a given **y** are as "different" as possible: here they are elements of different bins.

of the space of the source sequences into subsets (or bins) such that the vectors in the bins are as far apart (or as much "jointly nontypical") as possible (see Figure 6.2). In 1974 [39], Wyner suggested constructing the bins as cosets of a binary linear code and showed the optimality of this construction. More precisely, if the linear block code achieves the capacity of the binary symmetric channel (BSC) that models the correlation between the two sources, then this capacity-achieving channel code can be turned into a SW-achieving source code.

In order to present Wyner's algebraic binning scheme, we first review some notations on binary linear codes. A binary (n, k) code $\mathcal{C}$ is defined by an $(n-k)\times n$ parity-check matrix H and contains all the n-length vectors[1] $\mathbf{x}$ such that $\mathbf{x}H^T=\mathbf{0}$:

$$\mathcal{C}=\left\{\mathbf{x}:\mathbf{x}H^T=\mathbf{0}\right\}.$$

Notice that the code $\mathcal{C}$ is equivalently defined by a $k\times n$ generator matrix G such that $\mathcal{C}=\{\mathbf{x}:\mathbf{x}=\mathbf{u}G\}$, where $\mathbf{u}$ is a k-bit vector.

If H has full row rank, then the rate of the channel code is k/n. Moreover, for a good code, all the vectors of the code (called words or codewords) have maximum Hamming distance. The code partitions the space containing 2^n sequences into 2^{n-k} *cosets* of 2^k words with maximum Hamming distance. Each coset is indexed by an $(n-k)$-length *syndrome*, defined as $\mathbf{s}=\mathbf{x}H^T$. In other words, all sequences in a coset share the same syndrome: $\mathcal{C}_\mathbf{s}=\{\mathbf{x}:\mathbf{x}H^T=\mathbf{s}\}$. Moreover, as a consequence of the linearity of the code, a coset results from the translation of the code by any representative of the coset:

$$\forall\mathbf{v}\in\mathcal{C}_\mathbf{s},\ \mathcal{C}_\mathbf{s}=\mathcal{C}\oplus\mathbf{v}. \tag{6.1}$$

A geometric approach to binary coding visualizes binary sequences of length n as vertices of an n-dimensional cube. In Figure 6.3, the code $\mathcal{C}$ (the set of codewords or the set of vectors of syndrome 0 that is represented with •) is a subspace of $\{0, 1\}^3$, whereas the coset with syndrome 1, denoted $\mathcal{C}_1$ and represented with ∘, is an affine subspace parallel to $\mathcal{C}$. Notice that properties of a subspace and the linearity property (6.1) are satisfied (to verify this, all operations have to be performed over the finite field of order 2).

If a codeword $\mathbf{x}$ is sent over a BSC with crossover probability p and error sequence $\mathbf{z}$, the received sequence is $\mathbf{y}=\mathbf{x}+\mathbf{z}$. Maximum likelihood (ML) decoding over the BSC searches for the closest codeword to $\mathbf{y}$ with respect to the Hamming distance $d_H(.,.)$:

$$\hat{\mathbf{x}}=\arg\min_{\mathbf{x}:\mathbf{x}\in\mathcal{C}} d_H(\mathbf{x},\mathbf{y}).$$

This can be implemented with *syndrome decoding*. This decoding relies on a function $f:\{0, 1\}^{n-k}\rightarrow\{0, 1\}^n$, which computes for each syndrome a representative called the *coset leader* that has minimum Hamming weight. Thus, a syndrome decoder first computes the syndrome of the received sequence: $\mathbf{s}=\mathbf{y}H^T$, then the coset leader $f(\mathbf{s})$. This coset leader is the ML estimate of the error pattern $\mathbf{z}$. Finally, the ML estimate of $\mathbf{x}$ is given by $\hat{\mathbf{x}}=\mathbf{y}\oplus f(\mathbf{y}H^T)$.

In order to use such a code for the asymmetric SW problem, Wyner [39] suggests constructing bins as cosets of a capacity-achieving parity check code. Let $\mathbf{x}$ and

[1] All vectors are line vectors, and ^{T} denotes transposition. Moreover, $\oplus$ denotes the addition over the finite field of order 2 and + the addition over the real field.

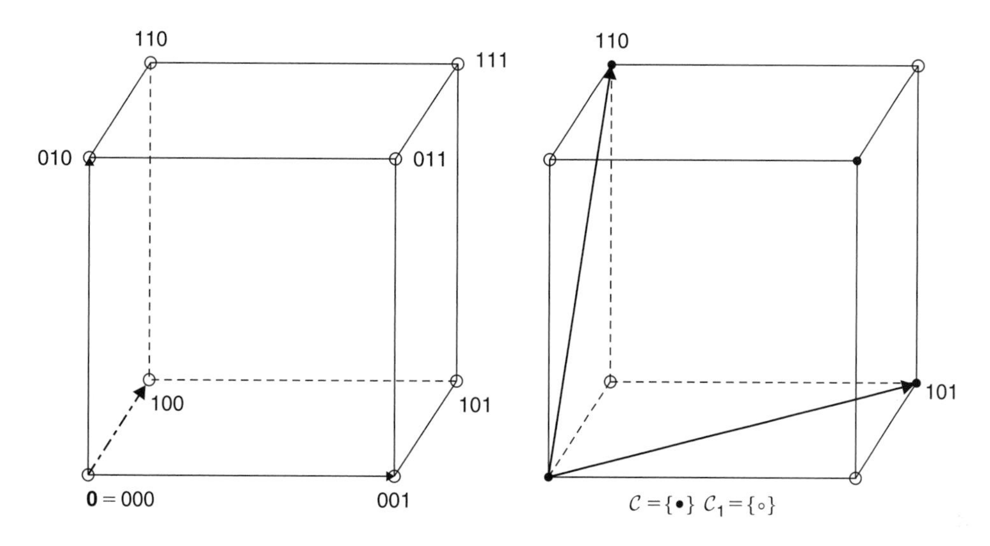

FIGURE 6.3

Construction of a binning scheme with a linear block code in 0, 1^3. The cube and the basis vector in 0, 1^3 (left); a code defined by the parity-check matrix $H = (111)$ or equivalently the generator matrix $G = \binom{110}{101}$ (right).

$\mathbf{y}$ be two correlated binary sequences of length n. These sequences are the realizations of the sources X and Y. The encoder computes and transmits the syndrome of $\mathbf{x}$: $\mathbf{s} = \mathbf{x}H^T$. The sequence $\mathbf{x}$ of n input bits is thus mapped into its corresponding $(n-k)$ syndrome bits, leading to a compression ratio of $n:(n-k)$. The decoder, given the correlation between the sources X and Y and the received coset index $\mathbf{s}$, searches for the sequence in the coset that is closest to $\mathbf{y}$. In other words, ML decoding is performed in order to retrieve the original sequence $\mathbf{x}$:

$$\hat{\mathbf{x}} = \arg \min_{\mathbf{x}:\mathbf{x} \in \mathcal{C}_{\mathbf{s}}} d_H(\mathbf{x}, \mathbf{y}). \tag{6.2}$$

Note that the maximization is performed in a set of vectors with syndrome that may not be $\mathbf{0}$. Therefore, the classical ML channel decoder has to be adapted in order to be able to enumerate all vectors in a given coset $\mathcal{C}_{\mathbf{s}}$. This adaptation is straightforward if syndrome decoding is performed. In this case, the decoder can first retrieve the error pattern (since for each syndrome, the corresponding coset leader is stored) and add it to the SI $\mathbf{y}$.

Syndrome decoding is used here as a mental representation rather than an efficient decoding algorithm, since it has complexity $O(n2^{n-k})$. There exist codes with linear decoding complexity, and we will detail these constructions in Section 6.2.2. In this section, we have presented a structured binning scheme for the lossless source coding problem with SI at the decoder. However, the principle is quite general and can be applied to a large variety of problems as shown in [40].

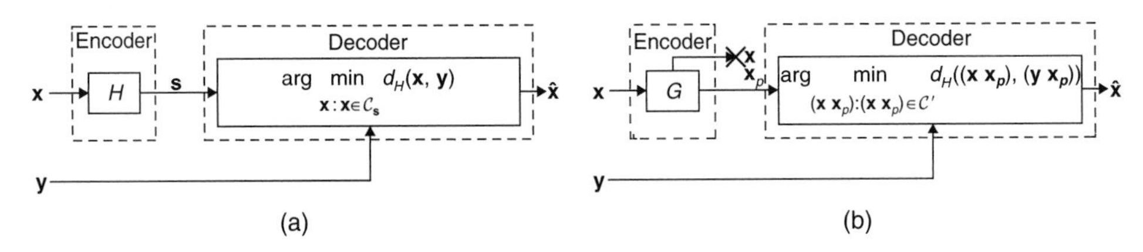

FIGURE 6.4

Asymmetric SW coding. The syndrome approach (left); the parity approach (right).

6.2.1.2 The Parity Approach

The syndrome approach detailed earlier is optimal [39], however, it may be difficult to construct rate-adaptive codes by puncturing the syndrome. Therefore, another approach called *parity approach* has been proposed [9],[2] [1].

Let $\mathcal{C}'$ be an $(n, 2n-k)$ systematic binary linear code, defined by its $(2n-k)\times n$ generator matrix $G=(I\ P)$:

$$\mathcal{C}' = \{\mathbf{x}G = (\mathbf{x}\ \mathbf{x}_p) : \mathbf{x} \in \{0,1\}^n\}.$$

The compression of the source X is achieved by transmitting only the parity bits $\mathbf{x}_p$ of the source sequence $\mathbf{x}$ (see Figure 6.4). The systematic bits $\mathbf{x}$ are not transmitted. This leads to a compression ratio $n:(n-k)$. Here again, the correlation between the source X and the SI Y is modeled as a "virtual" channel, where the pair $(\mathbf{y}\ \mathbf{x}_p)$ is regarded as a noisy version of $(\mathbf{x}\ \mathbf{x}_p)$. The channel is therefore a parallel combination of a BSC and a perfect channel. The decoder corrects the "virtual" channel noise and thus estimates $\mathbf{x}$ given the parity bits $\mathbf{x}_p$ and the SI $\mathbf{y}$ regarded as a noisy version of the original sequence $\mathbf{x}$. Therefore, the usual ML decoder must be adapted to take into account that some bits of the received sequence $(\mathbf{x}_p)$ are perfectly known. We will detail this in Section 6.2.2.

Interestingly, both approaches (syndrome or parity) are equivalent if the generator matrix (in the parity approach) is the concatenation of the identity matrix and the parity-check matrix used in the syndrome approach: $G=(I\ H)$. However, the two codes $\mathcal{C}$ and $\mathcal{C}'$ (defined by H and G, respectively) are not the same. They are not even defined in the same space ($\mathcal{C}$ is a subspace of $\{0,1\}^n$, whereas $\mathcal{C}'$ is a subspace of $\{0,1\}^{2n-k}$). Figure 6.4 compares the implementation of the syndrome and parity approaches.

6.2.2 Practical Code Design Based on Channel Codes

6.2.2.1 The Syndrome Approach

Practical solutions based on the syndrome approach first appeared in a scheme called DISCUS [20]. For block codes [20], syndrome decoding is performed. For convolutional

---

[2] Note that this paper introduces the principle of the parity approach in the general case of nonasymmetric coding.

codes, the authors in [20] propose to apply the Viterbi decoding on a modified trellis. In order to solve the problem (6.2), the decoder is modified so that it can enumerate all the codewords in a given coset. The method uses the linearity property (6.1). For systematic convolutional codes, a representative of the coset is the concatenation of the k-length zero vector and the $n-k$-length syndrome $\mathbf{s}$. This representative is then added to all the codewords labeling the edges of the trellis (see Figure 6.5). Note that the states of the trellis depend on the information bits, and thus on the systematic bits only. Thus, there exists one trellis section per syndrome value. The decoder, knowing the syndrome $\mathbf{s}$, searches for the sequence with that particular syndrome that is closest to $\mathbf{y}$. First, it builds the complete trellis. Each section of the trellis is determined by the syndrome. Once the whole trellis is built, the Viterbi algorithm chooses the closest sequence to the received $\mathbf{y}$.

A variation of the above method is proposed in [35], where the translation by a coset representative is performed outside the decoder (see Figure 6.6). First, the representative ($\mathbf{0}$ $\mathbf{s}$) is computed (this step is called inverse syndrome former—ISF) and added to $\mathbf{y}$. Then the decoding is performed (in the coset of syndrome $\mathbf{0}$), and finally the representative is retrieved from the output of the decoder. The advantage of this method is to use a conventional Viterbi decoder without a need to modify it. This method can also be applied to turbo codes [35]. In this case, two syndromes ($\mathbf{s}_1$, $\mathbf{s}_2$) are computed, one for each constituent code. A representative of the coset is ($\mathbf{0}$ $\mathbf{s}_1$ $\mathbf{s}_2$).

The authors in [24] propose a SW scheme based on convolutional codes and turbo codes that can be used for any code (not only systematic convolutional code). Rather than considering the usual trellis based on the generator matrix of the code, the

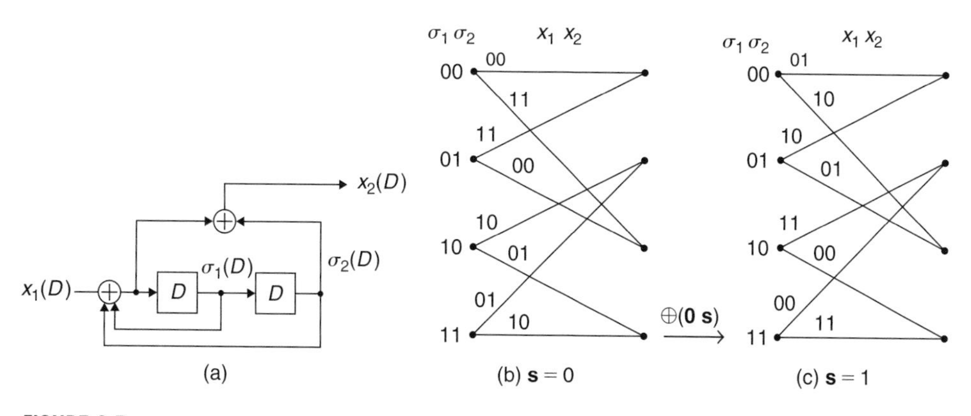

FIGURE 6.5

Block diagram (a) and trellises (b, c) for the systematic convolutional code $H=[1\,\frac{5}{7}]$. The diagram (a) defines the states $\sigma_1(D)\ \sigma_2(D)$ of the trellis, where D is a dummy variable representing a time offset. The trellis section in (b) corresponds to the syndrome value $\mathbf{s}=0$ and is called the *principal trellis*. The *complementary trellis* corresponds to $\mathbf{s}=1$ and is obtained by adding ($\mathbf{0}$ $\mathbf{s}$) to all the codewords labeling the edges of the principal trellis.

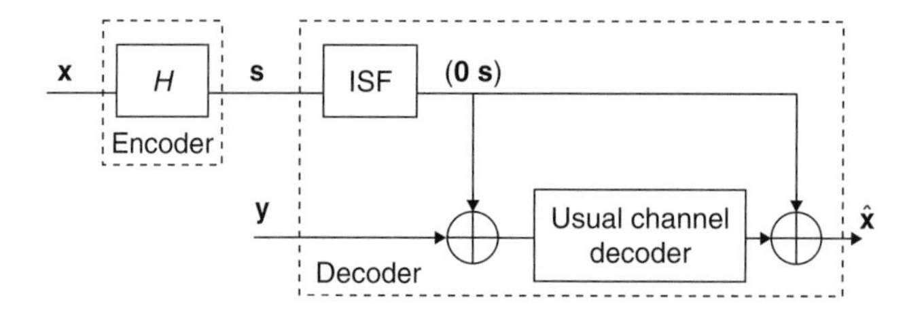

FIGURE 6.6

Implementation of the syndrome approach with a usual channel decoder. ISF stands for inverse syndrome former and computes a representative of the coset with syndrome **s**.

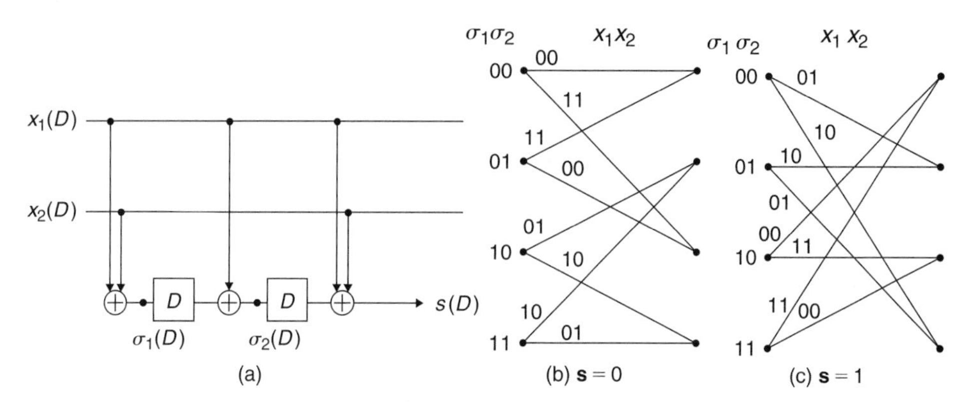

FIGURE 6.7

Block diagram (a) and syndrome trellises (b–c) for the rate 1/2 convolutional code $H = [5\ 7]$. The states $\sigma_1(D)\ \sigma_2(D)$ are defined in the block diagram (a). Two trellis sections are obtained: one for the syndrome value $\mathbf{s} = 0$ (b) and one for $\mathbf{s} = 1$ (c).

decoder is based on a syndrome trellis. This trellis was first proposed for binary linear block codes [38] and was then generalized to convolutional codes [28]. Such a trellis can enumerate a set of sequences in any coset. More precisely, a section corresponds to the reception of n bits of $\mathbf{x}$ and $n-k$ syndrome bits (see Figure 6.7). Thus, there exist 2^{n-k} possible trellis sections, one for each syndrome value. The decoder, knowing the syndrome $\mathbf{s}$, searches for the sequence with that particular syndrome that is closest to $\mathbf{y}$. First, it builds the complete trellis. Each section of the trellis is determined by the syndrome. Once the whole trellis is built, the Viterbi algorithm is performed and chooses the closest sequence to the received $\mathbf{y}$.

For low-density parity-check (LDPC) codes, the belief propagation decoder can be adapted to take into account the syndrome [16]. More precisely, the syndrome bits are added to the graph such that each syndrome bit is connected to the parity check equation to which it is related. The update rule at a check node is modified in order to take into account the value of the syndrome bit (known perfectly at the decoder).

6.2.2.2 *The Parity Approach*

The parity approach presented in Section 6.2.1.2 has been implemented using various channel codes, for instance, turbo codes [1, 5]. Consider a turbo encoder formed by the parallel concatenation of two recursive systematic convolutional (RSC) encoders of rate $(n-1)/n$ separated by an interleaver (see Figure 6.8). For each sequence $\mathbf{x}$ (of length n) to be compressed, two sequences of parity bits $\mathbf{x}_p = (\mathbf{x}_p^1\ \mathbf{x}_p^2)$ are computed and transmitted without any error to the decoder. The turbo code is composed of two symbol Maximum A Posteriori (MAP) decoders. Each decoder receives the (unaltered) parity bits together with the SI $\mathbf{y}$, seen as a noisy version of the sequence $\mathbf{x}$. The concatenation $(\mathbf{y}\ \mathbf{x}_p)$ can be seen as a noisy version of the coded sequence $(\mathbf{x}\ \mathbf{x}_p)$, where the channel is a parallel combination of a BSC and of a perfect channel. Therefore, the usual turbo decoder must be matched to this combined channel. More precisely, the channel transitions in the symbol MAP algorithm which do not correspond to the parity bits are eliminated. Such a scheme achieves an $n:2$ compression rate.

6.2.3 Rate Adaptation

Note that in all methods presented so far, in order to select the proper code and code rate, the correlation between the sources needs to be known or estimated at the transmitter before the compression starts. However, for many practical scenarios, this correlation may vary, and rate-adaptive schemes have to be designed. When the correlation decreases, the rate bound moves away from the estimate (see Figure 6.9). The rate of the code can then be controlled via a feedback channel. The decoder estimates the Bit Error Rate (BER) at the output of the decoder with the help of the

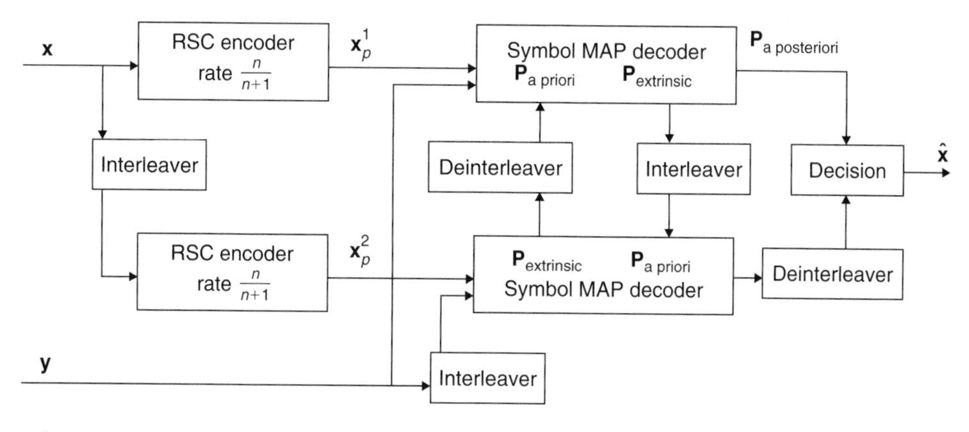

FIGURE 6.8

DSC system for two binary-correlated sources using punctured turbo codes.

log-likelihood ratios computed by the channel decoder. If the BER at the output of the decoder exceeds a given value, more bits are requested from the encoder. In this context of feedback-controlled schemes, the code should be *incremental* such that the encoder does not need to re-encode the data. The first bits are kept, and additional bits are only sent upon request. In the following, we present various rate-adaptive methods and specify whether or not they are incremental.

6.2.3.1 The Parity Approach

The parity approach was originally proposed to easily construct rate-adaptive schemes (see Section 6.2.2). At the encoder, the parity bits are punctured (some parity bits are not transmitted), and the decoder compensates for this puncturing, using standard techniques coming from channel coding. The source sequence $\mathbf{x}$ is compressed through some punctured parity bits $\tilde{\mathbf{x}}_p$, and the decoder retrieves the original sequence aided by the SI $\mathbf{y}$. The sequence $(\mathbf{y}\ \tilde{\mathbf{x}}_p)$ can be seen as the output of a channel, which is a combination of a perfect channel (unpunctured parity bits), an erasure channel (punctured parity bits), and a BSC channel (correlation between the sources $\mathbf{x}$ and $\mathbf{y}$). The method is by construction incremental.

6.2.3.2 The Syndrome Approach

Because of the optimality of the syndrome approach [39], a natural design of a rate-adaptive scheme consists of puncturing the syndrome. However, using puncturing mechanisms leads to performance degradation as reported in [15, 32, 36]. More precisely, it is shown in [32] that the syndrome approach is very sensitive to errors

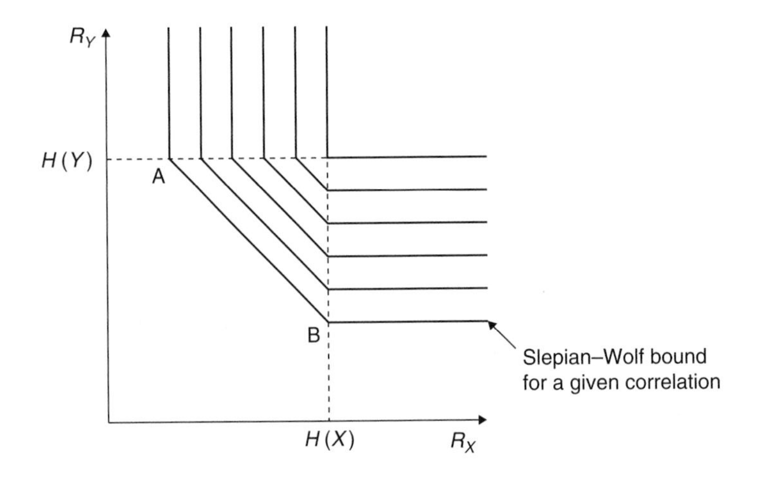

FIGURE 6.9

Evolution of the SW region, when the correlation between the sources varies.

or erasures, which indicates that the puncturing of syndromes will lead to performance degradation. Therefore, the first contributions avoid or compensate for this performance degradation.

A first method proposed in [15] punctures the parity bits rather than the syndrome bits. Each puncturing pattern defines a new code for which a parity check matrix can be computed. Then, one can construct a syndrome former and inverse syndrome former for the new punctured code (presented in Section 6.2.1.2). This method can be implemented using any code, in particular convolutional and turbo codes. However, the resulting code is not incremental.

A first incremental SW coding scheme based on Serially Concatenated Accumulate (SCA) codes is proposed in [7]. An inner encoder concatenates the first source bit with the modulo-2 sum of consecutive pairs of source bits. The output of the accumulator is interleaved and fed into a so-called base code, which is either an extended Hamming code or a product of single-parity check codes, which then compute the syndrome bits. The accumulator code is of rate 1, so that the compression rate is controlled by the base code. The base code is first decoded using a MAP decoder for extended Hamming codes or a belief propagation algorithm for product parity-check codes. The accumulator code is decoded with a symbol MAP decoder, A turbo-like algorithm iterates between both decoders.

The authors in [36] investigate the puncturing of LDPC (low-density parity-check code) encoded syndromes. To avoid degrading the performance of the LDPC code, the syndrome bits are first protected with an accumulator code before being punctured. Here also the accumulator code is of rate 1, so that the compression rate is not modified. The combined effect of the puncturing and of the accumulator code is equivalent to merging some rows of the parity-check matrix by adding them. This defines a set of parity-check matrices, one for each rate. Then, for each parity-check matrix, decoding is performed according to the modified sum-product algorithm [16] presented in Section 6.2.1.1. An interesting feature of this method is that it is incremental. If the merging of rows satisfies a regularity rule (one can only add two rows that have no "1" at the same column position), then the accumulated syndrome value of the new (smallest compression rate) matrix can be computed by adding the previous syndrome value (corresponding to the highest compression rate matrix) to the current one. The performance of the method is shown in Figure 6.10(a). Another method is proposed in [14]. Instead of protecting the syndrome bits, the authors combine the syndrome and the parity approaches. Given an input vector $\mathbf{x}$ of length n, the encoder transmits m syndrome bits plus l parity bits, defining an $(n+l, n-m)$ linear error-correcting code. Rate adaptation is then performed by puncturing the parity bits.

In [24], the optimal decoder for convolutional codes under syndrome puncturing is designed. It is shown that performance degradation due to syndrome puncturing can be avoided without a need to encode (or protect) the syndromes. When syndrome bits are punctured, the optimal decoder should in principle search for the closest sequence in a union of cosets. This union corresponds to all possible syndromes that equal the received syndrome in the unpunctured positions. Therefore, the number of cosets grows exponentially with the number of punctured bits. A brute force decoder consists

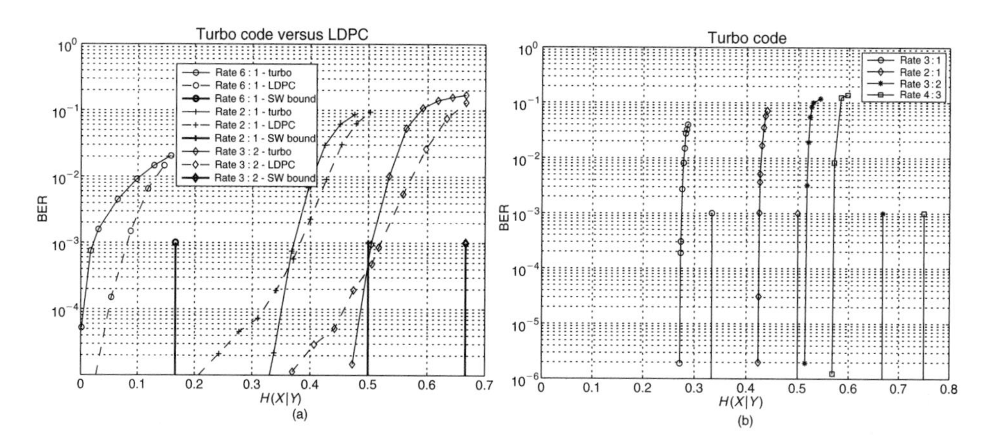

FIGURE 6.10

Syndrome puncturing of an LDPC code [36] and of a turbo code [24]: two rate-adaptive schemes. Performance versus the entropy rate H(X|Y) and comparison with the SW bounds. The 1/2-rate constituent code of the turbo code is defined by its parity-check matrix $H = (23/33, 35/33)$. The overall compression rate for both schemes is $1:1$, and puncturing leads to various compression rates. (a) Comparison of the syndrome punctured LDPC and turbo code for a blocklength of 2046. (b) Performance of the turbo code for an interleaver of size 10^5.

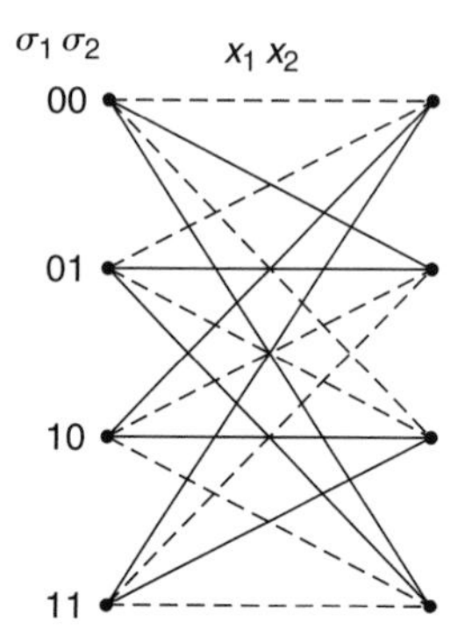

FIGURE 6.11

Super trellis for the rate 1/2 convolutional code defined by its polynomial parity-check matrix H(D) = [5 7]. This trellis is the union of the two trellis sections of Figure 6.7.

in performing an ML algorithm in each coset, but then the decoding complexity of such a method would grow exponentially with the number of punctured bits. On the other hand, if the decoder is not modified, it leads to systematic errors, since the search may be performed in the wrong coset. The method proposed in [24] is based on the syndrome trellis presented in Section 6.2.1.1. Whenever some syndrome bits are punctured, a new trellis section is constructed as the union of the trellis sections compatible with the received punctured syndrome. This enumerates the sequences in the union of cosets. The resulting trellis is called the *super trellis* (see Figure 6.11), and the Viterbi algorithm is performed on this new trellis. The complexity of the proposed algorithm grows only linearly with the code blocklength and with the number of punctured positions. The authors in [24] also show that this decoder can be applied to turbo codes. The performance of the method is shown in Figure 6.10.

6.3 NONASYMMETRIC SW CODING

For the sake of simplicity, we focus again on the two-user setting, and we now consider the case where both sources are compressed in order to reach any point of the segment between A and B in Figure 6.1(b). For some applications, it may be desirable to vary the rates of each encoder while keeping the total sum rate constant. This setup is said to be *nonasymmetric*. It will be said to be *symmetric* when both sources are compressed at the same rate (point C in Figure 6.1(b)). Note that both sources can be compressed at the same rate if and only if $\max(H(X|Y), H(Y|X)) \leqslant \frac{H(X,Y)}{2} \leqslant \min((H(X), H(Y))$. Several methods can be used to reach any point of the SW rate bound. The first approach called *time sharing* uses two asymmetric coder/decoder pairs alternatively. Other methods code each input sequence $\mathbf{x}$ and $\mathbf{y}$ with linear channel codes. For each source, part of the information bits plus syndrome or parity bits are transmitted. The decoder then needs to estimate the n-length sequences $\mathbf{x}$ and $\mathbf{y}$ knowing the partial information bits, the syndrome or parity bits and the correlation between $\mathbf{x}$ and $\mathbf{y}$.

6.3.1 Time Sharing

Let $\mathbf{x}$ and $\mathbf{y}$ again denote two random binary correlated sequences of source symbols of length n. It is assumed that the correlation is defined by a BSC of crossover probability p. All points of the segment between A and B of the SW rate bound are achievable by time sharing. More precisely, a fraction α of samples ($\alpha.n$ samples, where n is the sequence length) is coded at the vertex point A, that is, at rates $(H(Y), H(X|Y))$, with the methods described in Section 6.2, and a fraction $(1-\alpha)$ of samples is then coded at rates $(H(X), H(Y|X))$ corresponding to the corner point B of the SW rate region. This leads to the rates $R_X = \alpha H(X) + (1-\alpha)H(X|Y)$ and $R_Y = (1-\alpha)H(Y) + \alpha H(Y|X)$.

6.3.2 The Parity Approach

The sequences $\mathbf{x}$ and $\mathbf{y}$ of length n are fed into two linear channel coders (turbo coders [6] or LDPC coders [25]), which produce two subsequences $\mathbf{x}^h = (x_1 \dots x_l)$ and $\mathbf{y}^h = (y_{l+1} \dots y_n)$ of information bits and two sequences $x_1^p \dots x_a^p$ and $y_1^p \dots y_b^p$ of parity bits, where $a \geqslant (n-l)H(X|Y)$ and $b \geqslant lH(Y|X)$. The achievable rates for each source are then $R_X \geqslant \frac{l}{n}H(X) + \frac{a}{n} \geqslant \frac{l}{n}H(X) + \frac{n-l}{n}H(X|Y)$ and $R_Y \geqslant \frac{n-l}{n}H(Y) + \frac{b}{n} \geqslant$

$\frac{n-l}{n}H(Y)+\frac{l}{n}H(Y|X)$. The SW boundary can be approached by varying the ratio $\frac{l}{n}$ between 0 and 1. Unlike the time-sharing approach, the parity bits are computed on the entire input sequences of length n. The sequence $(\mathbf{x}^h\mathbf{x}^p)$ is regarded as a noisy version of the sequence $(\mathbf{y}^p\mathbf{y}^h)$. As in the asymmetric setup, the channel is a parallel combination of a BSC and of a perfect channel. The decoder then needs to estimate $\mathbf{x}$ and $\mathbf{y}$ given $(\mathbf{x}^h\mathbf{x}^p)$ and $(\mathbf{y}^p\mathbf{y}^h)$.

A first practical implementation of the parity approach is proposed in [9] with turbo codes and considers the particular case of symmetric SW coding. The solution is then extended to reach any point of the SW boundary in [6, 8]. In [8], the two information sequences of length n are encoded with a systematic punctured turbo code. The information bits of the second source are interleaved before being fed to the encoder. All the systematic bits as well as a subset of the parity bits are punctured in order to leave only a bits for the sequence $\mathbf{x}$ and b bits for the sequence $\mathbf{y}$. Each turbo decoder proceeds in the standard way, with the difference that part of the bits is received via an ideal channel. After each iteration, the measures obtained as a result of a turbo decoding on each input bit of one sequence, for example, $\mathbf{x}$, are passed as additional extrinsic information to the turbo decoder used to estimate the sequence $\mathbf{y}$, via the binary symmetric correlation channel (i.e., $\mathbb{P}(y_k=0)=(1-p)\mathbb{P}(x_k=0)+p\mathbb{P}(x_k=1)$).

Another implementation of the parity approach is described in [25] with LDPC codes and later studied in [26] for short to moderate-length codes. The decoder needs to determine an n-length sequence from the part of information bits and parity bits received for the sequence $\mathbf{x}$ and from the correlated sequence $\mathbf{y}$. For this, a message-passing algorithm is used by setting the LLR (log-likelihood ratio) of the bits received via the ideal channel to infinity. The LLR of the bits received via the correlation channel are set to $\pm\log(\frac{1-p}{p})$. The LLR for the punctured bits are set to zero. Two sets of degree distributions for the variables nodes are used to account for the fact that information bits are received via the correlation channel, whereas the parity bits are received via an ideal channel.

6.3.3 The Syndrome Approach

Let us consider an (n, k) linear channel code $\mathcal{C}$ defined by its generator matrix $G_{k\times n}$ and its parity-check matrix $H_{(n-k)\times n}$. A syndrome approach was first proposed in [19], based on the construction of two independent linear binary codes $\mathcal{C}^1$ and $\mathcal{C}^2$ with G_1 and G_2 as generator matrices, obtained from the main code $\mathcal{C}$. More precisely, the generator matrices G_1 and G_2 of the two subcodes are formed by extracting m_1 and m_2 lines, respectively, where $m_1+m_2=k$, from the matrix G of the code $\mathcal{C}$. The parity-check matrices H_1 and H_2 are then of size $(n-m_1)\times n$ and $(n-m_2)\times n$, respectively. A geometric interpretation of the code splitting is shown in Figure 6.12. Each coder sends a syndrome, defined as $\mathbf{s}_x=\mathbf{x}H_1^T$ and $\mathbf{s}_y=\mathbf{y}H_2^T$, respectively, to index the cosets of $\mathcal{C}^1$ and $\mathcal{C}^2$ containing $\mathbf{x}$ and $\mathbf{y}$. The total rate for encoding the input sequences of length n is then $n-m_1+n-m_2=2n-k$ bits. The cosets containing $\mathbf{x}$ and $\mathbf{y}$ are translated versions of the codes $\mathcal{C}^1$ and $\mathcal{C}^2$ (i.e., the cosets having null syndromes)

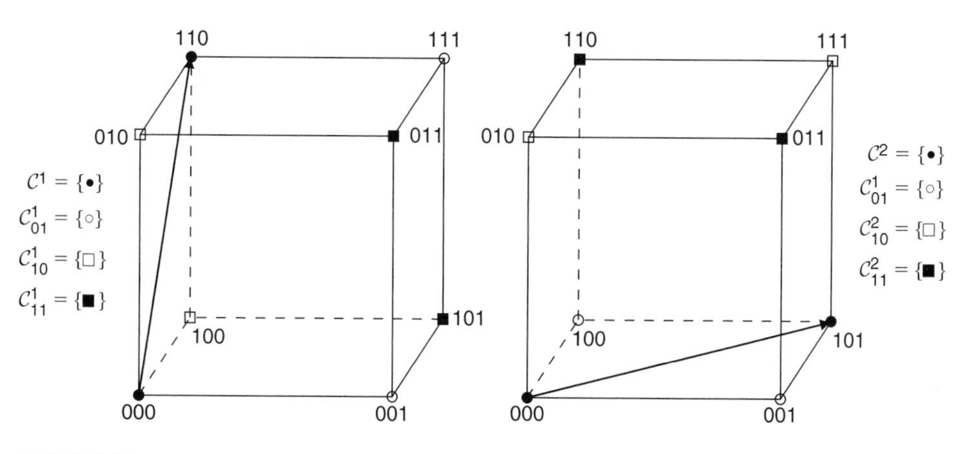

FIGURE 6.12

Splitting the code of Figure 6.3 into two subcodes: code $\mathcal{C}^1$ defined by $G_1 = (110)$ or $H_1 = \binom{110}{001}$ (left); code $\mathcal{C}^2$ defined by $G_2 = (101)$ or $H_2 = \binom{010}{101}$ (right).

by vectors $\mathbf{t}_x$ and $\mathbf{t}_y$. This translation is illustrated in Figure 6.12, where the addition by the translation vector is performed over the finite field $\{0, 1\}$. The codes $\mathcal{C}^1$ and $\mathcal{C}^2$ must be chosen such that all pairs $(\mathbf{x}, \mathbf{y})$ can be determined uniquely given the two coset indexes. We now explain how to determine a pair $(\mathbf{x}, \mathbf{y})$ from a usual ML decoder.

Let $\mathbf{u}_x G_1$ and $\mathbf{u}_y G_2$ define codewords in $\mathcal{C}^1$ and $\mathcal{C}^2$. By definition these vectors have a null syndrome. All possible codewords in a coset of syndrome $\mathbf{s}$ can be enumerated by translating a codeword belonging to the coset of null syndrome by a vector $\mathbf{t}$ belonging to the coset of syndrome $\mathbf{s}$ (see Equation (6.1)). If the code is systematic, a candidate for the translating vector is $\mathbf{t} = (\mathbf{0}\ \mathbf{s})$, as explained in Section 6.2.2. Thus, $\mathbf{x}$ and $\mathbf{y}$ can be seen as translated versions of codewords belonging to the cosets of null syndrome of the codes $\mathcal{C}^1$ and $\mathcal{C}^2$, respectively. The vectors $\mathbf{x}$ and $\mathbf{y}$ can thus be expressed as $\mathbf{x} = \mathbf{u}_x G_1 \oplus (\mathbf{0}\ \mathbf{s}_x)$, $\mathbf{y} = \mathbf{u}_y G_2 \oplus (\mathbf{0}\ \mathbf{s}_y)$, where $\mathbf{t}_x = (\mathbf{0}\ \mathbf{s}_x)$ and $\mathbf{t}_y = (\mathbf{0}\ \mathbf{s}_y)$ are the representatives of the cosets of $\mathcal{C}^1$ and $\mathcal{C}^2$ with syndromes $\mathbf{s}_x$ and $\mathbf{s}_y$. The decoder must search for a pair of codewords, one from each translated coset of $\mathcal{C}^1$ and $\mathcal{C}^2$. When receiving the syndromes $\mathbf{s}_x$ and $\mathbf{s}_y$, it first finds a codeword $\mathbf{c}$ of the main code $\mathcal{C}$ (which is actually the error pattern between $\mathbf{x}$ and $\mathbf{y}$) that is closest to the vector $\mathbf{t} = \mathbf{t}_x \oplus \mathbf{t}_y = (\mathbf{0}\ \mathbf{s}_x) \oplus (\mathbf{0}\ \mathbf{s}_y)$. The error pattern $\mathbf{x} \oplus \mathbf{y}$ is the minimum-weight codeword having the same syndrome as the vector $\mathbf{t}$. If the code is systematic, the sequences $\mathbf{x}$ and $\mathbf{y}$ are reconstructed as $\hat{\mathbf{x}} = \hat{\mathbf{u}}_x G_1 \oplus \mathbf{t}_x$ and $\hat{\mathbf{y}} = \hat{\mathbf{u}}_y G_2 \oplus \mathbf{t}_y$, where $\hat{\mathbf{u}}_x$ and $\hat{\mathbf{u}}_y$ are the systematic bits of the codeword $\mathbf{c}$. This method achieves the optimal SW compression sum rate if the global code $\mathcal{C}$ achieves the capacity of the equivalent correlation channel (between $\mathbf{x}$ and $\mathbf{y}$). The number of lines assigned to G_1 (and therefore to G_2) allows choosing any rate on the dominant face of the SW region (between points A and B in Figure 6.1(b)). The above ideas are further developed in [30, 31], for nonsystematic linear codes. The code construction has been extended in [27] to the case where the sources X and Y are binary but nonuniformly distributed.

In a second approach proposed in [10], the vectors $\mathbf{x}$ and $\mathbf{y}$ belong to the same coset of a unique code $\mathcal{C}$, which is the translated version of the coset of null syndrome by a vector $\mathbf{t}$. Each source vector is identified by a syndrome computed with the parity-check matrix H of the code $\mathcal{C}$ plus part of the source information bits [10]. The syndromes $\mathbf{s}_x^T = H\mathbf{x}^T$ and $\mathbf{s}_y^T = H\mathbf{y}^T$, of length $(n-k)$, are thus computed for both sequences and transmitted to the decoder. In addition, the k' first bits of $\mathbf{x}$ (denoted $\mathbf{x}_1^{k'}$) and the $k-k'$ next bits for the source $\mathbf{y}$ (denoted $\mathbf{y}_{k'+1}^{k}$) are transmitted as systematic bits, where k' is an integer so that $k' \in [0, k]$. The total rate for the sequences $\mathbf{x}$ and $\mathbf{y}$ of length n is, respectively, $n-k+k'$ and $n-k'$ bits. The structure of the coders is depicted in Figure 6.13. Note that $k'=k$ and $k'=0$ correspond to the two asymmetric setups with rates given by the corner points A and B of the SW region.

As in the asymmetric case, the representative of the coset is thus the concatenation of the transmitted information bits (k' bits for the sequence $\mathbf{x}$) and of the $(n-k)$-length syndrome vector. As in the first approach, when receiving the syndromes $\mathbf{s}_x$ and $\mathbf{s}_y$, the decoder must find a codeword $\mathbf{c}$ of the code $\mathcal{C}$ that is closest to the translating vector $\mathbf{t} = \mathbf{t}_x \oplus \mathbf{t}_y = (\mathbf{0}\ \mathbf{s}_x) \oplus (\mathbf{0}\ \mathbf{s}_y)$. For this, it first computes $\mathbf{s}_z = \mathbf{s}_x \oplus \mathbf{s}_y$, which is the syndrome of the error pattern $\mathbf{z}$ between $\mathbf{x}$ and $\mathbf{y}$ ($\mathbf{z} = \mathbf{x} \oplus \mathbf{y}$). In [10], a modified LDPC decoder estimates $\hat{\mathbf{z}} = \widehat{\mathbf{x} \oplus \mathbf{y}}$ from the syndrome $\mathbf{s}_z$ and the all-zero word of size n (see Figure 6.14) [16]. The error pattern $\mathbf{z}$ is the smallest weight codeword with syndrome $\mathbf{s}_z$. When convolutional and turbo codes are used, the estimation of the error pattern can be performed with a conventional Viterbi decoder and an inverse syndrome former (ISF) [33] or with a syndrome decoder [24] (see Section 6.2.2 for further details).

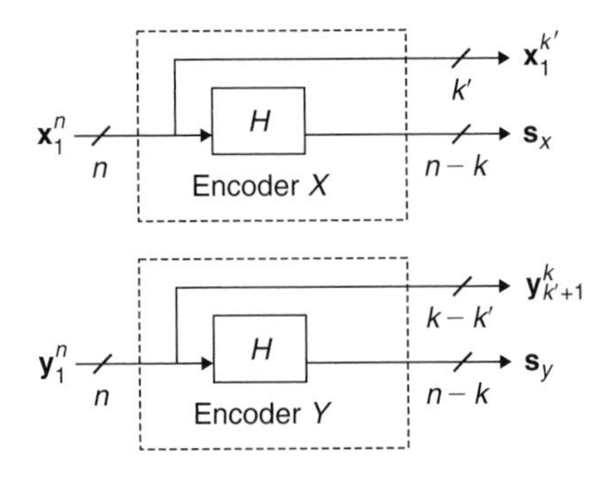

FIGURE 6.13

The nonasymmetric coders in the syndrome approach [10]. A single code $\mathcal{C}$ (determined by the parity check matrix H) is used for each sequence to be compressed.

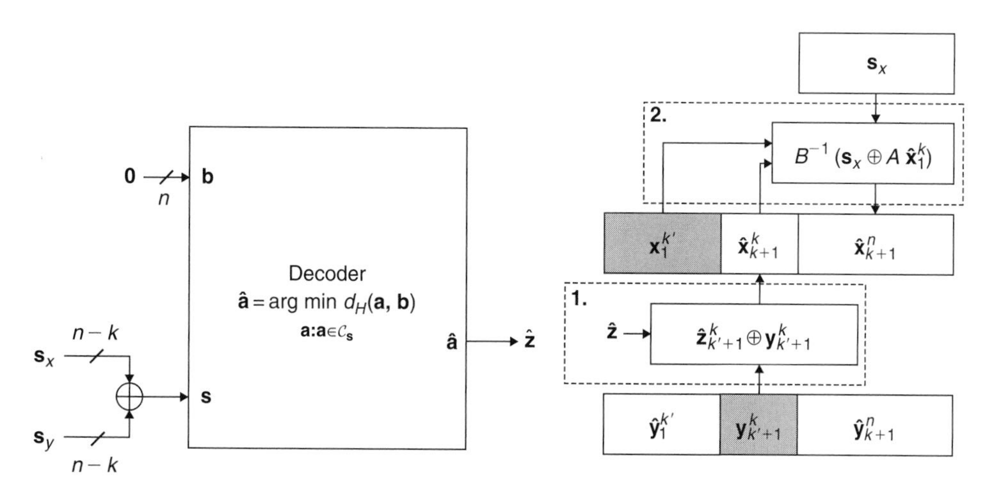

FIGURE 6.14

Nonasymmetric decoder: estimation of the error pattern $\mathbf{z} = \mathbf{x} \oplus \mathbf{y}$ (left); reconstruction of the source sequence $\mathbf{x}$ (right).

Once the error pattern is found, the subsequences of information bits $\mathbf{x}_{k'+1}^k$ and $\mathbf{y}_1^{k'}$ can be retrieved from the error pattern $\mathbf{z} = \mathbf{x} \oplus \mathbf{y}$ as (see Figure 6.14-right)

$$\hat{\mathbf{x}}_{k'+1}^k = \mathbf{y}_{k'+1}^k \oplus \hat{\mathbf{z}}_{k'+1}^k \quad \text{and} \quad \hat{\mathbf{y}}_1^{k'} = \mathbf{x}_1^{k'} \oplus \hat{\mathbf{z}}_1^{k'}. \tag{6.3}$$

Subsequences of $n-k$ bits (i.e., $\mathbf{x}_{k+1}^n$ and $\mathbf{y}_{k+1}^n$), remain to be computed for both sequences. Let us assume that $H = (A\ B)$, where B is an invertible square matrix of dimension $(n-k) \times (n-k)$. Note that for a rate k/n channel code, the parity-check matrix H has rank $n-k$. Therefore, one can always find a permutation to be applied on the columns of H so that the resulting parity-check matrix has the right form. Thus, $\mathbf{s}_x = H\mathbf{x} = (A\ B)\mathbf{x} = A\,\mathbf{x}_1^k \oplus B\,\mathbf{x}_{k+1}^n$, and the remaining $n-k$ unknown bits of the sequence $\mathbf{x}$ (and similarly for the sequence $\mathbf{y}$) can be computed as

$$\hat{\mathbf{x}}_{k+1}^n = B^{-1}\,(\mathbf{s}_x \oplus A\,\hat{\mathbf{x}}_1^k), \tag{6.4}$$

where B^{-1} denotes the inverse of the matrix B [10]. The authors in [10] have used LDPC codes.

If the error pattern is not perfectly estimated, the estimation of the remaining symbols $\mathbf{x}_{k+1}^n$ in Equation (6.4) may yield error propagation. Note that the unknown positions in the vector $\mathbf{x}$ are design parameters. These positions (or, equivalently, the columns of the matrix H to be extracted in order to build the matrix B) can be chosen so that the inverse B^{-1} is as sparse as possible. Figure 6.15 shows the BER of the error pattern $\mathbf{z}$ (continuous line) and its effect on the estimation of the source sequence $\mathbf{x}$. Performance is shown for a convolutional code. The dotted curve represents the BER, when Equation (6.4) is performed with an arbitrary invertible matrix B. This BER

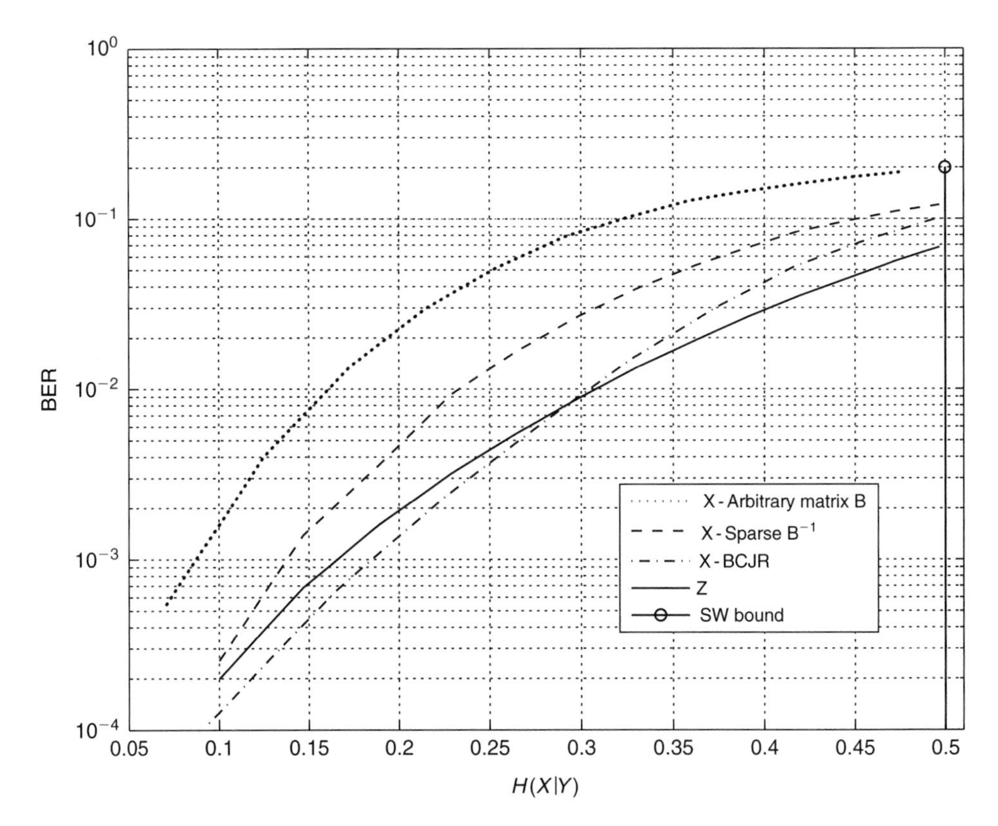

FIGURE 6.15

Error propagation in the estimation of the source X. The convolutional code is defined by its parity-check matrix $H = \begin{pmatrix} 11 & 15 & 06 \\ 15 & 12 & 17 \end{pmatrix}$ and is punctured over a period of four trellis sections in order to get a 2 : 1 compression rate for the source.

can be lowered to the dashed curve, if a matrix B with sparse inverse is chosen. This limits the error propagation. Finally, the error propagation can be further reduced, if one applies a modified decoder to solve Equation (6.4). This decoder (Viterbi for a convolutional code) is matched to a channel which combines a perfect channel (for $\mathbf{x}_1^{k'}$), a BSC (for $\hat{\mathbf{x}}_{k'+1}^{k}$ with crossover probability the BER of $\mathbf{z}$) and an erasure channel (for $\mathbf{x}_{k+1}^{n}$). Interestingly, with this decoder the BER of $\mathbf{x}$ remains almost the same as that of $\mathbf{z}$.

6.3.4 Source Splitting

Source splitting [21] is another approach to vary the rates between the two encoders. The approach is asymptotic and involves splitting the source into subsources having a lower entropy. More precisely, it transforms the two source problem into a three source problem where one source, for example, X is split into two i.i.d. discrete sources U and V as $U = XT$ and $V = Y(1-T)$. T is an i.i.d. binary source (which can be seen as a time-sharing sequence). By varying $\mathbb{P}(T=1)$, one can vary $H(Y|U)$ between $H(Y)$ and $H(Y|X)$. The sum rate becomes $R_X + R_Y = R_U + R_V + R_Y = H(U|T) + H(Y|U) + H(V|U, Y)$. A random coding argument is used to show that the rate triple is achievable [21]; however, no constructive codes are proposed.

6.3.5 Rate Adaptation

In practical applications, the correlation between the two sources may not be known at the time of code design. It may thus be necessary, as in the asymmetric setup, to have solutions allowing for flexible rate adaptation to any correlation between the two sources. Nonasymmetric rate-adaptive SW coding/decoding schemes are direct extensions of their asymmetric counterparts. Rate adaptation is performed by simply puncturing information, parity, or syndrome bits in the schemes described above.

In the parity approach, we have seen that for two correlated sequences of length n, the rate allocation between the two sequences is controlled by the number l of information bits transmitted for each. The change in the rate allocation between the two sources is thus performed by puncturing more or less information bits. For a given rate share between the two sources, rate adaptation to varying correlation is achieved by simply puncturing the parity bits. Standard decoding of punctured channel codes can then be applied.

For the syndrome approaches, as explained in Section 6.2.3, the application of the puncturing is less straightforward. Puncturing the syndrome bits may degrade the performance of the code. The approach considered in [34], similarly to [36] for the asymmetric setup, consists of first protecting the syndromes with an accumulator code. The effect of the accumulator code followed by the puncturing is equivalent to merging some rows of the parity-check matrix H by adding them, thus constructing a matrix H_i of dimension $(n-k_i) \times n$, of rank $n-k_i$. In the matrix H_i, with appropriate permutation, one may exhibit a submatrix B_i of dimension $(n-k_i) \times (n-k_i)$, which is invertible (see Equation (6.4)). The positions of the k' ($k' \in [0, k_i]$) and $k_i - k'$ information bits transmitted for the sequences $\mathbf{x}$ and $\mathbf{y}$, respectively, must then be chosen so that they correspond to the positions of the remaining nonfree k_i columns of H_i. If the correlation is not known to the encoder, the rate may be controlled via a feedback channel. In this case, once the error pattern $\hat{\mathbf{z}}$ has been estimated, the decoder verifies if the condition $\hat{\mathbf{z}} H^I = \mathbf{s}_z$ is satisfied. If this condition is not satisfied, it then requests for more syndrome bits. Note that, if the error pattern is not perfectly estimated, the last step of the decoding algorithm represented by Equation (6.4) may yield error propagation. The coding and decoding structures are depicted in Figure 6.16. Figure 6.17 illustrates the rate points achieved with a nonsymmetric SW coding/decoding scheme based on LDPC codes.

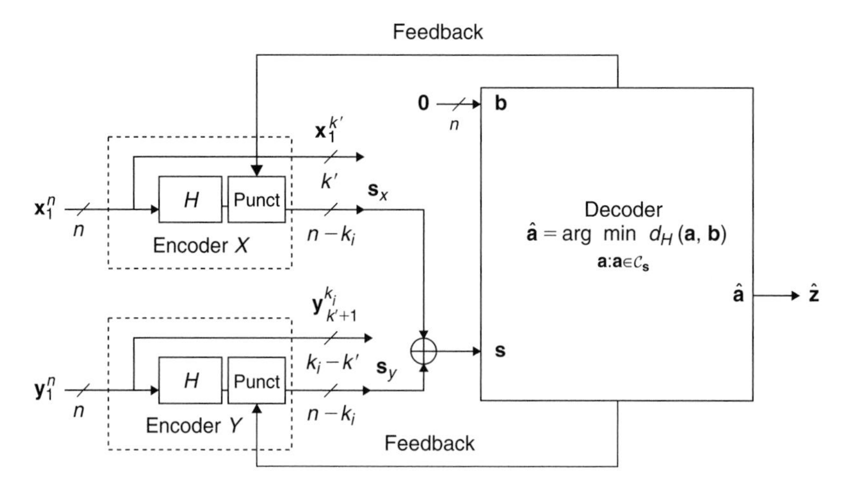

FIGURE 6.16

Rate-adaptive coding/decoding scheme.

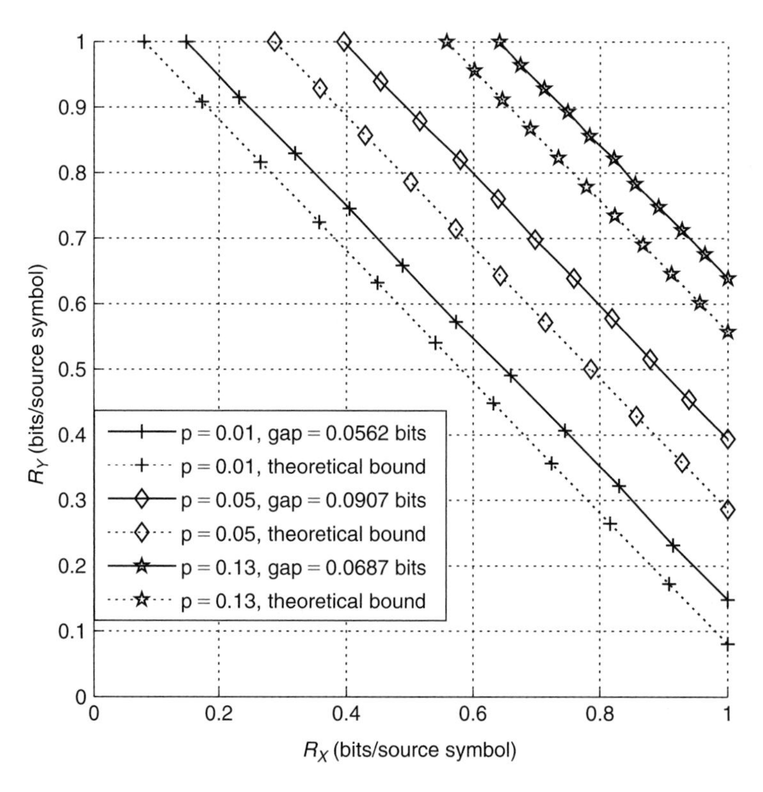

FIGURE 6.17

Rate points achieved for different parameters p of the binary symmetric correlation channel.

6.4 ADVANCED TOPICS

6.4.1 Practical Code Design Based on Source Codes

6.4.1.1 *Code Design Based on Source Alphabet Partitioning*

The use of source codes for SW coding was first been explored in [2] considering Huffman codes. The approach relies on partitioning the source alphabet into groups of symbols. One source Y is coded with a Huffman code designed according to its stationary probability $\mathbb{P}(Y)$. The alphabet of the second source X is partitioned so that two symbols x_i and x_j are not grouped together if there exists a symbol y such that $\mathbb{P}(x_i, y) > 0$ and $\mathbb{P}(x_j, y) > 0$. The joint probability mass function of (X, Y) may need to be slightly modified, for example, by thresholding the smallest values and then renormalizing, so that this condition is verified. In addition, the entropy of the group index must be minimum and approach at best $H(X|Y)$. Knowing the SI symbol y and the group index, the decoder can find the transmitted symbol x. The above code design based on alphabet partitioning has been formalized and generalized in [41] for arbitrary p.m.f. $\mathbb{P}(X, Y)$, assuming memoryless sources. The partition can be coded with a Huffman or an arithmetic code.

6.4.1.2 *Punctured and Overlapped Arithmetic Codes*

Different solutions based on overlapped arithmetic [11] and quasi-arithmetic (QA) [3] codes, as well as punctured (quasi-)arithmetic codes [17] have appeared recently for the case of binary sources. Arithmetic coding recursively divides the interval [0, 1) according to the source probabilities. Let us consider a binary memoryless source with probabilities denoted P_0 and P_1. At each symbol clock instant n, the current interval $[L_n, H_n)$ is divided into two subintervals whose widths are proportional to P_0 and $P_1 = 1 - P_0$, respectively. One of these subintervals is selected according to the value of the next symbol and becomes the current interval. Once the last symbol is encoded, the encoder produces a sequence of bits identifying the final subinterval. Practical implementations of arithmetic coding were first introduced in [18–22], and further developed in [23, 37].

The encoding and decoding processes can be modeled as state machines. But with optimal arithmetic coding, the number of states can grow exponentially with the sequence length. Controlled approximations can reduce the number of possible states without significantly degrading compression performances [13]. This fast, but reduced precision, implementation of arithmetic coding is called *quasi-arithmetic coding* [13]. Instead of using the real interval [0, 1], quasi-arithmetic coding is performed on an integer interval $[0, T]$. The value of T controls the trade-off between complexity, the number of states, and compression efficiency.

The average code length achieved per input source symbol is very close to the source entropy. To further reduce the rate down to the conditional entropy of two correlated sources, two strategies can be considered: puncturing the sequence of encoded bits [17] or overlapping the symbol probability intervals of the arithmetic [11] or quasi-arithmetic coder [3]. For the overlapped (quasi-)arithmetic codes, at a given instant n, the probability interval $[L_n, H_n)$ is partitioned into two subintervals $[L_n, P_0(H_n - L_n) + \rho T/2)$ associated with the symbol 0 and $[P_0(H_n - L_n) - \rho T/2, H_n)$ associated with the symbol 1. The parameter ρ controls the overlap between the subintervals. The larger the overlap, the lower the bit rate but the higher the uncertainty on the decoded sequence. In both cases, overlapped and punctured codes, the resulting code is no longer uniquely decodable. The ambiguity on the encoded sequence of symbols can be removed with the help of the SI (correlated source) and of a symbol MAP decoder.

To explain the decoding algorithm, let us first consider the punctured solution. Let $\mathbf{x}$ and $\mathbf{y}$ be two correlated sequences of length $L(\mathbf{x})$. The sequence $\mathbf{x}$ is coded with a QA code producing a sequence of bits $\mathbf{u}$ of length $L(\mathbf{u})$, which is then punctured. Given the received QA coded bitstream $u_1^{L(\mathbf{u})}$ and the correlated sequence $y_1^{L(\mathbf{x})}$ (SI), the symbol posterior marginals $\mathbb{P}(X_n = x_n \mid u_1^{L(\mathbf{u})}, y_1^{L(\mathbf{x})})$ are computed by running a BCJR algorithm [4] on the state model of the QA automaton. To cope with the fact that the QA automaton contains transitions corresponding to a varying number of symbols and a varying number of bits, the state model used in the decoding must keep track of the number of symbols being coded at a particular bit clock instant k. It is hence defined by the pairs of random variables $\mathcal{V}_k = (N_k, M_k)$, where N_k denotes the state of the QA automaton and M_k represents the symbol clock value at the bit clock instant k [12]. Since the variable M_k corresponds to the symbol clock, it accounts for the correlation between the encoded sequence and the SI. The transition probabilities on the state model $\mathcal{V}_k = (N_k, M_k)$ are equal to the transition probabilities between states N_k of the QA automaton, if the numbers of bits and symbols associated with this transition match the one defined by the automaton. The transition probabilities are equal to zero otherwise. The probabilities of transition between states N_k of the QA automaton depend on the source statistics and can account for source memory. Thus, for each state $\nu = (n, m)$, the BCJR algorithm computes the probabilities $\alpha_k(\nu) = \mathbb{P}(\mathcal{V}_k = \nu; \mathbf{U}_1^k)$ and $\beta_k(\nu) = \mathbb{P}(\mathbf{U}_{k+1}^{L(\mathbf{U})} \mid \mathcal{V}_k = \nu)$. The branch metric $\gamma_k(\nu'|\nu)$ used for the transition τ between two states $\nu = (n, m)$ and $\nu' = (n', m')$ is given by $\gamma_k(\nu'|\nu) = \mathbb{P}(\nu'|\nu) \times \mathbb{P}(\mathbf{b}_\tau) \times \mathbb{P}(\mathbf{x}_\tau|\mathbf{y}_\tau)$, where $\mathbf{b}_\tau$ and $\mathbf{x}_\tau$ denote the subsequences of bits and symbols associated with given transition τ on the QA automaton. The probability $\mathbb{P}(\mathbf{b}_\tau)$ is computed considering a probability of 0.5 for the punctured positions. The term $\mathbb{P}(\mathbf{x}_\tau|\mathbf{y}_\tau)$ represents the correlation channel between X and Y.

The decoding for the overlapped quasi-arithmetic codes proceeds in the same manner as for the punctured solution. The only difference is in the automaton, which in the case of overlapped codes depends not only on the distribution of the source X but also on the overlap factor ρ (for details see [17] and [3]). A second difference is in the branch metric $\gamma_k(\nu'|\nu)$ in which the term $\mathbb{P}(\mathbf{b}_\tau)$ disappears, since, in this case, all the coded bits are known to the decoder (and not punctured as in the previous method).

6.4.2 Generalization to Nonbinary Sources

Thus so far, we have considered only binary correlated sources. Practical applications often involve nonbinary (e.g., q-ary) sources. In practice, before being encoded, the q-ary sources are first binarized. Each resulting sequence of bits is then fed to the SW coder. For example, this is the case in [42] where the authors describe the principle using turbo codes to compress both sources X and Y. For each source, the sequences of bits at the output of the "symbol-to-bit" converter are interleaved and input to a punctured turbo coder. The turbo decoders act at a bit level in a standard fashion, but information exchange is performed between the two turbo decoders at the symbol level. The a posteriori probabilities computed by a turbo decoder for each input bit are converted into a measure on the corresponding q-ary symbol. The resulting symbol probabilities are then multiplied by the conditional probabilities characterizing the correlation channel and fed into the second turbo decoder as extrinsic information.

6.4.3 Generalization to *M* Sources

The asymmetric and nonasymmetric coding schemes presented earlier can be extended to more than two sources. The practical code design approach proposed in [19] has been extended in [30] and [31] for M sources (see Section 6.3.3). Each subcode is constructed by selecting a subset of rows from the generator matrix G of the starting code $\mathcal{C}$. Examples of design are given for irregular repeat-accumulate (IRA) codes. However, the method is shown to be suboptimal when the number of sources is greater (or equal) than three. The parity approach described in [25] has also been extended to M sources in [26]. The sequences $\mathbf{x}_i$, $i = 1, \ldots M$, of length n are fed into M LDPC coders, which produce subsequences of $a_i \times k$ information bits and $(1 - a_i) \times k$ parity bits. Joint decoding of the M sources is then performed.

6.5 CONCLUSIONS

This chapter has explained how near-capacity achieving channel codes can be used to approach the SW rate bound. Although the problem is posed as a communication problem, classical channel decoders need to be modified. This chapter has outlined the different ways to adapt channel decoding algorithms to the Slepian–Wolf decoding problem. In the approaches presented, it is assumed that syndrome or parity bits are transmitted over a perfect channel. As explained, the equivalent communication channel is thus the parallel combination of a BSC and a perfect channel. In some applications, this may not be the case: syndrome or parity bits may have to be transmitted on an erroneous communication channel. Distributed joint source-channel coding schemes, in which the SW code acts as a single code for both source coding and channel error correction, can be found in the literature to address this particular scenario. This last problem, however, remains a field of research. Finally, the use of SW codes is currently being explored for a number of applications such as low-power sensor networks, compression of hyperspectral images, monoview and multiview video compression, error-resilient video transmission, and secure fingerprint biometrics. These applications are presented in other chapters of Part 2.

REFERENCES

[1] A. Aaron and B. Girod. Compression with side information using turbo codes. *Proceedings of the IEEE International Data Compression Conference (DCC)*, 0(0):252–261, Apr. 2002.

[2] A. K. Al Jabri and S. Al-Issa. Zero-error codes for correlated information sources. *Proceedings of the IMA International Conference on Cryptography and Coding*, pp.17–22, Dec. 1997.

[3] X. Artigas, S. Malinowski, C. Guillemot, and L. Torres. Overlapped quasi-arithmetic codes for distributed video coding. *Proceedings of the IEEE International Conference on Image Processing (ICIP)*, pp. II.9–II.12, Sept. 2007

[4] L. R. Bahl, J. Cocke, F. Jelinek, and J. Raviv, Optimal decoding of linear codes for minimizing symbol error rate. *IEEE Transactions on Information Theory*, IT-20, pp. 284–287, Mar. 1974.

[5] J. Bajcsy and P. Mitran. Coding for the Slepian-Wolf problem with turbo codes. *Proceedings of the IEEE International Global Communications Conference (Globecom)*, pp. 1400–1404, Dec. 2001.

[6] F. Cabarcas and J. Garcia-Frias. Approaching the Slepian-Wolf boundary using practical channel codes. *Proceedings IEEE International Symposium on Information Theory*, p. 330, June 2004.

[7] J. Chen, A. Khisti, D. M. Malioutov, and J. S. Yedidia. Distributed source coding using serially-concatenated-accumulate codes, *Proceedings of the IEEE International Information Theory Workshop*, pp. 209–214, Oct. 2004.

[8] J. Garcia-Frias and F. Cabarcas. Approaching the Slepian-Wolf boundary using practical channel codes. *Signal Processing*, 86(11):3096–3101, 2006.

[9] J. Garcia-Frias and Y. Zhao. Compression of correlated binary sources using turbo codes. *IEEE Communications Letters*, 5(10):417–419, Oct. 2001.

[10] N. Gehrig and P. L. Dragotti. Symmetric and asymmetric Slepian–Wolf codes with systematic and non-systematic linear codes. *IEEE Communications Letters*, 9(1):61–63, Jan. 2005.

[11] M. Grangetto, E. Magli, and G. Olmo. Distributed arithmetic coding. *IEEE Communications Letters*, 11(11):883–885, Nov. 2007.

[12] T. Guionnet and C. Guillemot. Soft and joint source-channel decoding of quasi-arithmetic codes. *EURASIP Journal on Applied Signal Processing*, 2004(3):394–411, Mar. 2004.

[13] P. G. Howard and J. S. Vitter. *Image and Text Compression*. Kluwer Academic Publishers, pp. 85–112, 1992.

[14] J. Jiang, D. He, and A. Jagmohan. Rateless Slepian–Wolf coding based on rate adaptive low-density parity-check codes. *Proceedings of the IEEE International Symposium on Information Theory*, pp. 1316–1320, June 2007.

[15] J. Li and H. Alqamzi. An optimal distributed and adaptive source coding strategy using rate-compatible punctured convolutional codes. *Proceedings of the IEEE International Conference on Acoustics, Speech, and Signal Processing (ICASSP)*, 3:iii/685–iii/688, pp. 18–23 Mar. 2005.

[16] A. D. Liveris, Z. Xiong, and C. N. Georghiades. Compression of binary sources with side information at the decoder using LDPC codes. *IEEE Communications Letters*, 6(10):440–442, Oct. 2002.

[17] S. Malinowski, X. Artigas, C. Guillemot, and L. Torres. Distributed source coding using punctured quasi-arithmetic codes for memory and memoryless sources. Submitted to *IEEE International Workshop on Signal Processing Systems*, Oct. 2008.

[18] R. Pasco. Source coding algorithms for fast data compression. Ph.D thesis, Dept. of Electrical Engineering, Stanford Univ., Stanford, CA, 1976.

[19] S. S. Pradhan and K. Ramchandran. Distributed source coding: Symmetric rates and applications to sensor networks. *Proccedings of the IEEE International Data Compression Conference (DCC)*, pp. 363-372, Mar. 2000.

[20] S. S. Pradhan and K. Ramchandran. Distributed source coding using syndromes (DISCUS): Design and construction. *Proceedings of the IEEE International Data Compression Conference (DCC)*, pp. 158-167, Mar. 1999.

[21] B. Rimoldi and R. Urbanke. Asynchronous Slepian-Wolf coding via source-splitting. *Proceedings of the IEEE International Symposium on Information Theory*, p. 271, July 1997.

[22] J. J. Rissanen. Generalized Kraft inequality and arithmetic coding. *IBM J. Res. Develop.*, 20, pp. 198-203, May 1976.

[23] ——, Arithmetic codings as number representations. *Acta Polytech. Scand. Math.*, 31, pp. 44-51, Dec. 1979.

[24] A. Roumy, K. Lajnef, and C. Guillemot. Rate-adaptive turbo-syndrome scheme for Slepian-Wolf coding. *Proceedings of the IEEE Asilomar Conference on Signals, Systems, and Computers*, Nov. 2007.

[25] M. Sartipi and F. Fekri. Distributed source coding in wireless sensor networks using LDPC coding: The entire Slepian-Wolf rate region. *Proceedings of the IEEE Wireless Communications and Networking Conference*, 4(0):1939-1944, Mar. 2005.

[26] M. Sartipi and F. Fekri. Distributed source coding using short to moderate length rate-compatible LDPC codes: The entire Slepian-Wolf rate region. *IEEE Transactions on Communications*, 56(3):400-411, Mar. 2008.

[27] D. Schonberg, K. Ramchandran, and S. S. Pradhan. Distributed code constructions for the entire Slepian-Wolf rate region for arbitrarily correlated sources. *Proceedings of the IEEE International Data Compression Conference (DCC)*, pp. 292-301, Mar. 2004.

[28] V. Sidorenko and V. Zyablov. Decoding of convolutional codes using a syndrome trellis. *IEEE Transactions on Information Theory*, 40(5): 1663-1666, Sept. 1994.

[29] D. Slepian and J. K. Wolf. Noiseless coding of correlated information sources. *IEEE Transactions on Information Theory*, IT-19, pp. 471-480, July 1973.

[30] V. Stankovic, A. D. Liveris, Z. Xiong, and C. N. Georghiades. Design of Slepian-Wolf codes by channel code partitioning. *Proceedings of the IEEE International Data Compression Conference (DCC)*, pp. 302-311, 2004.

[31] V. Stankovic, A. D. Liveris, Z. Xiong, and C. N. Georghiades. On code design for the Slepian-Wolf problem and lossless multiterminal networks. *IEEE Transactions on Information Theory*, 52(4):1495-1507, Apr. 2006.

[32] P. Tan and J. Li. Enhancing the robustness of distributed compression using ideas from channel coding. *Proceedings of the IEEE International Global Telecommunications Conference (GLOBECOM)*, 4:5, Dec. 2005.

[33] P. Tan and J. Li. A practical and optimal symmetric Slepian-Wolf compression strategy using syndrome formers and inverse syndrome formers. *Proceeding of 43rd Annual Allerton Conference on Communication, Control and Computing*, Sept. 2005.

[34] V. Toto-Zarasoa, A. Roumy, and C. Guillemot. Rate-adaptive codes for the entire Slepian-Wolf region and arbitrarily correlated sources. *Proceedings of the IEEE International Conference on Acoustics, Speech and Signal Processing (ICASSP)*, Apr. 2008.

[35] Z. Tu, J. Li, and R. S. Blum. An efficient SF-ISF approach for the Slepian-Wolf source coding problem. *EURASIP Journal on Applied Signal Processing*, 2005(6):961–971, May 2005.

[36] D. Varodayan, A. Aaron, and B. Girod. Rate-adaptive codes for distributed source coding. *EURASIP Signal Processing*, 86(11):3123–3130, Nov. 2006.

[37] I. H. Witten, R. M. Neal, and J. G. Cleary. Arithmetic coding for data compression. *Communications of the ACM*, 30(6):520–540, June 1987.

[38] J. K. Wolf. Efficient maximum likelihood decoding of linear block codes using a trellis. *IEEE Transactions on Information Theory*, 24(1):76–80, Jan. 1978.

[39] A. Wyner. Recent results in the Shannon theory. *IEEE Transactions on Information Theory*, 20(1):2–10, Jan. 1974.

[40] R. Zamir, S. Shamai, and U. Erez. Nested linear/lattice codes for structured multiterminal binning. *IEEE Transactions on Information Theory*, 48(6):1250–1276, June 2002.

[41] Q. Zhao and M. Effros. Optimal code design for lossless and near lossless source coding in multiple access network. *Proceedings of the IEEE International Data Compression Conference (DCC)*, pp. 263–272, Mar. 2001.

[42] Y. Zhao and J. Garcia-Frias. Data compression of correlated non-binary sources using punctured turbo codes. *Proceedings of the IEEE International Data Compression Conference (DCC)*, pp. 242–251, Apr. 2002.

CHAPTER

Distributed Compression in Microphone Arrays

7

Olivier Roy, Thibaut Ajdler, and Robert L. Konsbruck

Audiovisual Communications Laboratory, School of Computer and Communication Sciences, Ecole Polytechnique Fédérale de Lausanne, Lausanne, Switzerland

Martin Vetterli

Audiovisual Communications Laboratory, School of Computer and Communication Sciences, Ecole Polytechnique Fédérale de Lausanne, Lausanne, Switzerland and Department of Electrical Engineering and Computer Sciences, University of California, Berkeley, CA

CHAPTER CONTENTS

Distributed Source Coding: Theory, Algorithms, and Applications

7.1 INTRODUCTION

Sensor networks have emerged as a powerful tool to acquire data distributed over a large area by means of self-powered and low-cost sensing units. They allow observation of physical fields at different time instants and space locations, thus acting as spatiotemporal sampling devices. Most envisioned deployments of such distributed infrastructures are tailored to a particular sensing task. Examples are environmental monitoring (temperature, humidity, or pressure measurements) [1], target tracking [2], or acoustic beamforming [3]. The design of these networked architectures usually involves a complex interplay of source and channel coding principles as a means of reproducing the sensed data within prescribed accuracy. In this context, a thorough understanding of the physical phenomenon under observation is crucial. It allows accommodating the design of sensing devices, sampling schemes, and transmission protocols to the targeted application, hence providing significant gains over blind communication strategies (e.g., see [4]).

This chapter addresses the problem of distributed coding for recording a spatiotemporal sound field using an array of microphones. As a means of designing efficient compression schemes that account for the physical properties of sound propagation, a large portion of our exposition is devoted to the study of the spatiotemporal characteristics of the sound field. This analysis is presented in Section 7.2. More precisely, in Section 7.2.1, we consider the evolution of the sound field in various recording environments. The spatiotemporal spectrum of the sound field recorded on a line is investigated in Section 7.2.2, under both the far-field and near-field assumptions. We then study, in Section 7.2.3, the trade-off that exists between the intermicrophone spacing and the reconstruction accuracy. Different sampling lattices along with their corresponding interpolation scheme are also discussed.

Next, we demonstrate how knowledge of the sound field characteristics can be beneficial for the design of efficient distributed compression schemes. Our analysis considers two limit configurations of interest. The first one, analyzed in Section 7.3, consists of an infinite number of sensors recording an acoustic field on a line and transmitting their observations to a base station. Under some simplifying assumptions, we show how the optimum rate-distortion trade-off can be achieved by judicious signal processing at the sensors, hence solving the multiterminal source coding problem for the considered scenario. As far as applications are concerned, a practical setup where such data gathering of audio signals could be of interest is a teleconferencing system made up of multiple cell phones. Typically, such devices are dispersed on a table, record the sound emitted by users talking around them, and transmit the signals to a central point by means of a wireless communication medium. This data may, for example, be conveyed to a remote location to re-create the spatial impression of the conference room. The second setup, studied in Section 7.4, involves only two sensors, namely, two digital hearing aids exchanging compressed audio signals over a wireless communication link in order to provide collaborative noise reduction. For a simple yet insightful scenario, we characterize the maximum achievable noise reduction as a function of the communication bit rate and provide optimal policies for the rate allocation between the two hearing devices. Numerical results obtained using data recorded in a realistic acoustic environment are also presented.

One of the main goals of distributed compression in acoustic sensor networks is to extend the concepts of multichannel audio coding [5, 6] to a scenario where the input channels cannot be encoded jointly. To the best of our knowledge, only a few research papers have addressed this topic. A practical compression scheme based on the DISCUS framework [7] has been proposed in [8]. More recently, the work in [9] considered a wavelet-based method in order to address the stringent power constraints of wireless sensor nodes. For the case of compression with side information at the decoder, a scheme capitalizing on perceptual aspects of human hearing was presented in [10]. Although the analysis presented in this chapter is more of a theoretical nature, it provides useful insights for the design of efficient distributed audio compression schemes.

7.2 SPATIOTEMPORAL EVOLUTION OF THE SOUND FIELD

We present a study on the spatial evolution of the sound field for different scenarios. In Section 7.2.1, we first describe a few recording setups that will be of interest throughout this discussion. Then, in Section 7.2.2, we investigate the spatiotemporal characteristics of a sound field recorded using a line of microphones. In particular, we show that the support of its spatiotemporal Fourier transform exhibits a butterfly shape. From this observation, sampling results are derived in Section 7.2.3. Different sampling lattices, along with their corresponding interpolation schemes, are also presented.

7.2.1 Recording Setups

The spatiotemporal characteristics of the signals recorded using an array of microphones are directly related to the characteristics of the emitting sources and the type of recording environment. Generally speaking, the effect of the environment on the sound emitted by a source can be described by means of the wave equation [11]. Unfortunately, analytical solutions to this equation can only be derived for very simple scenarios. In most cases, the solution needs to be approximated. This section presents four recording setups of interest. For the first two, exact solutions can be derived while, for the remaining ones, we must resort to approximations.

7.2.1.1 Source Emitting in Free Field and Recorded in One Point

Let us first consider the setup depicted in Figure 7.1(a). A sound $u(t)$ is emitted in free field by an omnidirectional source located at position (x_u, y_u, z_u). The sound recorded at a microphone located at position (x_v, y_v, z_v) is denoted by $v(t)$. Considering the acoustic channel between the source and the microphone as a linear and time-invariant system, we can define an impulse response between these two points. This impulse response, denoted by $h(t)$, can be expressed as [11]

$$h(t) = \frac{1}{4\pi d}\,\delta\left(t - \frac{d}{c}\right), \tag{7.1}$$

where $\delta(t)$ denotes the delta function, c the speed of sound propagation, and d the distance between the source and the microphone, which is given by $d = \sqrt{(x_v - x_u)^2 + (y_v - y_u)^2 + (z_v - z_u)^2}$. In the rest of the chapter, the speed of sound is set to $c = 340$ m/s. Under these considerations, the sound recorded at the microphone is obtained as the convolution of the source with the above impulse response, that is,

$$v(t) = h(t) * u(t), \tag{7.2}$$

where $*$ denotes the convolution operator. The signal $v(t)$ is thus simply a delayed and attenuated version of the source $u(t)$.

7.2.1.2 Source Emitting in Free Field and Recorded on a Line

We now consider the acoustic channels between a source emitting in free field and any point on a line at a distance d of the source. This setup is depicted in Figure 7.1(b). The impulse response now depends on both position and time, and is thus referred to as a spatiotemporal impulse response. It can be directly obtained by considering the evolution of (7.1) as a function of x_v and redefining d as $d = \sqrt{(y_v - y_u)^2 + (z_v - z_u)^2}$.

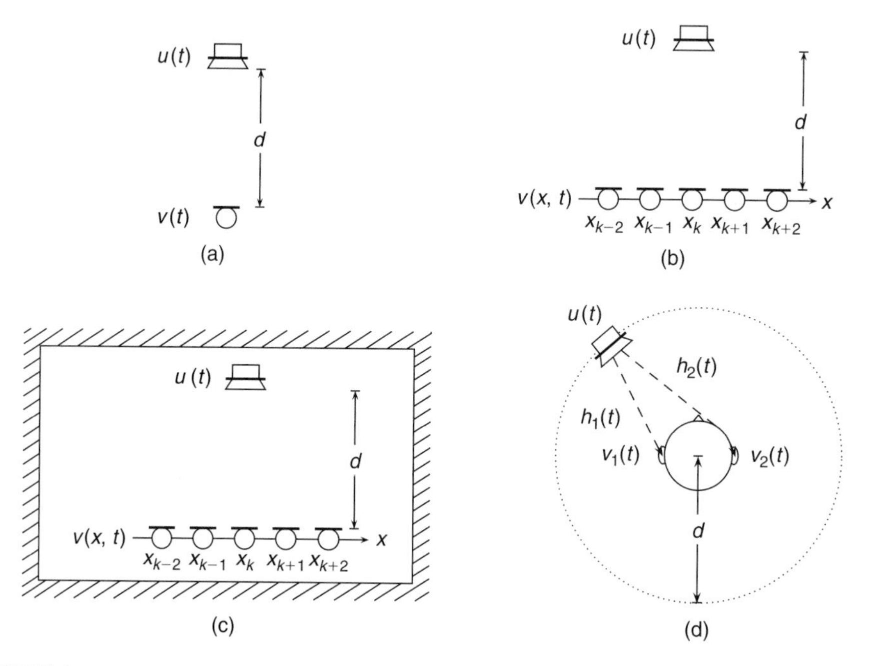

FIGURE 7.1

Setups for the study of the sound field. (a) One source and one microphone in free field. (b) One source and a line of microphones in free field. (c) One source and a line of microphones in a room. (d) One source and two microphones at the ears of a listener.

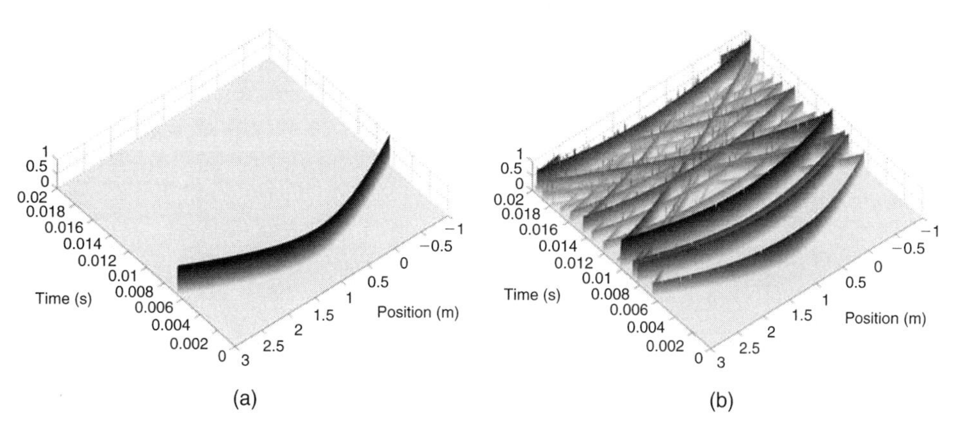

FIGURE 7.2

Theoretical spatiotemporal impulse response on a line. (a) In free field. (b) Inside a room. The spatiotemporal impulse response inside a room is obtained using the image method (see Section 7.2.1.3).

For simplicity, we denote the variable x_v by x such that the spatiotemporal impulse response can be expressed as

$$h(x,t) = \frac{1}{4\pi\sqrt{(x-x_u)^2+d^2}}\,\delta\left(t - \frac{\sqrt{(x-x_u)^2+d^2}}{c}\right). \tag{7.3}$$

The function $h(x,t)$ is plotted in Figure 7.2(a). The signal $v(x_0,t)$ recorded at a fixed position x_0 on the line can thus be obtained as the (one-dimensional) convolution of the impulse response at position x_0, given by $h(x_0,t)$, with the source $u(t)$. In this scenario, referred to as *near-field*, both the delay and the attenuation depend on the distance between the source and the microphone. To simplify matters, it is often assumed [12, 13] that the source is far enough away from the microphones such that the wave front impinging on the line of microphones appears as a plane wave. This assumption is known as the *far-field* assumption. In this case, the delay only depends on the direction of propagation of the wave. Moreover, the attenuation is assumed to be constant across microphones and is set to unity. The spatiotemporal impulse response (7.3) hence reduces to

$$h(x,t) = \delta\left(t - \frac{x\cos\alpha}{c}\right), \tag{7.4}$$

where α denotes the angle of arrival of the plane wave (see Figure 7.3(a)).

7.2.1.3 Source Emitting in a Room and Recorded on a Line

Let us now turn our attention to the recording setup illustrated in Figure 7.1(c). The source no longer emits in free field but instead emits inside a rectangular room. In room

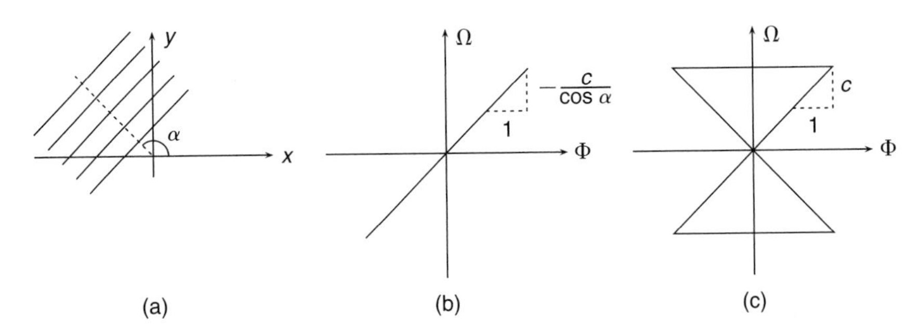

FIGURE 7.3

Far-field assumption for the spatiotemporal study of the sound field. (a) Plane wave impinging on a line of microphones with angle α. (b) Support of the 2D spectrum of the spatiotemporal impulse response. (c) If one considers all possible angles of arrival, the spectral support of the spatiotemporal impulse response has a butterfly shape.

acoustics, the impulse response is referred to as a *room impulse response* (RIR). It can be directly measured (e.g., see [14]) or estimated using a parametric model. The most common approach for simulating RIRs for simple geometries is the image method [15]. The idea is to model the reflections of the sound on the walls as new virtual sources located on the far side of these walls. Reverberation in a room is thus modeled by adding the effect of a large number of virtual free-field sources (theoretically an infinite number). The RIR is obtained by adding the impulse responses corresponding to each of these virtual sources. A spatiotemporal RIR on a line simulated using the image source model is shown in Figure 7.2(b).

7.2.1.4 Source Emitting in Free Field and Recorded at the Ears of a Listener

We assume now that the source is recorded at the left and right ears of a listener, as depicted in Figure 7.1(d). The direction-dependent transfer characteristics from a sound source to the ear is called *head-related impulse response* (HRIR) in the time-domain representation and *head-related transfer function* (HRTF) in the frequency-domain representation. These functions describe the acoustic filtering, due to the presence of a listener, of a sound wave propagating in free field. The HRTF is defined as the ratio between the Fourier transform of the sound pressure at the entrance of the ear canal and the one in the middle of the head in the absence of the listener [16]. HRTFs are therefore filters quantifying the effect of the shape of the head, body, and pinnae on the sound arriving at the entrance of the ear canal. A typical situation is described in Figure 7.1(d). The source emits the sound $u(t)$ in free field. The HRIRs describing the filters between the source and the left and right ears are denoted by $h_1(t)$ and $h_2(t)$, respectively. The sound signals recorded at the ears, denoted by $v_1(t)$ and $v_2(t)$, are again obtained by convolution. Many models exist for simulating HRTFs (e.g., see [17]). These models mostly consider the diffraction effect of the sound around the head. In this scenario, the head reduces the intensity of the sound arriving at the contralateral ear, a phenomenon commonly referred to as *head shadowing*.

7.2.2 Spectral Characteristics

We study the spectral characteristics of the spatiotemporal sound field recorded on a line. As observed in Section 7.2.1.2, the recorded signals can be obtained by the convolution of the source signals with the appropriate impulse responses. We will thus concentrate on the spectral characteristics of the spatiotemporal impulse response. A description of the spectrum under the far-field assumption is given in Section 7.2.2.1. This far-field assumption is then relaxed in Section 7.2.2.2, and an analytical expression of the two-dimensional (2D) Fourier transform of the spatiotemporal impulse response is derived. We further analyze the spatial frequency decay of the obtained spectrum.

7.2.2.1 *Far-field Assumption*

As illustrated in Figure 7.3(a), let us consider a plane wave impinging on a line of microphones with an angle α. The spatiotemporal impulse response is given by (7.4). Its 2D Fourier transform can be computed as

$$H(\Phi, \Omega) = 2\pi\delta\left(\Phi + \Omega\,\frac{\cos\alpha}{c}\right),$$

with Φ and Ω being, respectively, the spatial and temporal frequencies. It is nonzero on a line with slope $-c/\cos\alpha$, as shown in Figure 7.3(b). Consider now the case where plane waves arrive from every possible angle, that is, for $\alpha \in [0, 2\pi)$. In this case, the support of the spectrum has the butterfly shape depicted in Figure 7.3(c). This support corresponds to the region where

$$|\Phi| \leqslant \frac{|\Omega|}{c}. \tag{7.5}$$

Throughout this discussion, it will be referred to as the *butterfly support*. Outside this region, no energy is present. An intuitive explanation for the observed butterfly support is as follows. At low temporal frequencies, the sound wavelength is very large, and thus the wave varies slowly across space, which explains the small spatial support. For higher temporal frequencies, the wavelength is smaller such that the spatial variations are more significant. The spatial support is thus larger for higher temporal frequencies.

7.2.2.2 *Near-field Assumption*

The evolution of the impulse responses on a line in the near-field case has been studied in Section 7.2.1.2. The 2D Fourier transform of the spatiotemporal impulse response (7.3) can be computed as [18]

$$H(\Phi, \Omega) = -\frac{j}{4} H_{2,0}\left(d\sqrt{\left(\frac{\Omega}{c}\right)^2 - \Phi^2}\right), \tag{7.6}$$

where $H_{2,0}$ denotes the Hankel function of the second kind of order zero.

The region $|\Phi| > |\Omega|/c$ corresponds to the evanescent mode of the waves. The waves lose their propagating character to become exponentially fast-decaying waves

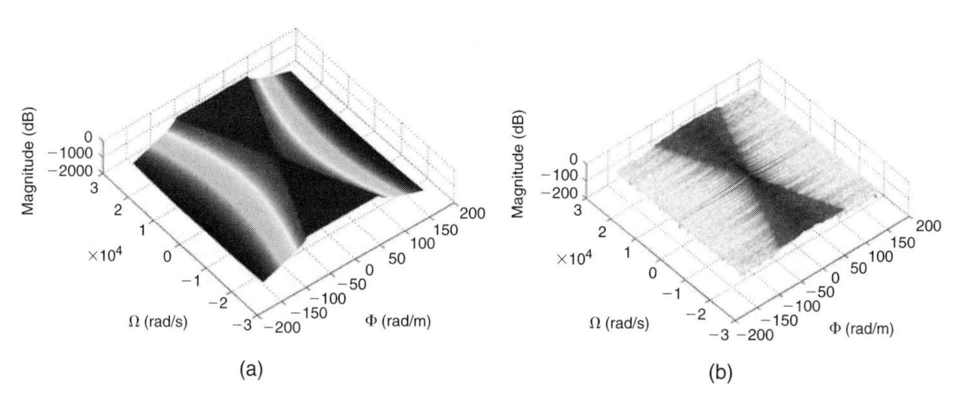

FIGURE 7.4

Magnitude of the spectrum of the spatiotemporal impulse response. (a) Theoretical result for a distance $d = 1$ m. (b) Experimental result obtained from lab measurements. Due to the finite length of the microphone array, the spectrum obtained experimentally decays more slowly outside the butterfly region compared to the theoretical solution.

[19]. Therefore, most of the energy is contained in the butterfly region (7.5). This observation will be used later in the context of spatial sampling. Note also that for $|\Phi| > |\Omega|/c$, the argument of the Hankel function in Equation (7.6) becomes imaginary, and the expression can be rewritten as

$$H(\Phi, \Omega) = \frac{1}{2\pi} K_0\left(d\sqrt{\Phi^2 - \left(\frac{\Omega}{c}\right)^2}\right), \tag{7.7}$$

where K_0 is the modified Bessel function of the second kind of order zero. It can be shown that, for a fixed Ω, the spectrum (7.7) asymptotically behaves as [18]

$$H(\Phi, \Omega) \sim \frac{1}{2\sqrt{\pi}} \frac{e^{-d\Phi}}{\sqrt{d\Phi}}.$$

The decay along the spatial frequency axis is thus faster than exponential. We plot the magnitude of (7.7) in Figure 7.4(a) for $d = 1$ m. In Figure 7.4(b), we also plot the spectrum obtained using 72 RIRs measured in a partially sound-insulated room using an intermicrophone spacing of 2 cm.

7.2.3 Spatiotemporal Sampling and Reconstruction

In Section 7.2.3.1, we first address the spatiotemporal sampling in the far-field case. The near-field case is then considered in Section 7.2.3.2. Two sampling strategies are discussed along with their corresponding ideal interpolation methods.

7.2.3.1 *Far-field Assumption*

In the far-field scenario, spatial sampling is rather simple and is usually referred to as the "half-wavelength rule" [13]. In fact, as mentioned in Section 7.2.2.1, the spectrum of the spatiotemporal impulse response is bandlimited to the region satisfying $|\Phi| \leqslant |\Omega|/c$. Hence, the sound field can be spatially sampled and perfectly reconstructed, provided that the spectral replicas do not overlap. For sampling the sound field on a line, we consider an infinite number of equally spaced microphones. Call Φ_S the spatial sampling frequency defined as $2\pi/\Delta x$, where Δx is the sampling interval, that is, an intermicrophone spacing. Besides spatial sampling, the spatiotemporal sound field also needs to be sampled at a temporal sampling rate that depends on the desired audio bandwidth. We call Ω_S the temporal sampling frequency, such that $\Omega_S = 2\pi/\Delta t$, where Δt is the sampling period. The values Δx and Δt define a *rectangular sampling* grid, as depicted in Figure 7.5(a). To allow perfect reconstruction, the spacing between

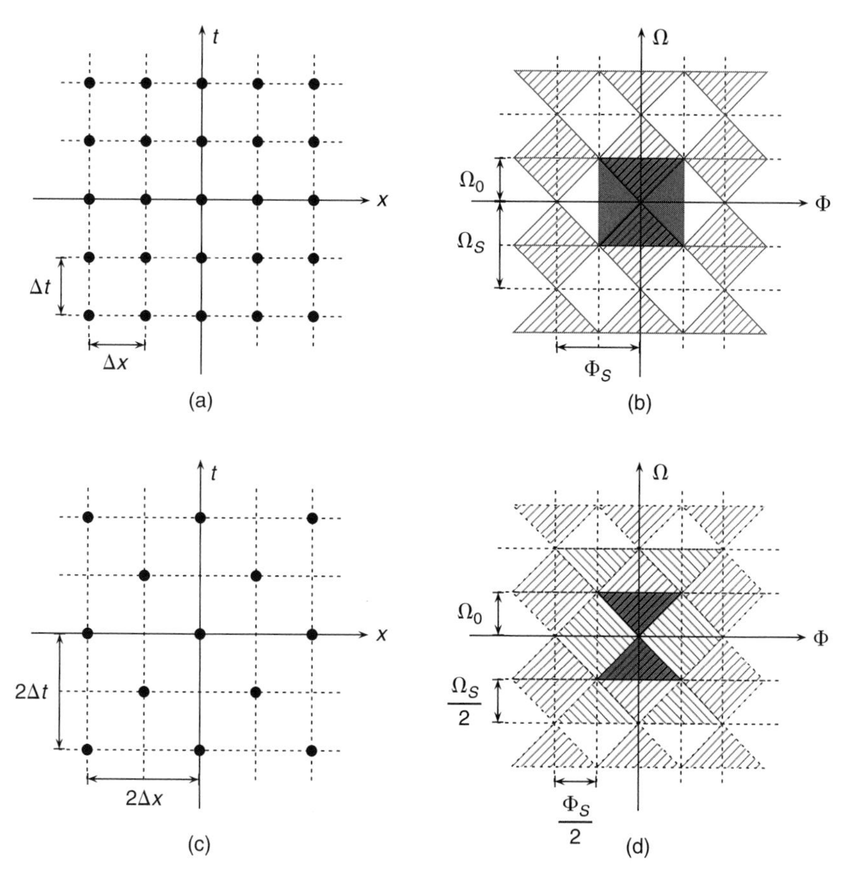

FIGURE 7.5

Sampling and interpolation of the sound field. (a) Rectangular sampling grid. (b) Support of the spectrum with its repetitions for a rectangular sampling grid. The support of the interpolation filter is given by the shaded area. (c) Quincunx sampling grid. (d) Support of the spectrum with its repetitions for a quincunx sampling grid. The support of the interpolation filter is given by the shaded area.

consecutive microphones must be at most equal to

$$\Delta x = \frac{\pi c}{\Omega_0},$$

where Ω_0 denotes the maximum temporal frequency. Equivalently, the minimum spatial sampling frequency is equal to

$$\Phi_S = \frac{2\pi}{\Delta x} = \frac{2\Omega_0}{c}.$$

It can be directly observed from Figure 7.5(b) that, under this condition, the spectral replicas do not overlap. The original sound field can thus be recovered using standard interpolation techniques (e.g., see [20]). The interpolation filter, in this case, is an ideal low-pass filter whose support is shown in bold in Figure 7.5(b).

A tighter packing of the spectrum can be achieved by using *quincunx sampling*. In time domain, the grid to be used is shown in Figure 7.5(c). In the corresponding spectrum, the spectral replicas are placed such that they fill the whole frequency plane as shown in Figure 7.5(d). With quincunx sampling, the temporal sampling frequency can be halved, while preserving the perfect reconstruction property. In this case, the even microphones are sampled at even times while the odd microphones are sampled at odd times. This leads to a gain factor of two in the processing. However, it does not reduce the necessary number of microphones. The ideal interpolation filter in this case is referred to as a "fan filter" [21] whose support is indicated in bold in Figure 7.5(d).

7.2.3.2 Near-field Assumption

In the near-field scenario, the situation is different. Let us consider the spectrum of the spatiotemporal impulse response given by (7.6) at a particular temporal frequency Ω_0. It has approximately the shape given in Figure 7.6(a). In particular, it is not bandlimited. When the sound field is sampled, repetitions of the spectrum occur as shown in Figure 7.6(b). As the spectrum is not perfectly bandlimited, the repetitions will affect the reconstruction; that is, aliasing will occur. To quantify this effect, we consider the signal-to-noise ratio (SNR) of the reconstructed sound field. Let $\text{SNR}(\Phi_S, \Omega_0)$ denote the SNR of the reconstructed sound field corresponding to a sinusoid emitted at frequency Ω_0 and where the spatial sampling frequency of the microphone array is Φ_S. We define this SNR as

$$\text{SNR}(\Phi_S, \Omega_0) = \frac{\int_{-\infty}^{\infty} |H(\Phi, \Omega_0)|^2 \, d\Phi}{4\int_{\Phi_S/2}^{\infty} |H(\Phi, \Omega_0)|^2 \, d\Phi}. \tag{7.8}$$

The numerator in Equation (7.8) corresponds to the energy of the spectrum at temporal frequency Ω_0. The denominator contains two different kinds of energy: the in-band energy and the out-of-band energy. The in-band energy corresponds to the energy of all the spectral replicas that contaminate the domain of interest, that is, $[-\Phi_S/2, \Phi_S/2]$. The out-of-band energy is the energy present in the original spectrum that is outside the domain of interest. It can be shown that these two energies are equal in the case of

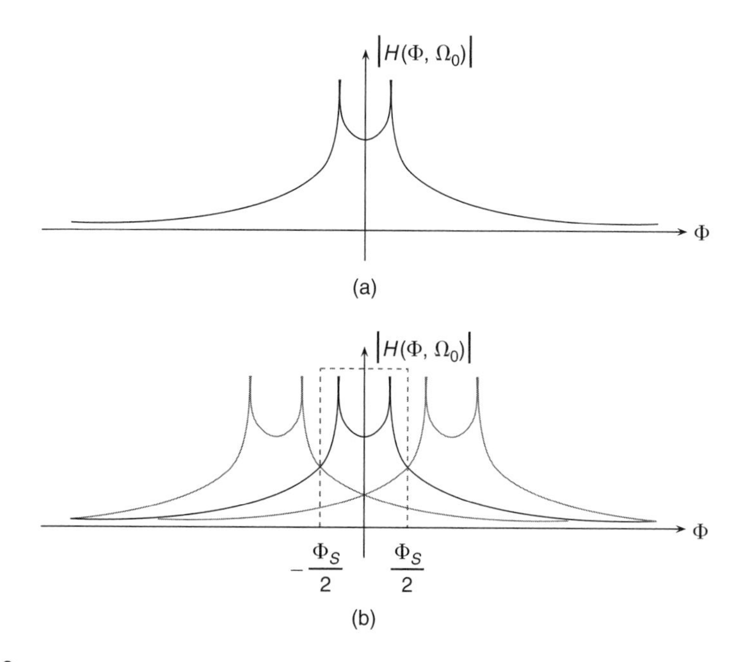

FIGURE 7.6

Magnitude of the spectrum of the spatiotemporal impulse response for a particular temporal frequency Ω_0. (a) Original spectrum. (b) Replicated spectrum after spatial sampling. We observe that the replicas overlap with the original spectrum, preventing perfect reconstruction.

an infinite line of microphones. This explains the factor 4 in the denominator of (7.8). The SNR evaluates as [18]

$$\mathrm{SNR}(\Phi_S, \Omega_0) = \frac{1}{2d \int_{\Phi_S/2}^{\infty} \left| H_{2,0}\left(d\sqrt{\left(\frac{\Omega_0}{c}\right)^2 - \Phi^2} \right) \right|^2 d\Phi}. \tag{7.9}$$

The computation of the SNR (7.9) is rather involved. For $\Phi_S > 2\Omega_0/c$, however, it can be lower bounded by [18]

$$\mathrm{SNR}(\Phi_S, \Omega_0) \geqslant \frac{\pi}{4E_i\left(2d\sqrt{\left(\frac{\Phi_S}{2}\right)^2 - \left(\frac{\Omega_0}{c}\right)^2}\right)}, \tag{7.10}$$

where E_i denotes the exponential integral function defined as $E_i(x) = \int_x^{\infty} \frac{e^{-t}}{t}\, dt$. As a numerical check of the tightness of the bound, the SNR (7.9) and its lower bound (7.10) are plotted in Figure 7.7 as a function of the normalized sampling frequency $\Phi_S\, d$, for different normalized temporal frequencies $\Omega_0\, d$. For a given normalized temporal frequency, it can be observed that the SNR increases as the normalized spatial sampling frequency increases. We also observe that the lower bound tightly follows the exact

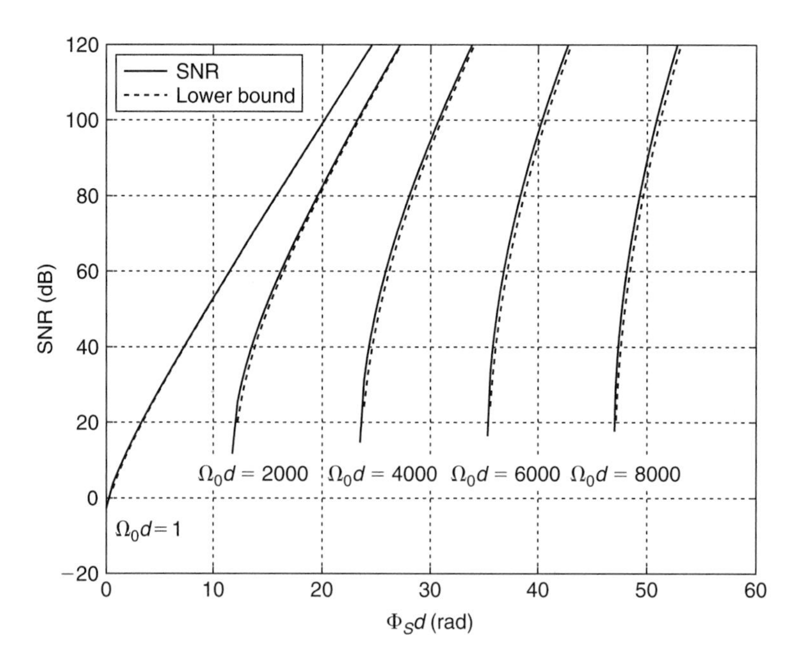

FIGURE 7.7

Reconstruction SNR and its lower bound as a function of the normalized spatial sampling frequency $\Phi_s d$. The different sets of curves correspond to different normalized temporal frequencies $\Omega_0 d$.

Table 7.1 Minimum Intermicrophone Spacing (in cm) for Different Temporal Frequencies Ω_0 and Different SNR Requirements. The Distance is $d = 1$ m

	Far Field	Near Field				
		20 dB	40 dB	60 dB	80 dB	100 dB
2000 rad/s (318 Hz)	53.41	51.30	45.31	38.14	31.84	26.86
8000 rad/s (1.3 kHz)	13.35	13.32	13.19	12.97	12.65	12.27
15,000 rad/s (2.4 kHz)	7.12	7.12	7.10	7.06	7.01	6.94
20,000 rad/s (3.2 kHz)	5.34	5.34	5.33	5.32	5.29	5.27

SNR computed using (7.9). It can be computed from (7.10) that to reconstruct the sound field at a temporal frequency of 8000 rad/s (1.3 kHz), a spatial distance of 12.35 cm is sufficient to achieve a reconstruction quality of 100 dB when a unit distance between the source and the line of microphones is considered. Under the far-field assumption, the required spacing to obtain perfect reconstruction would be 13.35 cm. Further numerical examples are provided in Table 7.1.

Similarly to the far-field scenario, quincunx sampling may be used to achieve a tighter packing of the spectrum, and an SNR formula similar to (7.9) may be derived.

7.3 HUYGENS'S CONFIGURATION

Let us now turn our attention to the distributed compression problem of audio signals. The first setup of interest consists of an infinite number of sensors equally spaced on a line. They record a continuous, stationary, Gaussian sound field that is emitted from a line parallel to the recording line. This setup is motivated by the following observation. According to Huygens's principle [22], a wave front generated by a sound source can be replaced by an infinite number of point sources placed on the wave front. This principle is at the basis of wave field synthesis [23]. If this sound source is sufficiently far away from the recording devices, the far-field assumption allows neglecting the curvature of the wave front such that it reduces to a line. For this reason, we refer to the considered setup as *Huygens's configuration*. Closed-form rate-distortion formulas are obtained for various sampling lattices and coding strategies. In particular, we show that, under a whiteness requirement, the best achievable rate distortion trade-off can be obtained by judicious signal processing at the sensors. We thus provide, for this particular example, the solution to the multiterminal source coding problem.

7.3.1 Setup

The recording setup is illustrated in Figure 7.8(a). Sound sources distributed on an infinite line $\mathcal{U}$ emit a spatiotemporal acoustic field $U(x, t)$. The sound field $U(x, t)$ is assumed to be a zero-mean Gaussian random field that is stationary across both space and time. We denote its power spectral density by $S_U(\Phi, \Omega)$. Recall that the variables Φ and Ω refer to the spatial and temporal frequencies, respectively. The field $U(x, t)$ induces a sound field $V(x, t)$ recorded on a line $\mathcal{V}$, parallel to and at a distance d from the line $\mathcal{U}$. The field $V(x, t)$ can be written as the convolution of $U(x, t)$ with

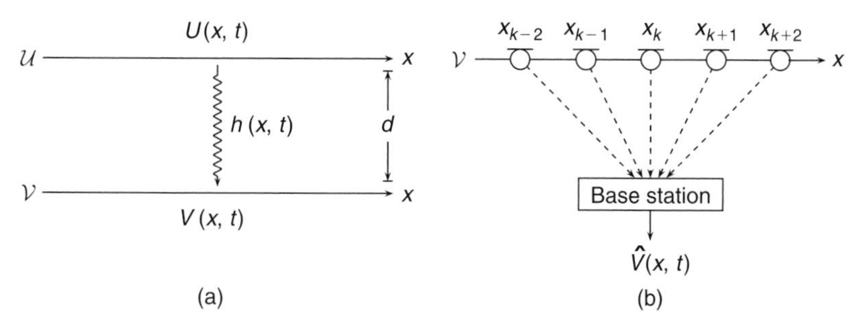

FIGURE 7.8

Huygens's configuration. (a) Sound field propagation setup. (b) Spatiotemporal sampling and distributed coding setup.

the space and time invariant filter $h(x, t)$, which we obtain from (7.3). The process $V(x, t)$ is thus also a stationary Gaussian random field with mean zero and a power spectral density given by [24]

$$S_V(\Phi, \Omega) = |H(\Phi, \Omega)|^2 S_U(\Phi, \Omega), \qquad (7.11)$$

where $H(\Phi, \Omega)$ is the 2D Fourier transform of $h(x, t)$ derived in (7.6). In particular, the spatiotemporal bandwidth of $S_V(\Phi, \Omega)$ is given by the intersection of the supports of $H(\Phi, \Omega)$ and $S_U(\Phi, \Omega)$.

As depicted in Figure 7.8(b), sensors located on the recording line $\mathcal{V}$ sample the sound field $V(x, t)$, encode their observation at a given rate, and transmit the quantized samples to a base station. More specifically, the sensor k samples the field $V(x, t)$ at position x_k and time instants $t_k(n)$, for $k, n \in \mathbb{Z}$. The values x_k and $t_k(n)$ define the sampling lattice. In the sequel, we will consider the rectangular and the quincunx lattices introduced in Section 7.2.3. The samples are then encoded into a bitstream that is transmitted to the base station. The goal of the base station is to provide, based on the received data, an estimate $\hat{V}(x, t)$ of the recorded sound field $V(x, t)$ for any position on the recording line and any time instant. The average rate R used by the sensors is measured in bits per meter per second, and the average distortion D is the mean-squared error (MSE) between $V(x, t)$ and $\hat{V}(x, t)$ per meter and per second. Our interest lies in the determination of rate-distortion functions under various simplifying assumptions on the spectral characteristics of the sound fields and under different constraints on the communication allowed between the sensors.

7.3.2 Coding Strategies

We consider the three coding strategies depicted in Figure 7.9. The first one, referred to as *centralized coding* (see Figure 7.9(a)), assumes that there is one encoder that has access to the measurements of all the sensors and can encode them jointly. For instance, we may assume the existence of a genie informing each sensor about the other sensors' measurements or, more realistically, the existence of unconstrained communication links between the sensors, which can be used without any cost (e.g., wired microphones). In the second coding strategy, referred to as *spatially independent coding* (see Figure 7.9(b)), each sensor encodes its samples, ignoring the correlation with the data measured by neighboring sensors. Similarly, the base station decodes the bitstream received from each sensor separately. In the last coding strategy, referred to as *multiterminal coding* (see Figure 7.9(c)), each sensor encodes its data separately but takes into account the intersensor correlation that is assumed to be known. The base station decodes the received bitstreams jointly. The rate-distortion function in this case is lower bounded by the rate-distortion function with centralized coding, because any scheme that works in the former scenario can also be applied in the latter. Similarly, it is upper bounded by the rate-distortion function with spatially independent coding since intersensor correlation is taken into account. The first two strategies thus provide bounds on the rate-distortion function of the multiterminal scenario whose general characterization remains unknown to date.

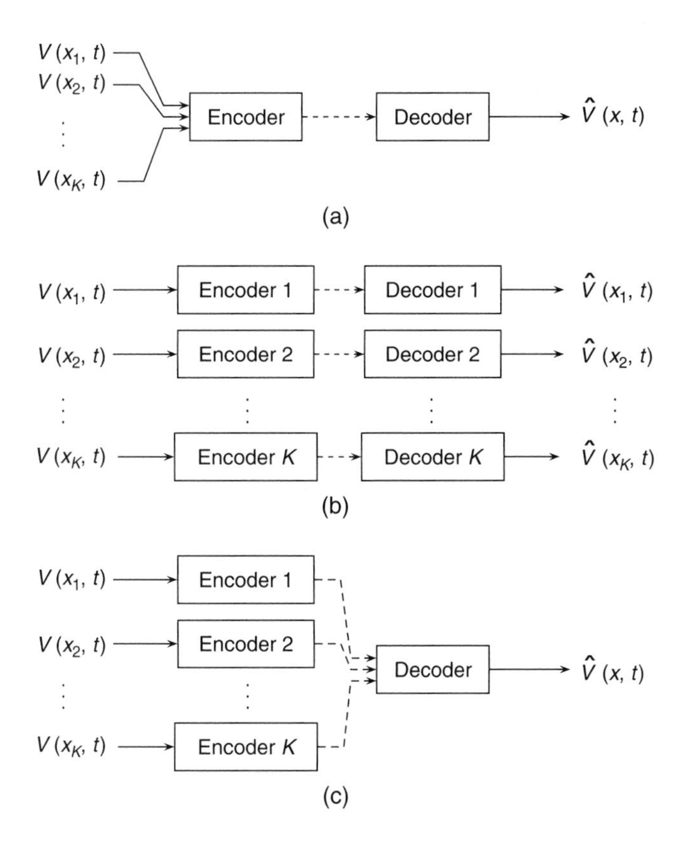

FIGURE 7.9

The three coding strategies considered for Huygens's configuration. (a) Centralized coding. (b) Spatially independent coding. (c) Multiterminal coding.

The assumption of known intersensor correlation is a strong one, for it may be difficult to compute such statistics in a practical setting. Coding methods that do not rely on such prior knowledge are thus of particular interest. In the sequel, we will show that, under simplifying assumptions, it is possible to achieve the lower bound given by centralized coding using an appropriate quincunx sampling lattice followed by independent coding.

7.3.3 Rate-distortion Trade-offs

In order to derive closed-form rate-distortion functions for the different coding strategies considered in Section 7.3.2, we will make two simplifying assumptions. First, we assume that $H(\Phi, \Omega)$ is zero outside the butterfly region. This corresponds to the far-field assumption (see Section 7.2.2.1) and is thus a good approximation of the true support when d is large. In this case, it follows from (7.11) that for a source $U(x, t)$

with maximum temporal frequency Ω_0, the maximum spatial frequency of $V(x,t)$ is given by $\Phi_0 = \Omega_0/c$, regardless of the actual spatial bandwidth of $U(x,t)$. The second assumption is that the power spectral density of the recorded sound field $V(x,t)$ is constant over its support and is given by σ_V^2. This assumption is a worst-case scenario in the sense that, for given bandwidth and total power, the Gaussian random field that requires the highest bit rate for centralized coding is the one with a constant power spectral density. Under these assumptions, we obtain the following optimal rate distortion trade-offs in closed form. For centralized coding, the rate-distortion function can be readily derived from standard results on compression of Gaussian random processes (e.g., see [25, 26]) as

$$R_c(D) = \frac{\Omega_0^2}{4\pi^2 c} \log\left(\frac{\sigma_V^2 \Omega_0^2}{2\pi^2 c D}\right). \tag{7.12}$$

For spatially independent coding and the rectangular sampling lattice, we obtain [27]

$$\begin{aligned} R_r(D) = {} & \frac{\Omega_0^2}{2\pi^2 c} \log\left[\frac{\sigma_V^2 \Omega_0^2}{e\pi^2 c D} \frac{1}{2}\left(1 + \sqrt{1 - 2\frac{\pi^2 c D}{\sigma_V^2 \Omega_0^2}}\right)\right] \\ & + \frac{\Omega_0^2}{2\pi^2 c}\left(1 - \sqrt{1 - 2\frac{\pi^2 c D}{\sigma_V^2 \Omega_0^2}}\right). \end{aligned} \tag{7.13}$$

For spatially independent coding and the quincunx sampling lattice, we find [27]

$$R_q(D) = \frac{\Omega_0^2}{4\pi^2 c} \log\left(\frac{\sigma_V^2 \Omega_0^2}{2\pi^2 c D}\right). \tag{7.14}$$

In the above equations, $D \in \left(0,\ (\sigma_V^2 \Omega_0^2)/(2\pi^2 c)\right)$. The important point regarding these rate-distortion functions is that the expressions (7.12) and (7.14) coincide. In other words, the strategy of sampling with a quincunx lattice followed by spatially independent coding achieves the lower bound given by the centralized scenario and hence determines the multiterminal rate-distortion function for this setup. This can be intuitively explained by the fact that the quincunx lattice allows for a perfect tiling of the frequency plane as shown in Figure 7.5(d). This results in a sampled spatiotemporal field whose power spectral density is flat. The processes sampled at each sensor are uncorrelated and can thus be encoded independently without any loss in terms of the rate-distortion trade-off. It is important to note that the sensor observations cannot be made independent using, for example, a transform coding approach [28, 29], because the measurements cannot be processed jointly. Under our assumptions, however, quincunx sampling achieves the same goal without the need for centralized processing. We plot the above rate-distortion functions in Figure 7.10. Note that the sampling lattice in this example is chosen as a function of the temporal and spatial bandwidth such as to obtain a perfect tiling of the frequency plane. If the sensor

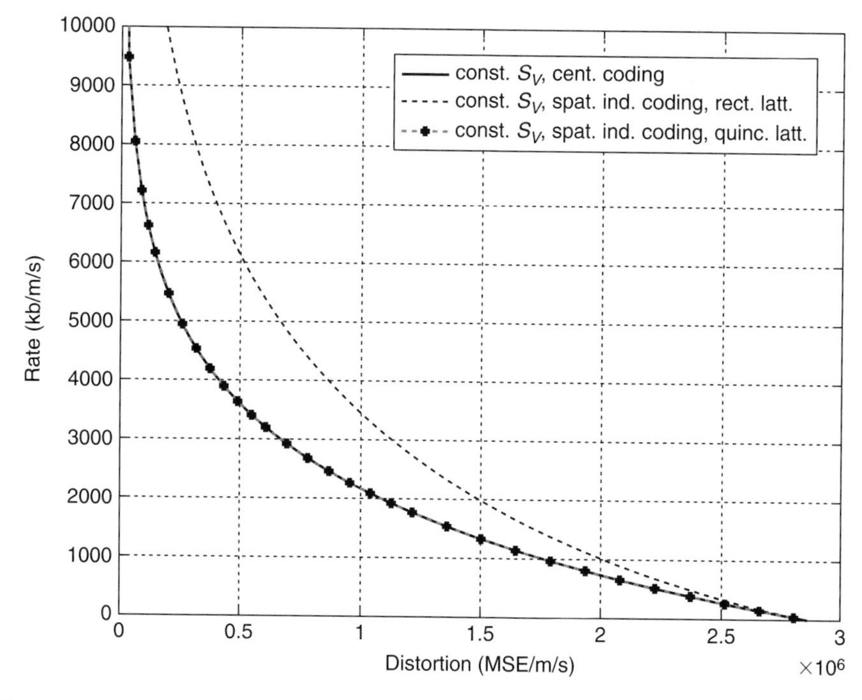

FIGURE 7.10

Rate-distortion functions for different sampling lattices and coding schemes in far field when the induced field's power spectral density $S_V(\Phi, \Omega)$ is constant on its support. The parameters are $\sigma_V^2 = 1$ and $f_0 = \Omega_0/(2\pi) = 22.05$ kHz.

density is not large enough, the replicas will overlap and independent coding will be strictly suboptimal. Similarly, if the sensor density is too large, the gap between consecutive replicas does not generally allow for the independent coding scheme to be optimal. Optimality can, however, be trivially preserved if the oversampling ratio is an integer L. In this case, only every other Lth sensor encodes its observation in order to re-obtain a perfect tiling of the frequency plane. It is important to mention that, although a desired sampling grid in time may be achieved using some sort of synchronization mechanism between the sensors, the design of a sampling pattern in space directly relates to the position of the sensors and is therefore less flexible. Nevertheless, with a uniform spacing of the sensors, the quincunx lattice always allows for a tighter packing than the rectangular one and is thus more suitable for our purpose.

To further study the influence of the wave kernel on the rate-distortion behavior of the sound field $V(x, t)$, we consider a source $U(x, t)$ whose power spectral density is a constant σ_U^2 on the butterfly region and is zero outside. In this case, the spectrum of

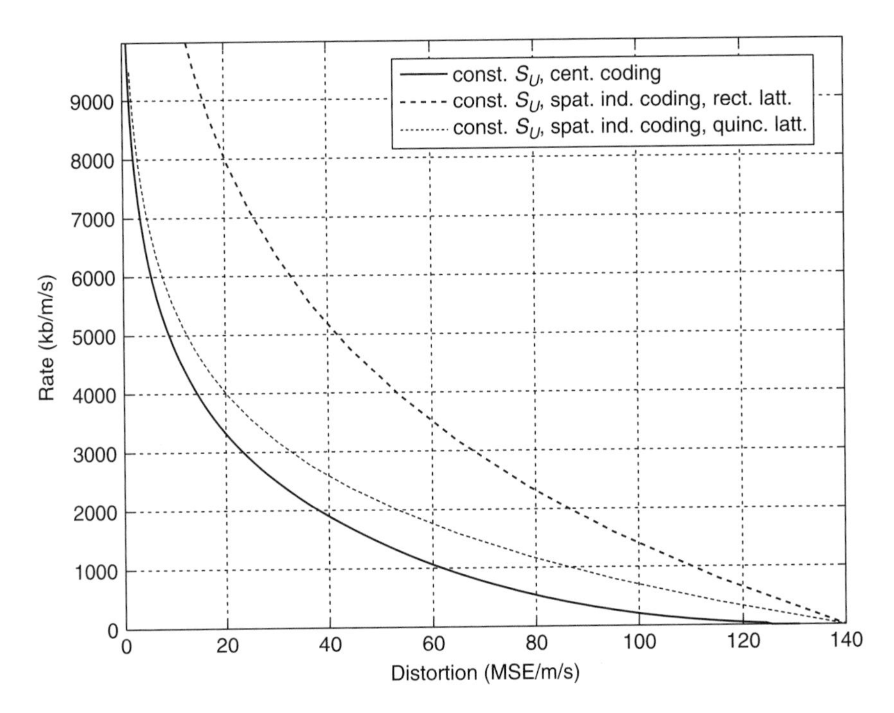

FIGURE 7.11

Rate-distortion functions for different sampling lattices and coding schemes in far field when the source field's power spectral density $S_U(\Phi, \Omega)$ is constant on its support. The parameters are $d = 100\,\text{m}$, $\sigma_U^2 = 1$ and $f_0 = \Omega_0/(2\pi) = 22.05\,\text{kHz}$.

the induced field $V(x, t)$ follows from (7.11) and is no longer constant on the butterfly support. We thus resort to numerical integration to compute the corresponding rate-distortion functions depicted in Figure 7.11. Unlike the previous case, where $S_V(\Phi, \Omega)$ is constant on the butterfly support, the spatially independent coding scheme with the quincunx sampling lattice does not achieve the performance of the centralized coding scheme, but it still outperforms the scheme with the rectangular sampling lattice by a significant margin. Hence, by adjusting the sampling lattice to the wave kernel's spectral properties, we achieve a good performance even with a coding scheme that treats the sensor measurements independently. This observation motivates the general advice that the communication schemes should be tailored to the physical properties of the problem under consideration.

Another way of getting a tighter packing of the replicas of $S_V(\Phi, \Omega)$, while preserving a rectangular lattice, consists in decomposing the sound field into M spectral subbands by filtering in the temporal domain, as shown in Figure 7.12. Each of the subband fields is sampled using a rectangular sampling lattice adjusted to the spatial and temporal bandwidths of the particular subband. This implementation generally

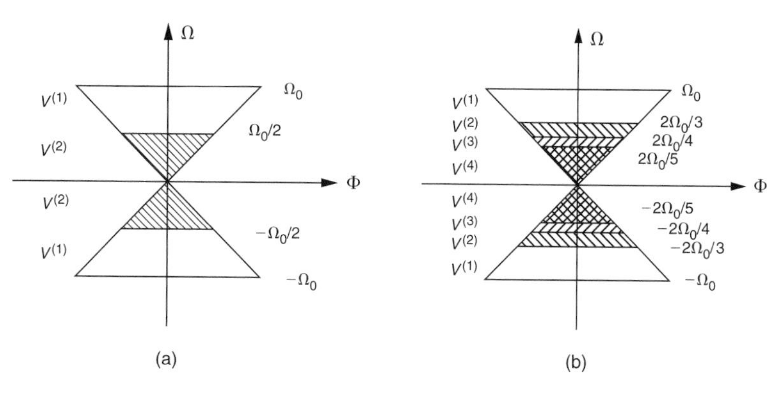

FIGURE 7.12

Spectral decomposition of the sound field by filtering in the temporal domain. (a) $L=1$ and $M=2$. (b) $L=2$ and $M=4$.

requires the deployment of additional sensors with appropriate intersensor spacings for each subband field. A uniform sensor deployment can, however, still be preserved by considering an L-times higher sensor density, for some integer L. More precisely, the intersensor spacing is chosen as $\Delta x/L$, where $\Delta x = \pi c/\Omega_0$ denotes the spacing at the Nyquist rate. We then define the subband field $V^{(m)}(x,t)$ as the field obtained by ideal bandpass filtering $V(x,t)$ on a pass-band given by

$$\begin{cases} \left[-\dfrac{L\Omega_0}{L+m-1}, -\dfrac{L\Omega_0}{L+m}\right] \cup \left[\dfrac{L\Omega_0}{L+m}, \dfrac{L\Omega_0}{L+m-1}\right] & \text{for } m=1,\ldots,M-1, \\ \left[-\dfrac{L\Omega_0}{L+M-1}, \dfrac{L\Omega_0}{L+M-1}\right] & \text{for } m=M. \end{cases}$$

Note that this scheme only requires temporal filtering (recall that spatial filtering is impossible in practice). Figure 7.12 shows the resulting decompositions for $L=1$ and $M=2$ (see Figure 7.12(a)) as well as for $L=2$ and $M=4$ (see Figure 7.12(b)). The spatial bandwidth of $V^{(m)}(x,t)$ is $2L\Omega_0/\big(c(L+m-1)\big)$, thus, $V^{(m)}(x,t)$ can be sampled with an intersensor spacing of $(L+m-1)\Delta x/L$. This sampling rate is obtained by having one sensor out of $L+m-1$ record the samples of the subband field $V^{(m)}(x,t)$. The sensors apply the spatially independent coding scheme defined in Section 7.3.2 to each sampled subband field. Since the spectral densities of the subband signals have nonoverlapping supports, these random fields are uncorrelated, and thus, they may be encoded independently without any penalty in terms of the rate-distortion trade-off. This is provided that the rate allocation among the subbands is performed according to the reverse water-pouring principle [25, 26].

To illustrate the performance of the subband coding scheme defined earlier, we first consider the decomposition of the sound field into two subbands of equal width, as

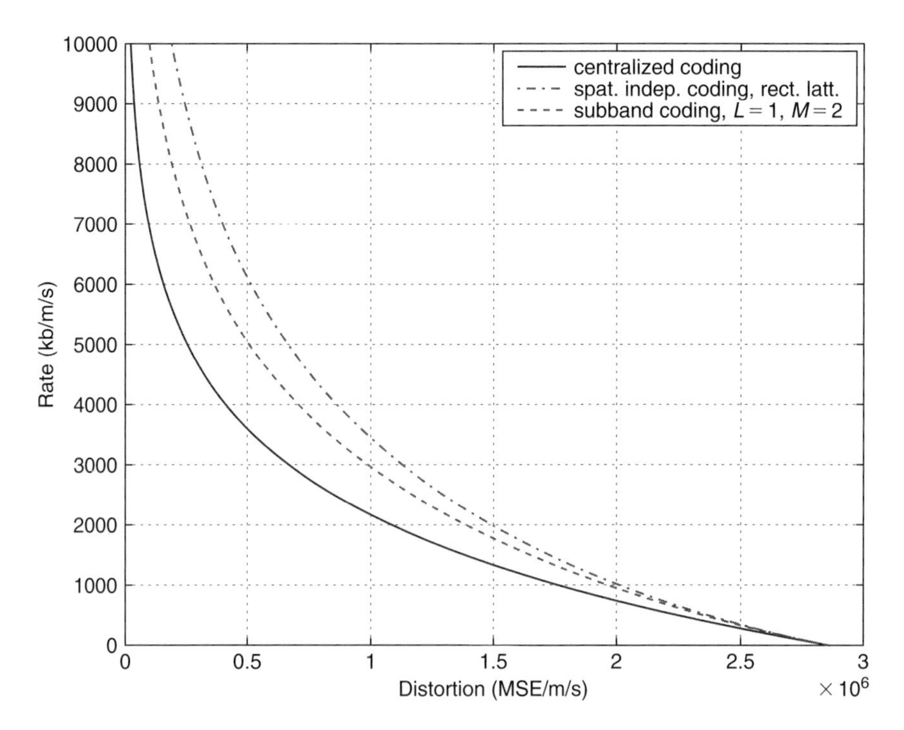

FIGURE 7.13

Rate-distortion function for the subband coding scheme with $L = 1$ and $M = 2$. The spectral density $S_V(\Phi, \Omega)$ is chosen constant on its support. The parameters are $\sigma_V^2 = 1$ and $f_0 = \Omega_0/(2\pi) = 22.05$ kHz.

shown in Figure 7.12(a). The sensors are deployed with an intersensor spacing of Δx, and the subband field $V^{(2)}(x, t)$ is sampled by every other sensor. Figure 7.13 shows the resulting rate-distortion curve in comparison with the ones for centralized coding and for spatially independent coding with the rectangular sampling lattice. Observe that, without deploying additional sensors, we can improve the performance of the latter scheme by combining it with the subband decomposition method. Next, we consider the spectral subdivision into multiple subbands. For a given value of the parameter L, we choose $M = 9L + 1$, so that the temporal bandwidth of the subband field $V^{(M)}(x, t)$ is equal to $\Omega_0/5$. Note that increasing the parameter L requires a proportional increase of the sensor density, whereas an increase of the number of subbands M only adds to the local processing complexity. Figure 7.14 shows the resulting rate-distortion curves for $L = 1, 2, 3$. Observe that, with a moderate increase of the sensor density, the performance of the spatially independent coding scheme with the rectangular sampling lattice combined with the subband decomposition method approaches the one of centralized coding.

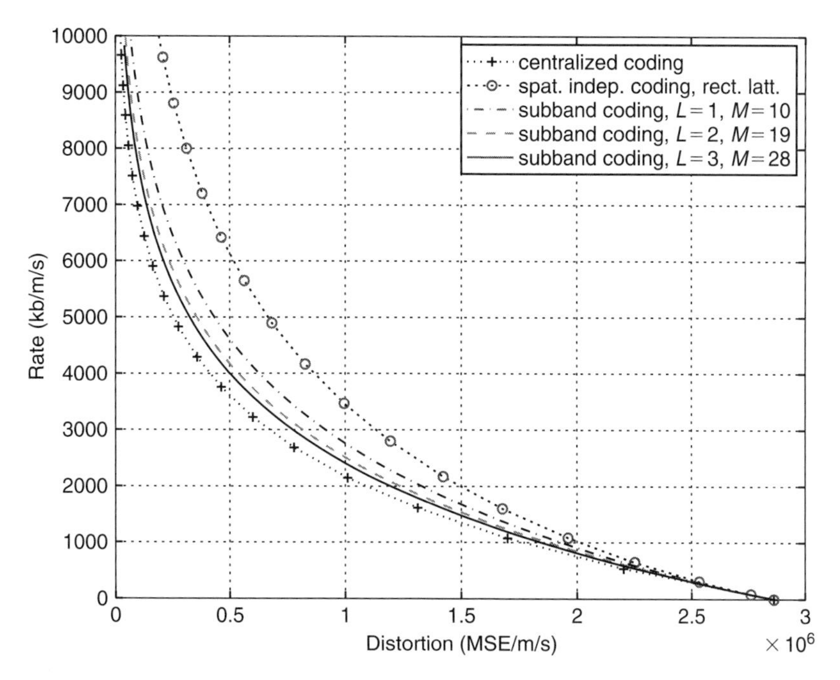

FIGURE 7.14

Rate-distortion functions for the subband coding scheme with $L = 1,2,3$ and $M = 9L+1$. The spectral density $S_V(\Phi, \Omega)$ is chosen constant on its support. The parameters are $\sigma_V^2 = 1$ and $f_0 = \Omega_0/(2\pi) = 22.05$ kHz.

7.4 BINAURAL HEARING AID CONFIGURATION

The distributed compression scenario considered in this section involves only two sensors, namely, two digital hearing aids (left and right) allowed to exchange compressed audio signals by means of a wireless link. The setup is referred to as the *binaural hearing aid configuration*. The study of optimal rate-distortion trade-offs in this context allows quantifying the noise reduction provided by the availability of a rate-constrained communication medium between the two hearing instruments. Closed-form rate-distortion formulas are derived in a simple yet insightful scenario, along with optimal rate-allocation strategies. More realistic acoustic environments are discussed, for which numerical results are presented.

7.4.1 Setup

Let us consider the binaural hearing aid configuration depicted in Figure 7.15. A user carries a left and a right hearing aid, both comprising a set of microphones, a processing

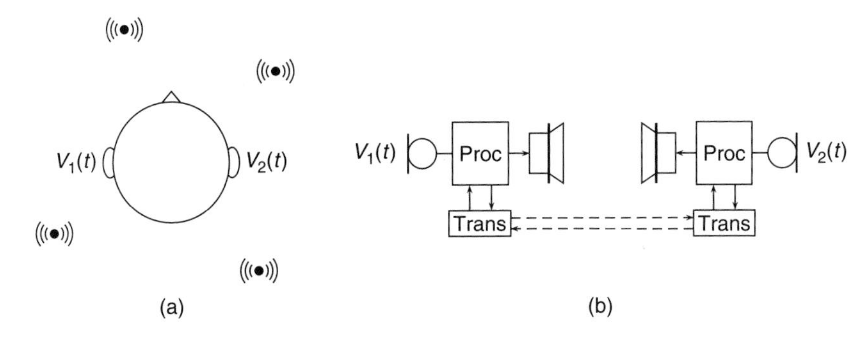

FIGURE 7.15

Binaural hearing aid configuration. (a) Recording setup. (b) Collaborative processing using wireless links.

unit with wireless communication capabilities, and a loudspeaker. For simplicity of exposure, we will consider that only one microphone is available at each hearing device, and we will denote the signal recorded at microphone k by $V_k(t)$ for $k = 1, 2$. We assume that $V_k(t)$ can be decomposed into a speech (or desired) component, denoted $V_k^s(t)$, and an additive noise (or undesired) component, denoted $V_k^n(t)$, as

$$V_k(t) = V_k^s(t) + V_k^n(t) \quad \text{for } k = 1, 2. \tag{7.15}$$

For the sake of mathematical tractability, the speech and noise signals are modeled as independent, zero-mean, stationary Gaussian random processes.

The goal of each hearing instrument is to recover the speech component of its microphone with minimum MSE. To make matters clearer, let us adopt the perspective of hearing aid 1. It aims at estimating the signal $V_1^s(t)$ with minimum MSE based on its own observation $V_1(t)$ and a compressed version of the signal $V_2(t)$ recorded at hearing aid 2. With the above considerations, the setup simply corresponds to a remote source coding problem with side information at the decoder. This problem, also referred to as indirect or noisy Wyner–Ziv coding in the literature, has been addressed by various researchers in the scalar case [30–32]. More generally, Witsenhausen [33] elegantly demonstrated how certain classes of remote rate-distortion problems can be reduced to direct ones, unifying earlier results by Dobrushin and Tsybakov [34], Sakrison [35], and Wolf and Ziv [36]. Extension to vector sources was investigated by Rebollo-Monedero et al. in the context of high-rate transform coding [37]. In the continuous-time case, we wish to encode $V_2(t)$ with rate R_1 such as to minimize the distortion D_1 between $V_1^s(t)$ and its reconstruction $\hat{V}_1^s(t)$, assuming the presence of the side information $V_1(t)$ at the decoder. The rate R_1 is measured in bits per second and the distortion D_1 in MSE per second. This procedure is illustrated for both devices in Figure 7.16. As in Section 7.3, our goal is to determine rate-distortion functions under two different coding strategies and various simplifying assumptions on the acoustic environment. The binaural hearing aid configuration, however, differs from the problem studied previously in two important points. First, the sound field is

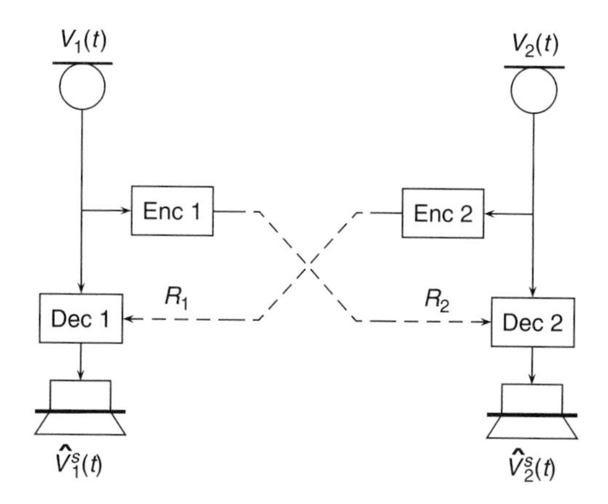

FIGURE 7.16

Collaborative noise reduction using rate-constrained communication links.

sampled only at two positions, namely, at the two microphones. Second, the signals of interest are not directly measured at the sensors but are recorded in a noisy (or remote) fashion [35].

In order to quantify the noise reduction provided by the wireless link, we define the *monaural* gain achieved at hearing aid k as

$$G_k(R_k) = \frac{D_k(0)}{D_k(R_k)} \quad \text{for } k = 1, 2, \tag{7.16}$$

where $D_k(R_k)$ denotes the corresponding optimal rate distortion trade-off, $D_k(0)$ being the distortion incurred when there is no collaboration ($R_k = 0$). Note that R_k refers to the rate at which data is delivered to hearing aid k. The quantity $G_k(R_k)$ actually corresponds to the signal-to-distortion improvement enabled by the wireless link when it operates at rate R_k. Similarly, the *binaural* gain is defined as

$$G(R) = \frac{D(0)}{D(R)}, \tag{7.17}$$

where $D(R)$ refers to the optimal trade-off between the sum distortion $D = D_1 + D_2$ and the total (bidirectional) transmission rate $R = R_1 + R_2$. The interesting question here is that of the rate allocation between the two devices.

7.4.2 Coding Strategies

We consider two coding strategies. The first one, referred to as *side information aware* (SIA), assumes that the correlation between the signals recorded at the hearing aids is known and can be used at the encoder for coding purposes. For example, hearing aid 2 records the signal $V_2(t)$ and encodes it taking into account that the signal $V_1(t)$ is

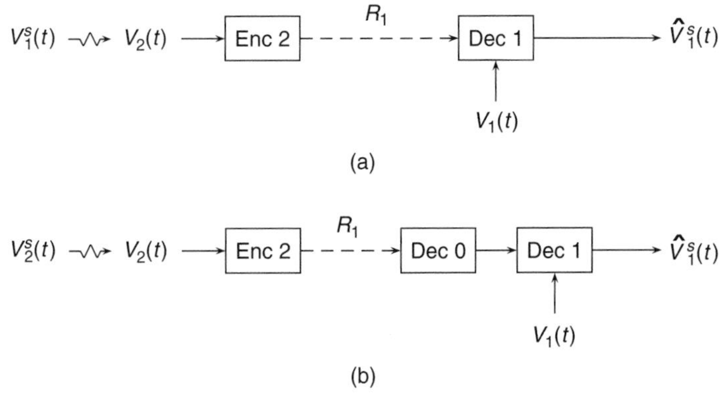

FIGURE 7.17

The two coding strategies considered for the binaural hearing aid configuration. (a) Side information aware coding. (b) Side information unaware coding.

available at the decoder, that is, at hearing aid 1. This coding scheme is schematically represented in Figure 7.17(a). The second class of coding methods, referred to as *side information unaware* (SIU), does not assume that such statistics can be computed at the encoder. In this case, hearing aid 2 encodes its observation $V_2(t)$ for a decoder (Dec 0) that does not have access to the side information $V_1(t)$. The transmitted data is decoded and subsequently used together with $V_1(t)$ in a second decoder (Dec 1) to estimate the desired speech component $V_1^s(t)$. This is depicted in Figure 7.17(b). In a practical setting, the statistics of the speech and noise components can be typically learned at the decoder by means of a voice activity detection mechanism [38]. Note also that, as the statistics between the signals recorded at the two hearing devices are not known at the encoder, this latter assumes that the process to be estimated at the other end is $V_2^s(t)$ and optimizes its coding strategy accordingly.

7.4.3 Rate-distortion Trade-offs

We first study a simple acoustic environment that allows us to derive optimal rate-distortion trade-offs and rate-allocation policies in closed form. Though clearly simplistic, the considered assumptions preserve many of the features observed in a realistic scenario, in particular for the rate-allocation problem, as observed in the sequel.

We consider a single speech source surrounded by ambient noise. The recorded signals (7.15) can be expressed as

$$V_k(t) = S(t) * h_k(t) + N_k(t) \quad \text{for } k = 1, 2,$$

where $S(t)$ denotes the source, $h_k(t)$ the acoustic impulse response from the source to microphone k, and $N_k(t)$ uncorrelated Gaussian noise signals. The desired component at hearing aid k is thus $V_k^s(t) = S(t) * h_k(t)$. We assume omnidirectional microphones, neglect the head-shadow effect (see Section 7.2.1.4) and work under the far-field assumption. In this case, $h_k(t)$ reduces to a simple delay proportional to the distance d_k between the source and the microphone, that is,

$$h_k(t) = \delta(t - d_k/c).$$

We further assume that the speech and noise sources have flat power spectral densities with maximal frequency Ω_0. The variance of the source is σ_S^2, and the SNR at microphone k is denoted γ_k. With these restrictions, the optimal rate-distortion trade-off at hearing aid 1 with the SIA coding scheme, denoted $D_{1,a}(R)$, can be expressed as [39]

$$D_{1,a}(R_1) = \frac{\Omega_0}{\pi} \frac{\sigma_S^2 \gamma_1 \gamma_2}{1 + \gamma_1 + \gamma_2} \left(1 + \frac{\gamma_2}{1 + \gamma_1} 2^{-\frac{2\pi R_1}{\Omega_0}}\right) \quad \text{for } R_1 \geqslant 0. \tag{7.18}$$

Similarly, the optimal rate-distortion trade-off obtained with the SIU coding scheme, denoted $D_{1,u}(R)$, is given by

$$D_{1,u}(R_1) = \frac{\Omega_0 \sigma_S^2}{\pi \gamma_1} \left(1 + \frac{1 + \gamma_2}{\gamma_1} \left(1 + \gamma_2 2^{-\frac{2\pi R_1}{\Omega_0}}\right)^{-1}\right)^{-1} \quad \text{for } R_1 \geqslant 0. \tag{7.19}$$

We observe that, in this scenario, the results depend neither on the position of the source nor on the geometrical properties of the binaural hearing aid configuration. This results from the far-field assumption and the fact that the noise is uncorrelated across microphones. From a more practical point of view, it can be shown [39] that an optimal encoding architecture with the SIA coding scheme can be broken into two stages. Call $\bar{V}_2(t)$ the minimum MSE estimate of $V_2(t)$ given $V_1(t)$. An optimal strategy amounts to (i) filtering $V_2(t)$ by the Wiener filter designed to estimate $V_1^s(t)$ using $\bar{V}_2(t)$ and (ii) encoding this estimate optimally, taking into account the side information $V_1(t)$ available at the decoder. Knowledge of the statistics between $V_1(t)$ and $V_2(t)$ is thus required at the encoder. Similarly, an optimal encoding architecture with the SIU strategy amounts to (i) computing the minimum MSE estimate of $V_2^s(t)$ using $V_2(t)$ and (ii) encoding this estimate optimally for a decoder that does not have access to the side information $V_1(t)$. In this case, however, the required statistics can be computed at the encoder by means of a voice activity detection mechanism [38].

To get some insights into the gain provided by the wireless link as a function of the communication bit rate, let us consider the case of equal SNRs ($\gamma_1 = \gamma_2 = \gamma$). In Figure 7.18, we plot the monaural gain rate function (7.16) corresponding to the two coding schemes for different values of γ. At 10 dB (see Figure 7.18(a)), we observe that the SIU coding strategy may lead to a significant loss in terms of noise reduction capability in comparison to the SIA scheme. However, as the input SNR decreases (see Figures 7.18(b) and 7.18(c)), the spatial correlation between the recorded signals decreases significantly and the gap between the SIA and SIU curves vanishes for all

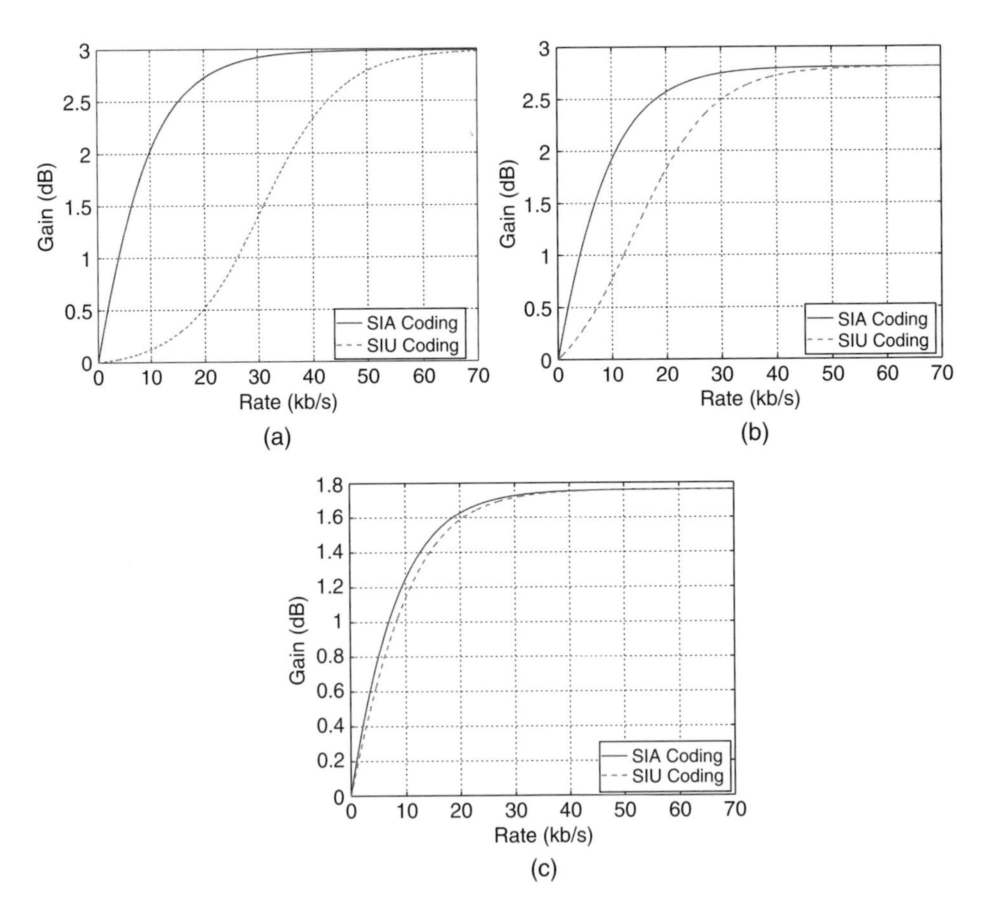

FIGURE 7.18

Monaural gain rate functions for different input signal-to-noise ratios γ and a maximum frequency $f_0 = \Omega_0/(2\pi) = 5$ kHz. (a) $\gamma = 20$ dB. (b) $\gamma = 10$ dB. (c) $\gamma = 0$ dB. We observe that the gain achieved by taking the side information into account may be significant at high SNR but vanishes in very noisy scenarios.

rates. The result suggests that the use of SIA coding strategies is uninteresting in very noisy scenarios.

Let us now turn our attention to the general binaural system depicted in Figure 7.16. Without loss of generality, we assume that $\gamma_1 \geqslant \gamma_2$. A natural question that arises in this context is that of the optimal rate allocation between the two hearing instruments. More precisely, if you assume that you are given a total bit budget of R bits per second, how should this be allocated to R_1 and R_2 to minimize the sum distortion $D = D_1 + D_2$, hence maximizing the binaural gain $G(R)$ given by (7.17)? Under the acoustic model we have considered, the optimal rate-allocation strategy for the SIA coding scheme follows as [39]

$$R_1^\star = \begin{cases} \frac{1}{2}(R-\bar{R}) & \text{if } R \geqslant \bar{R}_a, \\ 0 & \text{otherwise,} \end{cases} \quad \text{and} \quad R_2^\star = R - R_1^\star, \tag{7.20}$$

where the threshold rate is given by

$$\bar{R}_a = \bar{R} = \frac{\Omega_0}{2\pi} \log_2 \frac{\gamma_1(\gamma_1+1)}{\gamma_2(\gamma_2+1)}. \tag{7.21}$$

Similarly, the rate-allocation policy with the SIU coding method can be expressed as

$$R_1^\star = \begin{cases} \frac{1}{2}(R-\bar{R}) - \frac{\Omega_0}{2\pi} \log_2 \dfrac{1 - C2^{-\frac{\pi}{\Omega_0}(R+\bar{R})}}{1 - C2^{-\frac{\pi}{\Omega_0}(R-\bar{R})}} & \text{if } R \geqslant \bar{R}_u, \\ 0 & \text{otherwise,} \end{cases} \quad \text{and} \quad R_2^\star = R - R_1^\star, \tag{7.22}$$

where $\bar{R}$ is given by (7.21), the constant C is defined as

$$C = \frac{\gamma_1 \gamma_2}{\gamma_1 + \gamma_2 + 1},$$

and the threshold rate is given by

$$\bar{R}_u = \max \left\{ 0, \bar{R} + \frac{\Omega_0}{\pi} \log_2 \frac{C+1}{2} \left(1 + \sqrt{1 - \frac{4C}{(C+1)^2} 2^{-\frac{2\pi}{\Omega_0}\bar{R}}} \right) \right\}. \tag{7.23}$$

The rate-allocation strategies (7.22) and (7.23) suggest that the hearing device with smaller SNR does not transmit any data unless the total available bit rate is larger than a given threshold. Below this rate, the noisiest device benefits exclusively from the available bandwidth. At equal SNR ($\gamma_1 = \gamma_2 = \gamma$), the threshold rate $\bar{R}_a$ of the SIA coding scheme is equal to zero. In other words, the communication medium is evenly shared between the two hearing aids for all rates. By contrast, the threshold rate with SIU coding is greater than zero for large enough SNR. The surprising result is that, in this seemingly symmetric scenario, the communication bandwidth may not be equally shared. Figure 7.19 depicts the percentage of the total bit rate benefiting hearing aid 1 for different-input SNRs. With SIU coding at equal SNR, we observe a sharp transition between two policies, namely, unidirectional communication and equal rate allocation. In the former case, it can be checked that the hearing aid benefiting from the wireless link can be chosen arbitrarily. We also note that the SIU threshold rate is larger than that of the SIA coding scheme.

Finally, we compute the optimal rate-allocation strategy using data measured in a more realistic acoustic environment. To this end, two hearing aids, each equipped with two omnidirectional microphones at a distance of approximately 1 cm, have been mounted on a dummy head in a room with reverberation time $T_{60} \approx 120$ ms.

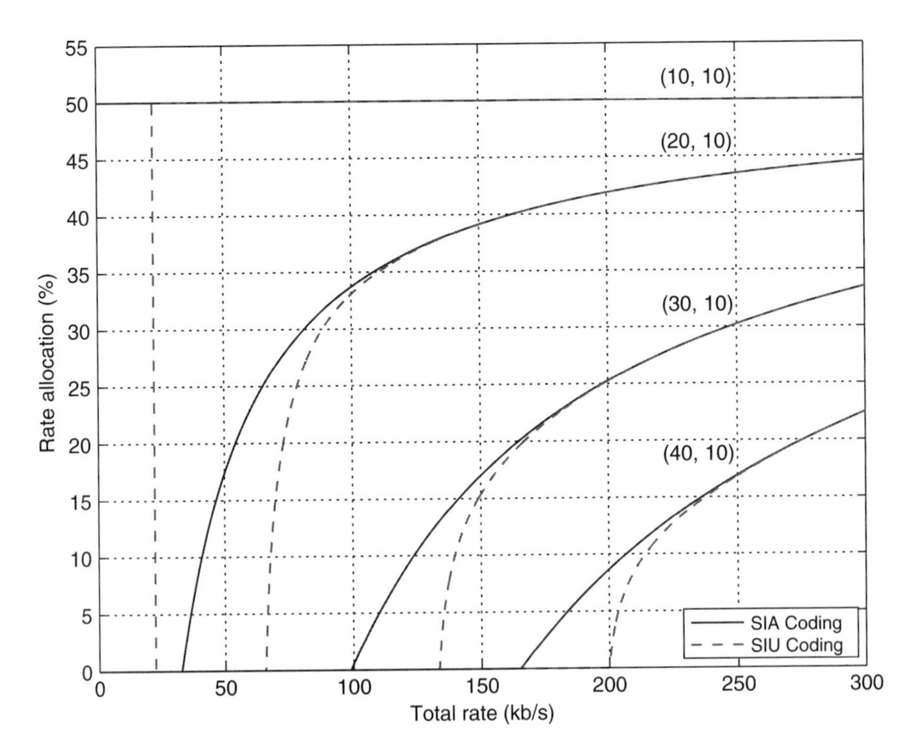

FIGURE 7.19

Percentage of the total bit rate benefiting hearing aid 1. The maximum frequency is $f_0 = \Omega_0/(2\pi) = 5$ kHz, and the different sets of curves correspond to various SNR pairs (γ_1, γ_2). We observe that the threshold rate for SIU coding is larger than that of SIA coding.

The acoustic transfer functions for the four microphones have been measured every 15° in the horizontal plane for a loudspeaker at a distance of 1 m. Here, the angles are measured clockwise, and the zero angle corresponds to the front. The sampling frequency is set to 20.48 kHz. The acoustic scene is synthesized using the measured transfer functions. The speech component in (7.15) is a single speech source at 0°. The noise component consists of a stationary white Gaussian ambient noise along with an interfering point source at 90°. The power spectral density of the speech component is estimated using 3 seconds of a sentence of the HINT database [40]. The power spectral density of the interferer is computed using a 3-second segment of multitalker babble noise available in the NOISEX-92 database [41]. The power of the involved sources is adjusted so as to meet a desired SNR and interference-to-noise ratio of 10 dB at the front microphone of hearing aid 1 that is, the signal-to-interference ratio is equal to 0 dB. The optimal rate allocation is obtained numerically and is plotted in Figure 7.20. Here, we only consider the front microphone of each hearing aid. Since the interfering point source is located on the right-hand side of the head, the SNR of hearing aid 1 is higher owing to the head-shadow effect. Up to a threshold bit rate, hearing aid 2

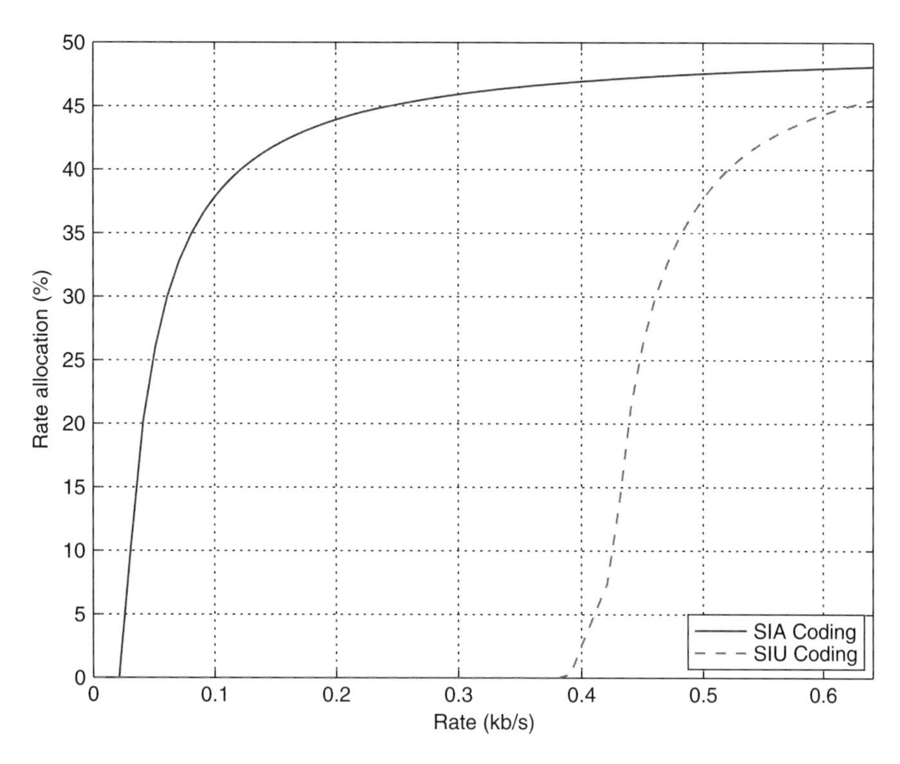

FIGURE 7.20

Percentage of the total bit rate benefiting hearing aid 1 in the presence of an interfering point source at 90°. Hearing aid 2 has the smallest SNR and thus benefits exclusively from the communication bandwidth up to a threshold rate that depends on the chosen coding strategy.

thus benefits exclusively from the available communication bandwidth. In this case, the fraction allocated to hearing aid 1 is zero. We observe that this threshold is higher with the SIU coding scheme, corroborating the analytical results obtained in the simple scenario considered previously.

7.5 CONCLUSIONS

The problem of distributed compression has been studied for the specific case of microphone arrays. A large portion of the exposition has been devoted to the study of the spatiotemporal characteristics of the sound field. In particular, it has been shown that the spatiotemporal impulse response is almost bandlimited to a region that has a butterfly shape. The insights gained from this analysis have been subsequently used for two distributed coding setups of practical interest. The first one has considered the gathering of audio signals recorded by remote sensors at a base station. It has been shown that, under simplifying assumptions, judicious signal processing at the sensors can achieve the best rate-distortion trade-off. We have thus provided the solution to the multiterminal source coding problem for this particular scenario. The second setup has involved two digital hearing aids allowed to exchange audio signals using a wireless communication medium as a way to provide collaborative noise reduction. Optimal gain rate trade-offs have been derived, and optimal rate-allocation strategies have been discussed.

ACKNOWLEDGMENT

This work was supported by the National Competence Center in Research on Mobile Information and Communication Systems (NCCR-MICS, http://www.mics.org), a center supported by the Swiss National Science Foundation under grant number 5005-67322.

REFERENCES

[1] G. Barrenetxea, F. Ingelrest, G. Schaefer, M. Vetterli, O. Couach, and M. Parlange, "SensorScope: Out-of-the-box environmental monitoring," *ACM/IEEE International Conference on Information Processing in Sensor Networks*, pp. 332–343, April 2008.

[2] D. Li, K. Wong, Y. Hu, and A. Sayeed, "Detection, classification, tracking of targets in micro-sensor networks," *IEEE Signal Processing Mag.*, vol. 19, no. 2, pp. 17–29, March 2002.

[3] J. C. Chen, K. Yao, and R. E. Hudson, "Source localization and beamforming," *IEEE Signal Processing Mag.*, vol. 19, no. 2, pp. 30–39, March 2002.

[4] B. Beferull-Lozano, R. L. Konsbruck, and M. Vetterli, "Rate-distortion problem for physics based distributed sensing," *ACM/IEEE International Conference on Information Processing in Sensor Networks*, pp. 330–339, April 2004.

[5] J. Breebaart and C. Faller, *Spatial Audio Processing*. Wiley, 2007.

[6] D. T. Yang, C. Kyriakakis, and C.-C. J. Kuo, *High-Fidelity Multichannel Audio Coding*, ser. EURASIP Book Series on Signal Processing and Communications. Hindawi Publishing, 2004, vol. 1.

[7] S. S. Pradhan and K. Ramchandran, "Distributed source coding using syndromes (DISCUS): Design and construction," *IEEE Trans. Inform. Theory*, vol. 49, no. 3, pp. 626–634, March 2003.

[8] A. Majumdar, K. Ramchandran, and I. Kozintsev, "Distributed coding for wireless audio sensors," *IEEE Workshop on Applications of Signal Processing to Audio and Acoustics*, pp. 209–212, October 2003.

[9] H. Dong, J. Lu, and Y. Sun, "Distributed audio coding in wireless sensor networks," *IEEE International Conference on Computational Intelligence and Security*, vol. 2, no. 1, pp. 1695–1699, November 2006.

[10] O. Roy and M. Vetterli, "Distributed spatial audio coding in wireless hearing aids," *IEEE Workshop on Applications of Signal Processing to Audio and Acoustics*, pp. 227–230, October 2007.

[11] P. M. Morse and K. U. Ingard, *Theoretical Acoustics*. McGraw-Hill, 1968.

[12] P. S. Naidu, *Sensor Array Signal Processing*. CRC Press, 2001.

[13] D. H. Johnson and D. E. Dudgeon, *Array Signal Processing: Concepts and Techniques*. Prentice-Hall, 1993.

[14] T. Adjler, L. Sbaiz, and M. Vetterli, "Dynamic measurement of room impulse responses using a moving microphone," *The Journal of the Acoustical Society of America*, vol. 122, no. 3, pp. 1636–1645, September 2007.

[15] J. B. Allen and D. A. Berkley, "Image method for efficiently simulating small-room acoustics," *The Journal of the Acoustical Society of America*, vol. 65, pp. 943–950, 1979.

[16] H. Moller, "Fundamentals of binaural technology," *Appl. Acoust.*, vol. 36, pp. 171–218, 1992.

[17] R. O. Duda and W. Martens, "Range dependence of the response of a spherical head model," *The Journal of the Acoustical Society of America*, vol. 104, pp. 3048–3058, 1998.

[18] T. Ajdler, L. Sbaiz, and M. Vetterli, "The plenacoustic function and its sampling," *IEEE Trans. Signal Processing*, vol. 54, no. 10, pp. 3790–3804, 2006.

[19] E. G. Williams, *Fourier Acoustics*. Academic Press, 1999.

[20] M. Vetterli and J. Kovačević, *Wavelets and Subband Coding*. Prentice Hall, 1995. [Online]. Available: http://www.waveletsandsubbandcoding.org.

[21] P. Vaidyanathan, *Multirate Systems and Filter Banks*. Prentice Hall, 1992.

[22] F. Bureau, "Le traité de la lumière de Christian Huygens," *Acad. Roy. Belgique. Bull. Cl. Sci.*, vol. 32, pp. 730–744, January 1946.

[23] A. J. Berkhout, D. de Vries, and P. Vogel, "Acoustic control by wave field synthesis," *The Journal of the Acoustical Society of America*, vol. 93, no. 5, p. 2764–2778, May 1993.

[24] E. Wong and B. Hajek, *Stochastic Processes in Engineering Systems*. Springer-Verlag, 1985.

[25] T. Berger, *Rate Distortion Theory: A Mathematical Basis for Data Compression*. Prentice-Hall, 1971.

[26] T. M. Cover and J. A. Thomas, *Elements of Information Theory*. Wiley, 1991.

[27] R. L. Konsbruck, E. Telatar, and M. Vetterli, "On the multiterminal rate distortion function for acoustic sensing," submitted to *IEEE Trans. Inform. Theory*, January 2008.

[28] V. Goyal, "Theoretical foundations of transform coding," *IEEE Signal Processing Mag.*, vol. 18, pp. 9–21, September 2001.

[29] M. Gastpar, P. L. Dragotti, and M. Vetterli, "The distributed Karhunen–Loève transform," *IEEE Trans. Inform. Theory*, no. 12, pp. 5177–5196, December 2006.

[30] T. Flynn and R. Gray, "Encoding of correlated observations," *IEEE Trans. Inform. Theory*, vol. 33, no. 6, pp. 773–787, November 1987.

[31] H. Yamamoto and K. Itoh, "Source coding theory for multiterminal communication systems with a remote source," *Trans. IECE Japan*, vol. E63, no. 10, pp. 700–706, October 1980.

[32] S. C. Draper, "Successive structuring of source coding algorithms for data fusion, buffering and distribution in networks," Ph.D. dissertation, Massachusetts Institute of Technology, June 2002.

[33] H. S. Witsenhausen, "Indirect rate-distortion problem," *IEEE Trans. Inform. Theory*, vol. 26, no. 5, pp. 518–521, September 1980.

[34] R. L. Dobrushin and B. S. Tsybakov, "Information transmission with additional noise," *IEEE Trans. Inform. Theory*, vol. 8, no. 5, pp. 293–304, September 1962.

[35] D. J. Sakrison, "Source encoding in the presence of random disturbance," *IEEE Trans. Inform. Theory*, vol. 14, no. 1, pp. 165–167, January 1968.

[36] J. K. Wolf and J. Ziv, "Transmission of noisy information to a noisy receiver with minimum distortion," *IEEE Trans. Inform. Theory*, vol. 16, no. 4, pp. 406–411, July 1970.

[37] D. Rebollo-Monedero, S. Rane, A. Aaron, and B. Girod, "High-rate quantization and transform coding with side information at the decoder," *EURASIP Journal on Signal Processing, Special Issue on Distributed Source Coding*, vol. 86, no. 11, pp. 3160–3179, 2006.

[38] L. R. Rabiner and R. W. Schafer, *Digital Processing of Speech Signals*. Prentice-Hall, 1978.

[39] O. Roy and M. Vetterli, "Rate-constrained collaborative noise reduction for wireless hearing aids," to appear in *IEEE Trans. Signal Processing*, 2009.

[40] M. Nilsson, S. D. Soli, and A. Sullivan, "Development of the Hearing in Noise Test for the measurement of speech reception thresholds in quiet and in noise," *The Journal of the Acoustical Society of America*, vol. 95, no. 2, pp. 1085–1099, 1994.

[41] A. Varga, H. Steeneken, M. Tomlinson, and D. Jones, "The NOISEX-92 study on the effect of additive noise on automatic speech recognition," Speech Research Unit, Defense Research Agency, Malvern, U.K., Tech. Rep., 1992.

CHAPTER

Distributed Video Coding: Basics, Codecs, and Performance

8

Fernando Pereira, Catarina Brites, and João Ascenso

Instituto Superior Técnico—Instituto de Telecomunicações, Lisbon, Portugal

CHAPTER CONTENTS

8.1 INTRODUCTION

Image, audio, and video coding are among the information technologies that have contributed more to change and improve the everyday life in recent years. Today, often without knowing, a large percentage of the world population, even in less developed countries, use image, video, and audio coding technologies on a regular basis. While users don't know much about these technologies, they know very well some of the most popular devices and services they made possible, such as digital cameras, digital television, DVDs, and MP3 players.

The key objective of digital audiovisual coding technologies is to compress the original audiovisual information into a much smaller number of bits, without adversely affecting the decoded signal quality, and following a set of requirements depending on the target application. Regarding video signals, the current coding paradigm is based mostly on four types of tools:

1. Motion-compensated temporal prediction between video frames to exploit the temporal redundancy
2. Transform coding, typically using the discrete cosine transform (DCT), to exploit the spatial redundancy
3. Quantization of the transform coefficients to exploit the irrelevancy intrinsic to the limitations of the human visual system limitations
4. Entropy coding to exploit the statistical redundancy of the created symbols

After a certain spatial resolution and frame rate have been adopted, the quality of the decoded video signal is controlled mainly through the quantization process. This process has to be adapted to the application needs, defined in terms of minimum quality or a target bit rate, depending on the network characteristics. Because the current video coding paradigm considers both the temporal (prediction) and frequency (DCT) domains, this type of coding architecture is known as hybrid or predictive coding.

Since the end of the 1980s, predictive coding schemes have been the solution adopted in most available video coding standards, notably the ITU-T H.26x and ISO/IEC MPEG-x families of standards. Today this coding paradigm is used in hundreds of millions of video encoders and decoders, especially in MPEG-2 Video set-top boxes and DVDs. In this video coding architecture, the correlation between (temporal) and within (spatial) the video frames is exploited at the encoder, leading to rather complex encoders and much simpler decoders. This scenario does not have much flexibility in terms of codec complexity budget allocation besides making the encoder less complex and thus less efficient; decoders are normative and basically much simpler than encoders. This approach fits some application scenarios very well, notably those that have been dominating the video coding developments in the past, such as broadcasting and video storage. These applications follow the so-called down-link model, in which a few encoders typically provide coded content for millions of decoders; in this case, the decoder complexity is the critical issue and thus has to be minimized. Moreover, the temporal prediction loop used to compute the residues to transmit, after the motion-compensated prediction of the current frame, requires the decoder to run the same loop in perfect synchronization with the encoder. This means that, when channel errors are present, errors may propagate in time, strongly affecting the video quality, typically until some (Intra) coding refreshment is performed.

The H.264/AVC (Advanced Video Coding) standard is the most efficient video coding standard available [1], typically spending about 50 percent of the rate for the same quality regarding previous video coding standards. It is being adopted for a wide range of applications from mobile videotelephony to HDTV and Blu-ray discs. Besides the technical novelty adopted in this standard (e.g., in terms of motion estimation and compensation, spatial transform, and entropy coding), the compression gains very much rely on additional significant encoder and decoder complexity. To provide coding solutions adapted to a wide range of applications, the standard defines a set of profiles and levels [1], which constrain the very flexible coding syntax in appropriate ways to allow successful deployment. Meanwhile, some of the most relevant compression gains are coming from control of the H.264/AVC codec through adequate encoder tools since, as usual, H.264/AVC encoders are not normative, thus leaving space for improvements in rate-distortion (RD) performance while still guaranteeing the same compatibility with much simpler decoders.

With the recent wide deployment of wireless networks, a growing number of applications do not fit well the typical down-link model but rather follow an up-link model in which many senders deliver data to a central receiver. Examples of these applications are wireless digital video cameras, low-power video sensor networks, and surveillance systems. Typically, these emerging applications require light encoding or a flexible distribution of the codec complexity, robustness to packet losses, high compression efficiency, and, often, low latency/delay as well. There is also a growing use of multiview video content, which means the data to be delivered regards many (correlated) views of the same scene. In many cases, to keep the sensing system simple, the associated cameras cannot communicate among themselves, preventing the usage of a predictive approach to exploit the interview redundancy.

In terms of video coding, these emerging applications are asking for a novel video coding paradigm that is better adapted to the specific characteristics of the new scenarios. Ideally, these applications would welcome a new video coding paradigm able to address all the requirements above with the same coding efficiency as the best predictive coding schemes available, as well as with an encoder complexity and error robustness similar to the current best Intra coding solutions, which are the simplest and most error-robust solutions.

To address these needs, around 2002, some research groups decided to revisit the video coding problem in light of two information theory results from the 1970s: the Slepian–Wolf theorem [2] and the Wyner–Ziv theorem [3]. These efforts gave birth to what is now known as distributed video coding (DVC), and Wyner–Ziv (WZ) video coding, as a particular case of distributed video coding. This chapter presents the developments that have been produced in DVC following the early research initiatives.

To achieve this purpose, this chapter is organized as follows: building on previous chapters, Section 8.2 will briefly review the basic concepts and theorems underpinning DVC. Next, Section 8.3 will present the first distributed video codecs developed, while Section 8.4 will review some of the most relevant developments following the early DVC codecs. To get a more precise understanding of how a DVC codec works, Section 8.5 will present in detail perhaps the most efficient DVC codec available, the DISCOVER WZ video codec [4, 5]. Section 8.6 will propose a detailed performance evaluation of the DISCOVER WZ codec, which may be used as benchmarking. Finally, Section 8.7 will summarize the past developments and project the future in the DVC arena.

8.2 BASICS ON DISTRIBUTED VIDEO CODING

As mentioned in a previous chapter, the Slepian–Wolf theorem addresses the case where two statistically dependent discrete random sequences, independently and identically distributed (i.i.d.), X and Y, are independently encoded, and thus not jointly encoded as in the largely deployed predictive coding solution. The Slepian–Wolf theorem states that the minimum rate to encode the two (correlated) sources is the same as the minimum rate for joint encoding, with an arbitrarily small error probability. This distributed source coding (DSC) paradigm is an important result in the context of the emerging application challenges presented earlier, since it opens the doors to new coding solutions where, at least in theory, separate encoding and joint decoding does not induce any compression efficiency loss when compared to the joint encoding and decoding used in the traditional predictive coding paradigm. In theory, the rate bounds for a vanishing error probability considering two sources are

$$\begin{aligned} &R_X \geqslant H(X|Y) \\ &R_Y \geqslant H(Y|X) \\ &(R_X + R_Y) \geqslant H(X,Y), \end{aligned} \tag{8.1}$$

which corresponds to the area identified in Figure 8.1. This basically means that the minimum coding rate is the same as for joint encoding (i.e., the joint entropy), provided that the individual rates for both sources are higher than the respective conditional entropies.

Slepian–Wolf coding is the term generally used to characterize lossless coding architectures that follow this independent encoding approach. Slepian–Wolf coding is also referred to in the literature as lossless distributed source coding since it considers that the two statistically dependent sequences are perfectly reconstructed at a joint decoder (neglecting the arbitrarily small probability of decoding error), thus approaching the lossless case.

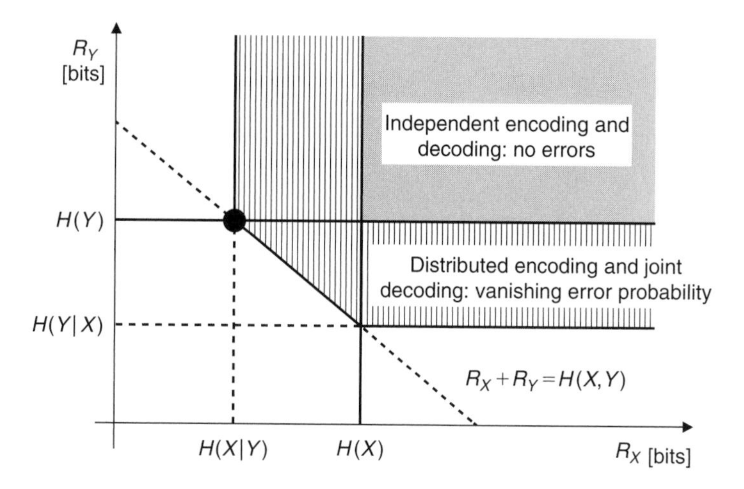

FIGURE 8.1

Rate boundaries defined by the Slepian–Wolf theorem for the independent encoding and joint decoding of two statistically dependent discrete random i.i.d. sources.

Slepian-Wolf coding has a deep relationship with channel coding, although it is possible to perform Slepian-Wolf coding in other ways, for example, using source codes [6]. Since the sequence X is correlated with the sequence Y, it can be considered that a virtual "dependence channel" exists between sequences X and Y. The Y sequence is, therefore, a "noisy" or "erroneous" version of the original uncorrupted X sequence, which will be corrected by Slepian-Wolf codes. Thus, the "errors" between the X and Y sequences can be corrected applying a channel coding technique to encode X, which will be conditionally (jointly) decoded using Y; this relationship was studied in the 1970s by Wyner [7]. Thus, channel coding tools typically play a main role in the novel distributed source coding paradigm.

In 1976, A. Wyner and J. Ziv studied a particular case of distributed source coding that deals with lossy compression of source X associated with the availability of the Y source at the decoder but not at the encoder. In these conditions, Y (or a derivation of Y) is known as side information. This case is well known as asymmetric coding since Y is independently encoded and decoded, while X is independently encoded but conditionally decoded. Wyner and Ziv performed a theoretical study for lossy coding to conclude that, typically, there is a rate loss incurred when the side information is not available at the encoder. They also derived the so-called Wyner-Ziv theorem [3], which states that when performing independent encoding with side information under certain conditions—that is, when X and Y are jointly Gaussian, memoryless sequences and a mean-squared error distortion measure is considered—there is no coding efficiency loss with respect to the case when joint encoding is performed, even if the coding process is lossy (and not "lossless" anymore as for Slepian-Wolf coding). For general statistics and a mean-squared error-distortion measure, Zamir [8] also proved that the rate loss is less than 0.5 bit/sample. Later, it was shown that, for no rate loss, only the innovation, this means the $X - Y$ difference needs to be Gaussian [9].

Together, the Slepian–Wolf and the Wyner–Ziv theorems suggest that it is possible to compress two statistically dependent signals in a distributed way (separate encoding, joint decoding), approaching the coding efficiency of conventional predictive coding schemes (joint encoding and decoding). Based on these theorems, a new video coding paradigm, well known as distributed video coding, has emerged having Wyner–Ziv coding as its lossy particular case.

DVC does not rely on joint encoding, and thus, when applied to video coding, it typically results in the absence of the temporal prediction loop (always used in predictive schemes) and lower complexity encoders. DVC-based architectures may provide the following functional benefits, which are important for many emerging applications:

1. Flexible allocation of the global video codec complexity
2. Improved error resilience
3. Codec independent scalability (since upper layers do not have to rely on precise lower layers)
4. Exploitation of multiview correlation without cameras/encoders communicating among them

These functional benefits can be relevant for a large range of emerging application scenarios such as wireless video cameras, low-power surveillance, video conferencing with mobile devices, disposable video cameras, visual sensor networks, distributed video streaming, multiview video systems, and wireless capsule endoscopy [10]. While this chapter will concentrate on monoview video coding, multiview video coding using a DVC approach is also a very hot research topic; check [11] for a review on this subject.

Practical design of WZ video codecs started around 2002, following important advances in channel coding technology, especially error-correction codes with a capacity close to the Shannon limit, for example, turbo and low-density parity-check (LDPC) codes. While theory suggests that DVC solutions may be as efficient as joint encoding solutions, practical developments did not yet achieve that performance for all types of content, especially if low-complexity encoding is also targeted. Part of this gap may be filled with more research to make the technology more mature, but there are also some practical limitations that may be difficult to overcome. For example, while the theory assumes that the encoder knows the statistical correlation between the two sources, X and Y, and that the innovation $X - Y$ is Gaussian, in practical conditions this is often not true. Naturally, the better the encoder knows the statistical correlation between X and Y, the higher the compression efficiency. This highlights a main DVC coding paradox: although encoders may be rather simple, they may also need to become more complex to increase the compression efficiency and reach the limits set by the theory. This is one of the DVC research challenges being faced today.

8.3 THE EARLY WYNER–ZIV VIDEO CODING ARCHITECTURES

The first practical WZ video coding solutions emerged around 2002, notably from Stanford University [12–14] and the University of California (UC), Berkeley [15, 16]. The Stanford architecture is mainly characterized by frame-based Slepian–Wolf coding, typically using turbo codes, and a feedback channel to perform rate control at the decoder; the Berkeley WZ video coding solution is mainly characterized by block-based coding with decoder motion estimation.

The Stanford architecture was later adopted and improved by many research groups around the world. However, while there are many papers published with changes and improvements to this architecture, the precise and detailed evaluation of its performance, targeting its deep understanding for later advances, was not presented until recently. Available performance results were mostly partial (e.g., typically only few RD performance results), under unclear and incompatible conditions (e.g., different sequences, different sets of frames for each sequence, different key frames coding), using vaguely defined and also sometimes architecturally unrealistic codecs (e.g., assuming the original frames available at the decoder), using side information at the encoder, or a vague side information creation process. This chapter will present an extensive performance evaluation of one of the most representative (and best performing) DVC codecs, the DISCOVER WZ codec. However, considering the impact they had in the DVC research arena, first this section will review and compare the early Stanford and Berkeley Wyner–Ziv video coding solutions.

8.3.1 The Stanford WZ Video Codec

The Stanford WZ video coding architecture was first proposed in 2002 for the pixel domain [12] and later extended to the transform domain [13, 14]. In this later approach, transform coefficients are WZ encoded to exploit, with a rather small encoder complexity, the spatial redundancy. In summary, the more efficient transform domain Stanford WZ video codec, shown in Figure 8.2, works as follows:

At the encoder:

1. **Frame Classification** The video sequence is divided into WZ frames and key frames, corresponding to the X and Y sequences mentioned in Section 8.2; the key frames are periodically inserted, determining the GOP (Group of Pictures) size. The key frames are Intra encoded; this means without exploiting temporal redundancy, notably without performing any motion estimation, for example, using the H.263+ Intra or H.264/AVC Intra standards.

2. **Spatial Transform** A block-based transform, typically a DCT, is applied to each WZ frame. The DCT coefficients of the entire WZ frame are then grouped together, according to the position occupied by each DCT coefficient within a block, forming DCT coefficient bands.

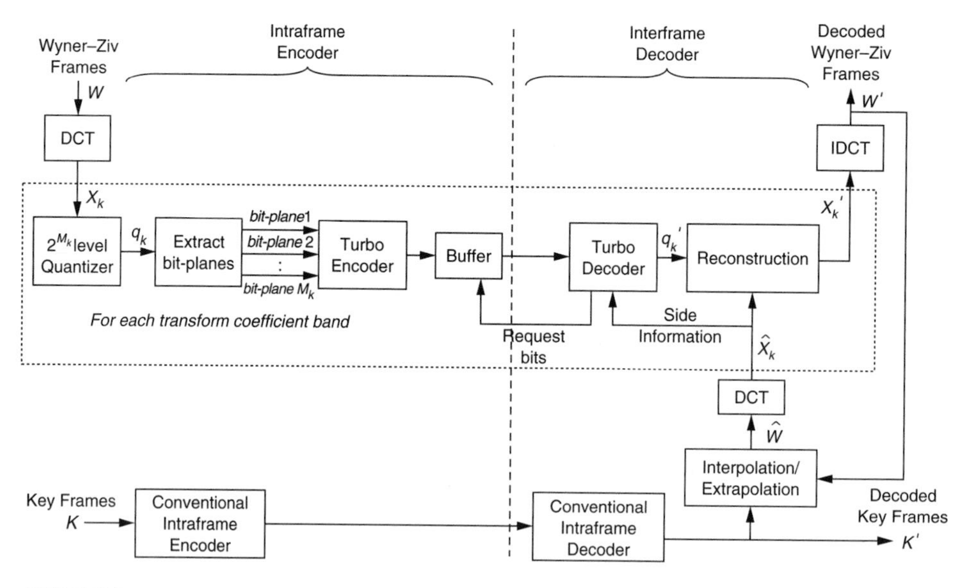

FIGURE 8.2

Stanford WZ video coding architecture [13].

3. **Quantization** Each DCT band is uniformly quantized with a number of levels that depend on the target quality [13]. For a given band, bits of the quantized symbols are grouped together, forming bit-planes, which are then independently turbo encoded.

4. **Turbo Encoding** The turbo encoding of each DCT band starts with the most significant bit-plane (MSB). The parity information generated for each bit-plane is then stored in the buffer and sent in chunks/packets upon decoder requests, made through the feedback channel.

At the decoder:

5. **Side Information Creation** The decoder creates the side information (SI) for each WZ frame by performing a motion-compensated frame interpolation (or extrapolation) using the closest already decoded frames. The side information for each WZ frame is taken as an estimate (noisy version) of the original WZ frame. The better the quality of the estimation, the smaller the number of "errors" the turbo decoder has to correct and thus the number of parity bits (or bit rate) needed.

6. **Correlation Noise Modeling** The residual statistics between corresponding DCT coefficients in the WZ frame and the side information are assumed to be modeled by a Laplacian distribution whose parameter was initially estimated using an off-line training phase. This is an unrealistic approach because

it either assumes the original data is available at the decoder or that the side information is available at the encoder. Recently, solutions have been developed to overcome this problem (see Section 8.4.1.3).

7. **Turbo Decoding** Once the side information DCT coefficients and the residual statistics for a given DCT coefficients band are known, each bit-plane is turbo decoded (starting from the MSB one). The turbo decoder receives from the encoder successive chunks of parity bits following the requests made through the feedback channel. To decide whether or not more parity bits are needed for successful decoding of a certain bit-plane, the decoder uses a request stopping criterion. Initially, this criterion assumed the availability of the originals at the decoder, but once again this unrealistic assumption was overcome (see Section 8.4.2). After successfully turbo decoding the MSB bit-plane of a DCT band, the turbo decoder proceeds in an analogous way with the remaining bit-planes associated with the same band. Once all the bit-planes of a DCT band are successfully turbo decoded, the turbo decoder starts decoding the next band.

8. **Reconstruction** After turbo decoding, all the bit-planes associated with each DCT band are grouped together to form the decoded quantized symbol stream associated with each band. Once all decoded quantized symbols are obtained, it is possible to reconstruct all the DCT coefficients with the help of the corresponding side information coefficients. The DCT coefficients bands for which no WZ bits were transmitted are replaced by the corresponding DCT bands of the side information.

9. **Inverse Transform** After all DCT bands are reconstructed, an inverse discrete cosine transform (IDCT) is performed and the decoded WZ frame is obtained.

10. **Frame Remixing** Finally, to get the decoded video sequence, decoded key frames and WZ frames are conveniently mixed.

Because, the side information for the WZ frames is here estimated or "guessed," at the decoder, based on the independently coded key frames, some authors label this type of DVC approach as a "Guess" DVC solution.

Over the last few years, many improvements have been proposed for most of the modules in the initial Stanford WZ video codec (e.g., new source/channel codes instead of turbo codes, better side information estimation, dynamic correlation noise modeling, enhanced reconstruction, realistic and efficient request stopping criteria). Other proposed solutions required revisiting the original architecture by introducing major changes, such as selective Intra coding of blocks in the WZ frame, selective transmission of hash signatures by the encoder, removal of the feedback channel, and provision of scalability and error resilience features. Section 8.4 will address these and other developments that have emerged since the early DVC activities.

8.3.2 The Berkeley WZ Video Codec

Almost at the same time as the Stanford WZ video coding solution, another WZ video coding approach was proposed at UC Berkeley, known in the literature as PRISM—from Power-efficient, Robust, High-compression, Syndrome-based Multimedia coding [15, 16]. The PRISM codec works at block level and does not require a feedback channel. In summary, the PRISM codec is shown in Figure 8.3 and works as follows:

At the encoder:

1. **Transform** Each video frame is divided into 8×8 samples blocks, and a DCT is applied over each block.
2. **Quantization** A scalar quantizer is applied to the DCT coefficients corresponding to a certain target quality.
3. **Block Classification** Before encoding, each block is classified into one of several predefined classes depending on the correlation between the current block and the predictor block in the reference frame. Depending on the allowed complexity at the encoder, such a predictor can be either the co-located block, or a motion-compensated block [16]; in the latter case, a fast motion estimation

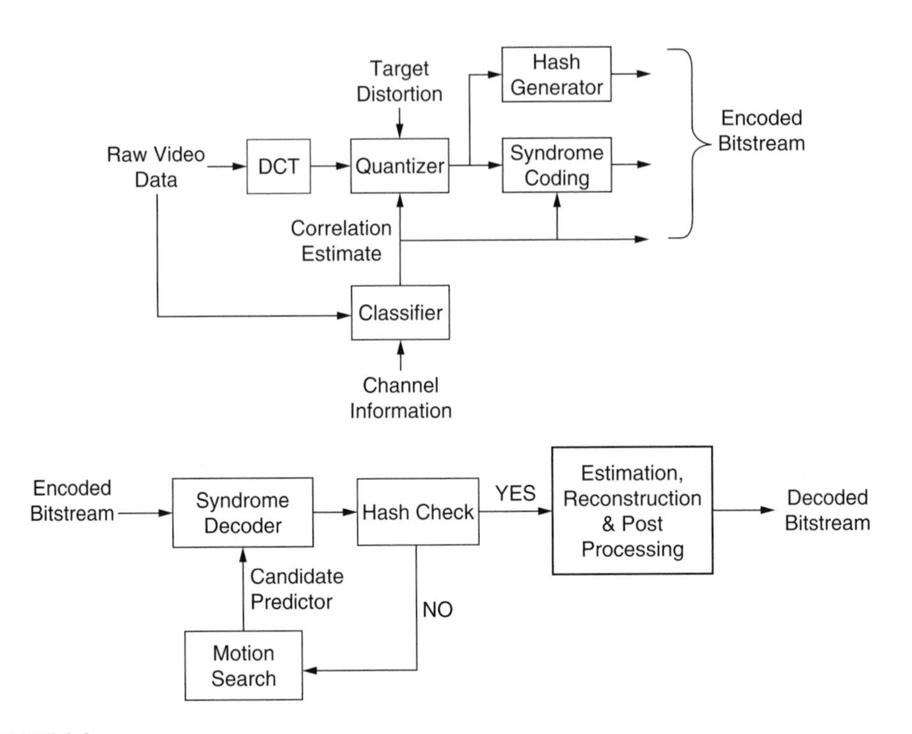

FIGURE 8.3

PRISM encoder and decoder architectures [16].

technique is suggested. The classification stage decides the coding mode for each block of the current frame: no coding (skip class), traditional Intraframe coding (entropy coding class), or syndrome coding (several syndrome coding classes), depending on the estimated temporal correlation. The blocks classified in the syndrome coding classes are coded using a WZ coding approach as described below. The coding modes are then transmitted to the decoder as header information.

4. **Quantization** A scalar quantizer is applied to the DCT coefficients corresponding to a certain target quality. For syndrome coded blocks, the quantizer step size depends on the coding class of the block, that is, on the estimated correlation between the current block and a predictor block (a coarse representation of the side information).

5. **Syndrome Coding** For those blocks that fall in the syndrome coding classes, only the least significant bits of the quantized DCT coefficients in a block are syndrome encoded, since it is assumed that the most significant bits can be inferred from the side information. The number of least significant bits to be sent to the decoder depends on the syndrome class to which the block belongs. Within the least significant bits, the lower part is encoded using a (run, depth, path, last) 4-tuple-based entropy codec. The upper part of the least significant bits is coded using a coset channel code, in this case a Bose, Ray-Chaudhuri, and Hocquenghem (BCH) code, since it works well for small blocklengths as is the case here.

6. **Hash Generation** In addition, for each block, the encoder sends a 16-bit cyclic redundancy check (CRC) checksum as a signature of the quantized DCT coefficients. This is needed in order to select the best candidate block (SI) at the decoder, as explained below.

At the decoder:

7. **Motion Search** The decoder generates side information candidate blocks, which correspond to all half-pixel displaced blocks in the reference frame, in a window positioned around the center of the block to decode.

8. **Syndrome Decoder** Each of the candidate blocks plays the role of side information for syndrome decoding, which consists of two steps: one step deals with the coset channel coded bit-planes and is performed for each candidate block; the other step deals with the entropy decoding of the least significant bit-planes and is performed only for the selected block by hash matching.

9. **Hash Checking** Each candidate block leads to a decoded block, from which a hash signature is generated. In order to select one of the candidate blocks and to detect successful decoding (i.e., blocks with a small error probability), the latter is compared with the CRC hash received from the encoder. Candidate blocks are visited until decoding leads to hash matching.

10. **Reconstruction and IDCT** Once the quantized sequence is recovered, it is used along with the corresponding side information to get the best reconstructed block. A minimum mean-squared estimate is computed from the side information and the quantized block.

Because the side information for the WZ frames is created here, at the decoder, in a tentative way, based on the previously decoded WZ frames, some authors label this type of DVC approach as a "Try" DVC solution.

8.3.3 Comparing the Early WZ Video Codecs

From the technical point of view, the following main functional differences between the two early WZ video codecs may be highlighted (Stanford versus Berkeley):

1. Frame-based versus block-based coding; in the latter approach, it is easier to accommodate coding adaptability to address the highly nonstationary statistics of video signals.
2. Decoder rate control versus encoder rate control; in the former case, a feedback channel is needed, restricting the scope to real-time applications, while making the rate control problem simpler.
3. Very simple encoder versus smarter and likely more complex encoder; enabling limited interframe operations at the encoder allows incorporating spatially varying coding mode decisions—for example, acknowledging that it is useless to adopt a WZ coding approach when the correlation is too weak or inexistent.
4. More sophisticated channel codes, notably turbo codes and later LDPC codes, versus simpler channel codes (e.g., BCH codes).
5. No auxiliary data versus hash codes sent by the encoder to help the decoder in the motion estimation process.
6. Less intrinsically robust to error corruption versus higher resilience to error corruption due to the PRISM motion search like approach performed at the decoder, which allows finding less corrupted side information, thus reducing the residual noise associated to the errors.

With time, some of the differences above between the two early WZ video codecs have been smoothed: for example, there are nowadays Stanford-based codecs with selective block-based Intra coding, encoder-transmitted hash signatures, and without feedback channel. Section 8.4 will address these and other developments that emerged after the early WZ video coding solutions. However, after a few years, the performance gap between the two early solutions seems to be rather substantial, at least in terms of error-free RD performance. In November 2007, the European project DISCOVER published error-free RD performance results for a Stanford-based WZ video codec, which is able to outperform H.264/AVC Intra and, sometimes, even H.264/AVC "no motion" with a lower encoding complexity [4,5]. In October 2007, the Berkeley team published error-free RD performance results that slightly outperform H.263+ coding, with both encoders performing encoder motion estimation [16]. However, for wireless error-prone conditions, notably packet errors according to the CDMA2000 1X standard, PRISM performs better than H.263+ as well as better than H.263+ with Reed-Solomon forward error correction (FEC) and H.263+ with Intra refreshing, providing graceful error degradation.

8.4 FURTHER DEVELOPMENTS ON WYNER–ZIV VIDEO CODING

In recent years, a significant number of research groups around the world have been developing research activities in the distributed video coding field, many of them building on and improving the previously presented early WZ video codecs. The main research target has clearly been to improve the RD performance, but other objectives, such as providing error resilience, and scalability or even removing the feedback channel in the Stanford architecture, have been addressed. This section briefly reviews some of the main developments; space considerations necessarily limit this review, as well as the associated list of references.

8.4.1 Improving RD Performance

Because the initial RD performance was rather poor in comparison with the alternative solutions provided by the available standards (e.g., H.263+, MPEG-4 Visual and H.264/AVC), most of the DVC research has focused on improving the coding efficiency, especially in the context of low-complexity encoding.

8.4.1.1 Slepian–Wolf Coding

Since Slepian–Wolf coding is the core of WZ coding, and channel coding plays a central role in Slepian–Wolf coding, channel coding developments play an important role not only in terms of RD performance but also codec complexity budget. Channel capacity-achieving codes, such as the block codes used in PRISM [15–17], the turbo codes used in the Stanford WZ solution [12–14, 18–20], or LDPC codes [21, 22], have been shown to reach the performance corresponding to the corner points of the Slepian–Wolf region in Figure 8.1. Beside turbo codes, the most frequently used codes for WZ video coding are clearly the LDPC codes due to their capacity approaching performance. The literature states that LDPC codes can better approach the capacity of a variety of communication channels than turbo codes [22]. This has led naturally to the consideration of LDPC codes for distributed source coding and, in particular, for distributed video coding and WZ coding. LDPC codes present a trade-off between latency, complexity, and system performance. In fact, the complexity of LDPC decoding, based on a sum-product algorithm, is lower than that of turbo codes. Moreover, similar encoding complexity can be achieved by a careful LDPC code design. Typically, the performance of these codes also depends on the blocklength. Classes of rate-compatible LDPC codes for distributed source coding (LDPC Accumulate (LDPCA) codes and Sum

LDPC Accumulate (SLDPCA) codes) have been introduced in [22]. The performances of LDPCA and SLDPCA codes are shown to be better than those of turbo codes [20] for moderate and high rates.

Recent interesting developments regard the use of raptor codes for distributed source-channel coding, targeting scalable video transmission over wireless networks [23]. Raptor codes are the latest addition to a family of low-complexity digital fountain codes [24], capable of achieving near-capacity erasure protection. They may be used as precoded Luby Transform (LT) codes—the first class of practical fountain codes that are near optimal erasure correcting codes—in combination with LDPC precoding. Fountain codes (also known as rateless erasure codes) are a class of erasure codes with the property that a potentially limitless sequence of encoding symbols can be generated from a given set of source symbols such that the original source symbols can be recovered from any subset of the encoding symbols of size equal to or only slightly larger than the number of source symbols. In [25], a receiver-driven layered multicast system over the Internet and 3G wireless networks based on layered Wyner–Ziv video coding and digital fountain codes is proposed.

8.4.1.2 Advanced Side Information Creation

The quality of the side information plays a central role in the WZ codec's overall RD performance. Without a powerful side information creation mechanism, no "decent" RD performance can be achieved. This fact motivated the development of many improvements in the simple and inefficient early side information methods such as computing the side information as the average of the two nearest key frames for the Stanford WZ solution. For example, Ascenso et al. developed a largely used solution in the literature [26, 27] where the side information is generated based on block matching using a modified mean absolute difference (MAD) to regularize the motion vector field; then, a hierarchical coarse-to-fine bidirectional motion estimation is performed (with half-pixel precision); spatial motion smoothing based on a weighted vector median filter is applied afterward to the obtained motion field to remove outliers before motion compensation is finally performed. If low delay is required, then side information may be created through extrapolation and not interpolation anymore [13, 28]. This type of solution corresponds to the so-called DVC Guess approach in which side information is estimated (guessed) based on independently coded key frames. A recent interesting approach is unsupervised learning; in [29], Varodayan et al. propose an expectation maximization (EM) algorithm to perform unsupervised learning of disparity at the decoder for the WZ coding of stereo images. This type of learning approach was also applied for video coding by exchanging soft information between an LDPC decoder and a probabilistic motion estimator in an iterative way; the side information is successively refined after syndromes are incrementally received by the LDPC decoder. Because the side information for the WZ frames is created or improved at the decoder, using some learning tools, some label this type of DVC approach a "Learn" DVC solution. This DVC approach can be typically combined with the popular Guess approach, described previously.

8.4.1.3 Correlation Noise Modeling

Since WZ video coding targets the lossy coding of the difference between the original data and its corresponding side information, it is essential for an efficient RD performance that the encoder and decoder are aware of the statistical correlation between the original and side information data. For the Stanford architecture with pure decoder rate control, only the decoder needs to be aware of the correlation noise model (CNM) between the original and its side information. This CNM has to be estimated in a realistic way, which means at the decoder, of course without having access to the original data. The correlation noise modeling may be performed at various levels of granularity (e.g., band or coefficient levels), allowing a dynamic adaptation of the model to the varying temporal and/or spatial correlation. In [30–32], Brites et al. propose CNM solutions for the pixel domain and transform domain Stanford-based WZ codecs, showing there are benefits in modeling this noise with a finer granularity.

8.4.1.4 Reconstruction

The last module in WZ coding architectures is typically the reconstruction module whose target is to convert the decoded quantized symbols or bins into a real value, either luminance value for pixel domain codecs or transform coefficient for transform domain WZ codecs. The initial solution presented in [12], where the decoded value is the side information if it falls within the decoded bin, or clips to the bin limit closer to the side information if it falls outside, has been used for a large number of WZ codecs. Recently, a novel reconstruction solution has been developed by Kubasov et al. in which the decoded values are reconstructed using an optimal MSE-based approach using closed-form expressions derived for a Laplacian correlation model [33]. Since this solution has been developed and adopted by the DISCOVER WZ codec, further details will be presented in Section 8.5.5.

8.4.1.5 Selective Block-based Encoding

Somehow inspired by the PRISM approach and its block-based approach, the addition of a block classification module to a Stanford-based WZ video codec has been proposed in [34], allowing the selection of one of two coding modes (Intra or WZ modes), depending on the available temporal correlation. This approach results from the observation that the correlation noise statistics describing the relationship between the original frame and its corresponding side information available at the decoder is not spatially stationary. A mode decision scheme (applied either at the encoder or at the decoder) works in such a way that (i) when the estimated correlation is weak, Intra coding is performed on a block-by-block basis and (ii) when it is strong, the more efficient WZ coding mode is used. Both spatial and temporal criteria are used to determine whether or not a block is better Intra coded. Large gains are reported in [34]: up to 5 dB with respect to the case where all the blocks are WZ encoded, for the News sequence coded at QCIF, 30 Hz resolution.

8.4.1.6 Encoder Auxiliary Data

Another way to overcome the "blind" frame-based approach adopted by the early Stanford WZ video coding solution, following the observation that the temporal correlation is not uniform within the frames, consists in incorporating the capability for the encoder to send some hash signatures to help the decoder generating better side information [35, 36]. In [35], the hash code for an image block simply consists of a small subset of coarsely quantized DCT coefficients of the block. Since the hash requires fewer bits than the original data, the encoder is allowed to keep the hash codewords for the previous frame in a small hash store. Strictly speaking, the encoder is no longer Intra frame due to the hash store. In [35], significant gains over conventional DCT-based Intra frame coding are reported, while having comparable encoding complexity. In [36], a novel bidirectional hash motion estimation framework is proposed, which enables the decoder to select the best block by hash matching from candidate blocks obtained from the past and/or future reference frames. New features include the coding of DCT hashes with zero motion, combination of the usual trajectory-based motion interpolation ("Guess") with hash-based motion estimation, and adaptive selection of the number of DCT bands (starting from the low-frequency DC band) to be sent to the decoder in order to have reliable estimation of the side information (using a motion-estimation hash-based procedure). Gains up to 1.2 dB compared to previous traditional motion interpolation approaches [27] may be reached.

Because the side information for the WZ frames is here created or improved, at the decoder, using the help of some auxiliary data, or hints, sent by the encoder, this type of DVC approach is labeled by some authors as a "Hint" DVC solution. A few DVC solutions are, in practice, a combination of the DVC "Guess" and "Hint" approaches.

8.4.2 Removing the Feedback Channel

The feedback channel is very likely the most controversial architectural element in the Stanford WZ video coding solution because it implies not only the presence of the feedback channel itself but also requires that the application has to work in real time; the application and the video codec must also be able to accommodate the delay associated with the feedback channel. On the other hand, usage of the feedback channel simplifies the rate control problem since the decoder equipped with realistic error-detection criteria and knowing the available side information can easily adjust the necessary bit rate. To allow the Stanford solution to be applicable to other applications that do not fulfill the conditions above (e.g., storage applications), a WZ codec with encoder rate control (without feedback channel) is proposed by Brites and Pereira in [37]. This paper reports a loss of up to 1.2 dB, especially for the highest bit rates, between the pure encoder and decoder rate control solutions.

An interesting improvement on decoder rate control without completely giving up on the feedback channel is hybrid rate control. For this case, rate control processing is made at both encoder and decoder [38, 39]. Although the encoder has to make a conservative estimation of the rate needed, the decoder has the task of complementing this rate using the feedback channel, if necessary. The main hybrid rate control advantage regards the reduction of the decoding complexity since Slepian–Wolf decoding will have to be run a substantially lower number of times. While encoder rate overestimation is paid with losses in RD performance regarding the pure decoder rate control, encoder rate underestimation is paid with increased decoding complexity and delay regarding a perfect encoder rate control.

For both decoder and hybrid rate control, it is essential to have efficient request stopping criteria that allow stopping the parity rate requests after estimating that a certain error probability (e.g., typically 10^{-3} for each DCT bit-plane) has been reached [38, 40].

8.4.3 Improving Error Resilience

Since predictive codecs are extremely sensitive to transmission errors, distributed video coding principles have been extensively applied in the field of robust video transmission over unreliable channels. To improve the final decoded quality in the presence of errors, predictive coding schemes have two main solutions: (1) add some FEC parity bits; and (2) perform postprocessing in the form of error concealment. The two solutions are not incompatible and can be used together. WZ video codecs are characterized by in-built error robustness, due to the lack of the prediction loop that characterizes conventional motion-compensated predictive codecs. When errors occur, the side information can be corrupted, and thus the WZ coding will operate as a joint source-channel codec and not only as a source codec.

In [41], the notion of systematic lossy source-channel coding is first introduced; in systematic transmission, the decoder has access to a noisy version of the original data, analog or digital. Then a coded version of the same data is used to reduce the average decoded distortion at the cost of an increase in the rate of information transmission. Wyner–Ziv coding has a very close relation with systematic lossy source-channel coding since it may efficiently play the role of the auxiliary coded channel. Most of the WZ video coding schemes that focus on error resilience try to increase the robustness of standard encoded video by adding an auxiliary channel with redundant information encoded according to WZ coding principles. In one of the first works with this focus, Sehgal et al. [42] use auxiliary WZ encoded data sent only for some frames, to stop drift propagation at the decoder. The proposed video coding algorithm mitigates the propagation of errors in predictive video coding by periodically transmitting a small amount of additional information, termed coset information, to the decoder, instead of periodically sending Intra coded frames. Following the distributed coding approach, the coset information is able to correct the errors, without the encoder having precise knowledge of the packets or information that was lost. In [14], Girod et al. protect an MPEG-2 Video coded bitstream with an independent WZ bitstream, achieving graceful degradation with increasing channel error rate without using a scalable representation. This systematic coding approach is compatible with systems already deployed, such as MPEG-2 digital TV broadcasting since the WZ bitstreams can be ignored by legacy systems while they would be processed by the new WZ enabled receivers. By protecting the original signal with one or more WZ bitstreams, rather than applying forward error correction to the standard compressed bitstream, graceful degradation with deteriorating channel conditions can be achieved without a scalable signal representation, overcoming a major limitation of available predictive video coding schemes. This systematic lossy error protection (SLEP) framework has been extended for the H.264/AVC standard [43]. In [44], Wang et al. model the correlation noise for the overall, source correlation plus network loss SI distortion channel; an auxiliary channel with a subset of the conventional stream transform coefficients is WZ coded to improve error robustness.

The error resilience performance of the Stanford DISCOVER WZ video codec [4, 5] has also been studied in [45]. The results confirm the intrinsic error resilience capability of the WZ codec, due mostly to the usage of turbo coding. The most interesting result is that, for the adopted test conditions, the DISCOVER WZ codec performs better than the H.264/AVC (Inter) standard considering an error-prone channel and small GOP sizes.

8.4.4 Providing Scalability

Scalability is a very important functionality when video streaming for heterogeneous environments, for example, in terms of networks, or terminals, has to be provided. In current scalable codecs, there is typically a predictive approach from lower layers to upper layers, requiring the encoder to use as reference the decoded frames from the previous layers in order to create the successive layers (e.g., with SNR or spatial resolution enhancements). However, the WZ prediction loop-free approach between the scalable layers no longer requires deterministic knowledge of the previous layers (just a correlation model), which means the layers may be generated by various different and rather unknown codecs. In this case, only the correlation between one layer and the side information created from the previous layer has to be known.

In 2004, Sehgal et al. [46] proposed a solution for the scalable predictive video coding problem using a WZ coding framework. For that, the predictor for each video frame is modeled as the result of passing the video frame through a hypothetical communication channel. In this way, the coding of a video frame is recast as the problem of correcting the errors introduced by this hypothetical communication channel. In conclusion, the authors propose a compression algorithm based on motion vectors to generate the side information at the decoder based on the previous decoded frames and employing a bank of LDPC codes applied to the original frame-quantized DCT coefficients to correct the decoder prediction. The authors claim results demonstrating that the proposed approach is approximately 4 dB superior to a H.26L-based conventional scalable video coding approach. In [47], Tagliasacchi et al. study a scalable version of PRISM addressing both spatial and temporal scalability. The proposed codec inherits the main PRISM codec attributes while also providing scalability. Experiments show that the proposed scalable PRISM codec is more resilient to packet losses than conventional predictive codecs and even outperforms, by a significant margin, predictive coded bitstreams protected with FEC codes under reasonable latency constraints. In [48], Wang et al. propose a FGS (fine granularity scalability)

WZ codec where refinement bit-planes are encoded with a hybrid approach, using either LDPC codes and interlayer prediction or conventional FGS variable-length coding (VLC) tools. The experimental results show coding efficiency gains of 3–4.5 dB for the FGS WZ codec over MPEG-4 FGS, for video sequences with high temporal correlation. A layered WZ coding architecture is proposed in [49], achieving both scalability and error resilience. There, Xu and Xiong propose a layered WZ video codec using as base layer, a standard scalable video codec (e.g., MPEG-4 FGS or H.263+), and LDPC coding (with irregular LDPC codes) of the DCT bit-planes of the enhancement layers. While for error-free conditions, the WZ codec performs worse than conventional FGS coding, the opposite is true for CDMA2000 1X standard error channel characteristics. Finally, in [50], Ouaret et al. propose several WZ coding-based scalable architectures providing different types of scalability, notably temporal, quality, and spatial, on top of the DISCOVER WZ codec.

8.5 THE DISCOVER WYNER–ZIV VIDEO CODEC

The WZ video codec evaluated in this chapter has been improved by the DISCOVER project team [4] based on a first codec developed and implemented at Instituto Superior Técnico in Lisbon [26]. The DISCOVER WZ video codec architecture, illustrated in Figure 8.4, is based on the early Stanford WZ video coding architecture proposed in [13, 14]; further information may be obtained at [4, 5]. However, the early Stanford codec has been much improved, for example, the architectural limitations regarding using originals at the decoder have been removed, side information generation

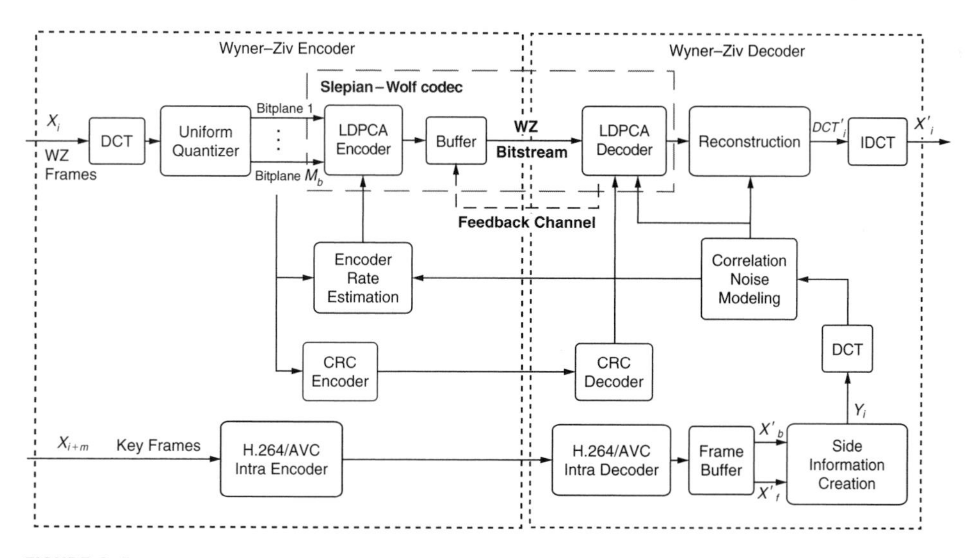

FIGURE 8.4

DISCOVER Wyner–Ziv video codec architecture.

and reconstruction have been improved, and an LDPCA code has been added for Slepian–Wolf coding and a CRC code for error detection. In practice, the tools for most of the WZ architecture modules are different (and globally much more efficient).

The DISCOVER codec is probably the most efficient WZ video codec now available. Its performance is reported in detail with the corresponding test conditions in [4]; moreover, executable code may be downloaded, allowing all researchers to compare performances for other sequences and conditions as well.

The DISCOVER WZ video codec works as follows:

At the encoder:

1. **Frame Classification** First, a video sequence is divided into WZ frames; this means the frames that will be coded using a WZ approach, and key frames that will be coded as Intra frames (e.g., using the H.264/AVC Intra codec [1]). Typically key frames are periodically inserted with a certain GOP size. An adaptive GOP size selection process may also be used, meaning that the key frames are inserted depending on the amount of temporal correlation in the video sequence [27]. Most results available in the literature use a GOP size of 2, which means that odd and even frames are key frames and WZ frames, respectively.

2. **Discrete Cosine Transform** Over each Wyner–Ziv frame, an integer 4×4 block-based DCT is applied. The DCT coefficients of the entire WZ frame are then grouped together, according to the position occupied by each DCT coefficient within the 4×4 blocks, forming the DCT coefficients bands.

3. **Quantization** After the transform coding operation, each DCT coefficients band b_k is uniformly quantized with 2^{M_k} levels (where the number of levels 2^{M_k} depends on the DCT coefficients band b_k). Over the resulting quantized symbol stream (associated with the DCT coefficients band b_k), bit-plane extraction is performed. For a given band, the quantized symbols bits of the same significance (e.g., the most significant bit) are grouped together, forming the corresponding bit-plane array, which is then independently LDPC (or turbo) encoded [22].

4. **LDPCA Encoding** The LDPCA (or turbo) encoding procedure for the DCT coefficients band b_k starts with the most significant bit-plane array, which corresponds to the most significant bits of the b_k band quantized symbols. The parity information generated by the LDPCA encoder for each bit-plane is then stored in the buffer and sent in chunks/packets upon decoder request, through the feedback channel.

5. **Encoder Rate Estimation** In order to limit the number of requests to be made by the decoder, and thus the decoding complexity and transmission delay (since each request corresponds to several LDPC decoder iterations), the encoder estimates for each bit-plane the initial number of bits to be sent

before any request is made [5]. This number should be an underestimation of the final number of bits needed if no RD performance losses associated with this step are welcome (regarding the pure decoder rate control case). If the rate is underestimated, the decoder will complement it by making one or more requests.

At the decoder:

6. **Side Information Creation** The decoder creates the side information for each WZ coded frame with a motion-compensated frame interpolation framework, using the previous and next temporally closer reference frames to generate an estimate of the WZ frame [26, 27]. The side information for each WZ frame corresponds to an estimation of the original WZ frame. The better is the quality of this estimation, the smaller are the number of "errors" the WZ LDPC (or turbo) decoder has to correct and the bit rate necessary for successful decoding (i.e., with a small error probability).

7. **DCT Estimation** A block-based 4×4 DCT is then carried out over the side information in order to obtain the DCT coefficients, which are an estimate of the WZ frame DCT coefficients.

8. **Correlation Noise Modeling** The residual statistics between corresponding WZ frame DCT coefficients and the side information DCT coefficients are assumed to be modeled by a Laplacian distribution. The Laplacian parameter is estimated on-line and at different granularity levels, notably at band and coefficient levels [31, 32].

9. **LDPCA Decoding** Once the DCT-transformed side information and the residual statistics for a given DCT coefficients band b_k are known, the decoded quantized symbol stream associated with the DCT band b_k can be obtained through the LDPCA decoding procedure. The LDPCA (or turbo) decoder receives from the encoder successive chunks of parity bits following the requests made through the feedback channel [22].

10. **Request Stopping Criterion** To decide whether or not more bits are needed for the successful decoding of a certain bit-plane, the decoder uses a simple request stopping criterion, that is, checks that all LDPC code parity-check equations are fulfilled for the decoded (hard decision) codeword. If no more bits are needed to decode the bit-plane, the decoding of the next bit-plane or band can start; otherwise, the bit-plane LDPC decoding task has to proceed with another request and receive another chunk of parity bits.

11. **Further LDPCA Decoding** After successfully LDPCA (or turbo) decoding the most significant bit-plane array of the b_k band, the LDPCA (or turbo) decoder proceeds in an analogous way to the remaining M_{k-1} bit-planes associated with that band. Once all the bit-plane arrays of the DCT coefficients band b_k are successfully LDPCA (or turbo) decoded, the LDPCA (or turbo) decoder

starts decoding the b_{k+1} band. This procedure is repeated until all the DCT coefficients bands for which WZ bits are transmitted are LDPCA (or turbo) decoded.

12. **CRC Checking** Because some residual errors are left even when all parity-check equations are fulfilled (step 10) and may have a rather negative subjective impact on the decoded frame quality, a CRC checksum is transmitted to help the decoder detect and correct the remaining errors in each bit-plane. Since this CRC is combined with the developed request stopping criterion, it does not have to be very strong in order to guarantee a vanishing error probability (≈ 0) for each decoded bit-plane. As a consequence, a CRC-8 checksum for each bit-plane was found to be strong enough for this purpose, which only adds minimal extra rate (8 bits).
13. **Symbol Assembling** After LDPCA (or turbo) decoding, the M_k bit-planes associated with the DCT band b_k, the bit-planes are grouped together to form the decoded quantized symbol stream associated with the b_k band. This procedure is performed over all the DCT coefficients bands to which WZ bits are transmitted. The DCT coefficients bands for which no WZ bits were transmitted are replaced by the corresponding DCT bands from the DCT side information.
14. **Reconstruction** Once all quantized symbol streams are obtained, it is possible to reconstruct the matrix of decoded DCT coefficients for each block.
15. **IDCT** After a block-based 4×4 IDCT is performed, the reconstructed pixel domain WZ frame is obtained.
16. **Frame Remixing** To finally get the decoded video sequence, decoded key frames and WZ frames are mixed conveniently.

It is important to stress that the DISCOVER WZ video codec does not include any of the limitations that are often still present in WZ video coding papers, notably those adopting the Stanford WZ architecture. This means, for example, that no original frames are used at the decoder to create the side information, to estimate the bit-plane error probability or to estimate the correlation noise model parameters for LDPCA (or turbo) decoding. The DISCOVER WZ video codec is a fully practical video codec for which a realistic performance evaluation was made [4]. For a better understanding of the DISCOVER WZ video codec, the next sections will provide some detail for the main coding tools. For more details, the readers should consult the papers listed in [4].

8.5.1 Transform and Quantization

In the DISCOVER architecture, illustrated in Figure 8.4, the first stage toward encoding a Wyner–Ziv frame is transform coding (corresponding to the DCT module). The transform employed in the DISCOVER WZ video codec relies on an integer 4×4

block-based DCT transform,[1] as defined by the H.264/AVC standard [1]. The DCT transform is applied to all 4×4 nonoverlapping blocks of the WZ frame, from left to right and top to bottom.

Afterward, quantization is applied to each DCT coefficient band b_k to exploit irrelevance and achieve the target quality. Two different quantization approaches are used in the DISCOVER solution.

1. **DC Coefficients** The DC coefficients band is quantized using a uniform scalar quantizer without a symmetric quantization interval around the zero amplitude. Typically, the DC coefficients band is characterized by high-amplitude positive values since each DC transform coefficient expresses the average energy of the corresponding 4×4 samples block.

2. **AC Coefficients** The remaining AC coefficients bands are quantized using a uniform scalar quantizer with quantized bins evenly distributed around zero in order to reduce the block artifacts effect. The zero bin size is twice the remaining bins size. Since most of the AC coefficients are concentrated around zero, by doubling the zero bin size, the matching probability between corresponding quantized bins of the WZ and SI frames increases, bringing bit rate savings. Some distortion loss is, however, expected since the larger the quantization bin, the worse the decoded frame quality is, but overall the RD performance improves.

Moreover, the DISCOVER codec considers a varying quantization step size for the AC coefficients. The decoder is informed by the encoder of the dynamic range for each AC band of each WZ frame. This allows a quantization bin width (step size) to be adjusted to the dynamic range of each AC band. Since the dynamic range of a given AC band may be lower than a fixed selected value, a smaller quantization bin width is used in the DISCOVER codec, for the same number of quantization levels. The smaller the quantization step size, the lower the distortion is at the decoder. In summary, the AC coefficients can be more efficiently encoded varying the quantization step size according to the dynamic range of each AC coefficients band. In the DISCOVER WZ codec, the quantization step size, W, for the AC bands $b_k (k = 2, \ldots, 16)$ is obtained from $W = 2|V_k|_{\max}/(2^{M_k} - 1)$ where $|V_k|_{\max}$ stands for the highest absolute value within b_k. For 4×4 samples blocks and 8-bit accuracy video data, the DC band dynamic range is 1024. This value is maintained fixed for all the video frames.

8.5.2 Slepian–Wolf Coding

The DISCOVER WZ video codec has adopted a class of LDPC codes for the Slepian–Wolf part of the WZ codec after using the popular turbo codes for a long time. The DISCOVER codec uses an LDPC Accumulate (LDPCA) codec, which consists of

[1] To be more precise, a separable integer transform with similar properties as a 4×4 DCT is used.

an LDPC syndrome code concatenated with an accumulator [22] to achieve a rate-compatible operation. For each bit-plane, syndrome bits are created using the LDPC syndrome code and accumulated modulo 2 to produce the accumulated syndrome. Basically, the LDPC syndrome encoder calculates for the input bit-plane x the syndrome $s = Hx$, where H corresponds to the parity-check matrix. The parity-check matrix H of the LDPC code represents the connections (i.e., edges) of a bipartite graph with two node types: the variable nodes and the check nodes. It also has a sparse representation, that is, with a low density of 1's, which guarantees a low encoder complexity. When an LDPC syndrome code is concatenated with an accumulator, the resulting concatenated code is rate adaptive and incremental, allowing the use of the request-decode strategy already employed for the turbo codes. Thus, after the calculation of the accumulated syndromes, they are stored in a buffer and transmitted to the decoder in chunks, upon decoder request.

Making use of the side information created at the decoder (see Section 8.5.3), the LDPC syndrome decoder attempts to reconstruct the original data, with the help of the correlation noise model, using a soft-bit decoding belief propagation algorithm that operates in bipartite graph (or equivalently matrix H) with high efficiency. In this case, the popular log-domain sum-product algorithm (SPA) is used with the adaptations described in [21] for LDPC syndrome codes with side information available at the decoder. The SPA algorithm uses an iterative approach and consists of exchanging messages along the edges, from variable nodes initialized with the correlation noise model soft-input information, to check nodes, which are affected by the received syndromes. At each node type, different calculations are performed. Moreover, they always exchange extrinsic information, that is, outgoing messages at an edge do not consider incoming information on the same edge.

If the number of syndrome bits is not enough, WZ decoding fails, the encoder is notified by the feedback channel and more accumulated syndromes are sent from the encoder buffer. To enable this type of rate control, the LDPCA decoder tests the convergence by computing the syndrome check error—that is, by counting the number of parity-check equations that are not satisfied, assuming a certain decoded bit-plane by the SPA. In fact, the syndrome check error is calculated after each LDPC decoding iteration. If the syndrome check error is different from zero and the maximum number of iterations was not reached, another LDPC decoding iteration will start: in this case up to 100 iterations are allowed. After 100 iterations, if the syndrome check error remains different from zero, the decoded bit-plane is assumed to be erroneously decoded, and the LDPCA decoder requests more accumulated syndromes. If at any time the syndrome check error is equal to zero, a CRC mechanism is used to detect if there are remaining errors in the decoded bit-plane. The CRC-8 checksum is computed by the encoder for every encoded bit-plane and transmitted to the decoder. If the decoded bit-plane has the same CRC checksum that the encoder CRC checksum has, the decoding is declared to be successful and the decoding of another band/bit-plane can start. Note also that the inverse of the H matrix is used when all syndrome bits are received (compression ratio of 1:1), allowing recovery of the source, losslessly, since H is full rank and the LDPC syndrome code is linear. This is a significant advantage when the correlation between the side information and original data is low, especially when compared to the turbo codes parity solution, which does guarantee compression ratios lower than 1.

The performance of the LDPC codes depends on several factors, for example, the irregular or regular nature of the graph, but one of the most important factors is the length of cycles in the bipartite graph since decoding algorithms such as SPA can only achieve optimal decoding in cycle-free graphs. In [51], advances were made with a new LDPC code for DVC, which was designed with powerful graph conditioning techniques and a new check node merging technique to obtain an alternative rate compatible strategy (i.e., lower rate codes are obtained from a high-rate base code).

8.5.3 Side Information Creation

This section presents a summary of the side information creation process. This process is rather complex and central to the performance of the Wyner–Ziv codec since it determines how many "errors" have to be corrected through LDPC (or turbo) syndrome (parity) bits. Thus, the efficiency of the techniques developed can significantly influence the rate-distortion performance of the Wyner–Ziv video codec, in the same way as efficient motion estimation and compensation tools have been establishing compression advances for block-based hybrid video coding. In the DISCOVER codec, an advanced motion compensation interpolation (MCI) framework, as proposed in [26, 27], is used to generate the side information. The techniques used work at the block level, as in predictive video coding, in order to capture the local motion in a robust and accurate way. The main objective of the MCI framework is to calculate a piecewise smooth motion field, that is, a motion field that simultaneously captures the local motion caused by moving objects, and the global motion of the scene.

The MCI framework generates the side information, an estimate of the WZ frame, based on two references—one temporally in the past (X_B) and another in the future (X_F), as follows (see Figure 8.5):

1. For a GOP length of 2, X_B and X_F are the previous and the next temporally adjacent decoded key frames to the WZ frame being decoded. For other GOP lengths, the frame interpolation structure definition algorithm proposed in [27]

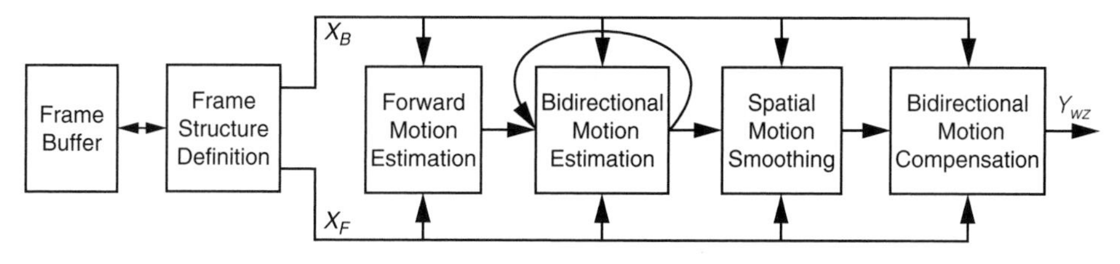

FIGURE 8.5

Architecture of the frame interpolation framework for side information creation.

indicates the decoding order and the reference frames to be used by the frame interpolation algorithm.

2. Both reference frames, X_B and X_F, are first low-pass filtered to improve the reliability of the motion vectors. Then, a block-matching algorithm is used to estimate the motion between the X_B and X_F frames. In this step, full-search motion estimation with modified matching criteria is performed [26]; the criteria include a regularized term that favors motion vectors that are closer to the origin.
3. The bidirectional motion estimation module refines the motion vectors obtained in the previous step with an additional constraint: the motion vector selected for each block has a linear trajectory between the next and previous reference frames and crosses the interpolated frame at the center of the blocks. This technique combines a hierarchical block-size technique with a new adaptive search range strategy in a very efficient way [27]. The hierarchical coarse-to-fine approach tracks fast motion and handles large GOP sizes in the first iteration (block size 16×16) and then achieves finer detail by using smaller block sizes (block size 8×8).
4. Next, a spatial motion smoothing algorithm based on weighted vector median filters [26] is used to make the final motion vector field smoother, except at object boundaries and uncovered regions.
5. Once the final motion vector field is obtained, the interpolated frame can be filled by simply using bidirectional motion compensation as defined in standard video coding schemes.

This type of advanced frame interpolation solution has a major contribution to the good RD performance of the DISCOVER WZ video codec since the quality of the generated side information has a crucial role in the overall codec RD performance.

8.5.4 Correlation Noise Modeling

To make good usage of the side information obtained through the frame interpolation framework, the decoder needs to have reliable knowledge of the model that characterizes the correlation noise between corresponding DCT bands of the WZ and side information frames. The Laplacian distribution is widely used to model the residual statistics between correspondent coefficients in X_{WZ}^{DCT} and Y_{WZ}^{DCT}, for example, [14], and, thus, it was also adopted by the DISCOVER WZ video codec. The Laplacian distribution parameter α is estimated on-line at the decoder, for each DCT coefficient, using (2) below, based on the residual, R, between the reference frames X_B and X_F used to create the side information after motion compensation [32]. Basically, after applying the 4×4 DCT transform over the R frame, each DCT coefficient is classified into one of two classes: (1) *inlier* coefficients corresponding to those whose absolute value (magnitude) is close to the corresponding *R* frame DCT band magnitude average value $\hat{\mu}_{|b|}$; and (2) *outlier* coefficients corresponding to those whose magnitude is far away from $\hat{\mu}_{|b|}$. To determine how close a certain coefficient is to the corresponding *R* frame DCT band magnitude average value, the distance between that coefficient magnitude and $\hat{\mu}_{|b|}$ with the DCT band magnitude variance $\hat{\sigma}^2_{|b|}$ is compared since the variance is a measure of how spread the coefficient values are regarding its average value. The α parameter estimation in (2) leads to a finer adaptation of the correlation noise model, both spatially (within a frame) and temporally (along the video sequence):

$$\hat{\alpha}_b(u,v) = \begin{cases} \sqrt{\frac{2}{\hat{\sigma}^2_b}} & , \ [D_{|b|}(u,v)]^2 \leqslant \hat{\sigma}^2_{|b|} \\ \sqrt{\frac{2}{[D_{|b|}(u,v)]^2}} & , \ [D_{|b|}(u,v)]^2 > \hat{\sigma}^2_{|b|} \end{cases}. \tag{8.2}$$

In Equation (8.1), $\hat{\alpha}_b(u,v)$ is the α parameter estimate for the DCT coefficient located at (u,v) position, $\hat{\sigma}^2_{|b|}$ is the estimate of the variance for the DCT band to which the DCT coefficient belongs, and $D_{|b|}(u,v)$ represents the distance between the (u,v) coefficient magnitude and the DCT coefficients band b magnitude average value. The distance $D_{|b|}(u,v)$ measures how spread the coefficient's values are regarding its corresponding DCT band average value [32]. The first branch in (8.2) corresponds to a block/region well interpolated. Thus, the estimate of the α parameter for the *R* frame DCT band is a more reliable estimation in this case. The second branch in (8.2) corresponds to a block where the residual error is high, which means that the SI generation process failed for that block. Thus, the α parameter estimate for the *R* frame DCT band is not the best approach in this case, since it corresponds to an average value for all DCT coefficients within the band b. In this case, the solution adopted by the DISCOVER codec is to estimate $\hat{\alpha}_b(u,v)$ using $[D_{|b|}(u,v)]^2$ since this enables giving less confidence to the turbo decoder on the DCT coefficients that belong to blocks where the frame interpolation algorithm essentially failed.

8.5.5 Reconstruction

The LDPC (or turbo) decoded bit-planes, together with the side information and the residual statistics for each DCT coefficient band, are used by the reconstruction to obtain the decoded DCT coefficients matrix, ${X'_{WZ}}^{DCT}$ as in [33]. Consider that the M_k bit-planes associated with each DCT coefficients band, for which WZ bits were received, are successfully decoded. For each band, the bit-planes are grouped and a decoded quantization symbol (bin) q' is obtained for each DCT coefficient, guiding the decoder about where the original DCT coefficient value lies (an interval). The decoded quantization bin q' corresponds to the true quantization bin q, obtained at the encoder before bit-plane extraction, if all errors in the decoded bit-planes were corrected (however, a very small error probability is allowed). The reconstruction function is optimal in the sense that it minimizes the mean-squared error (MSE) of the reconstructed value, for each DCT coefficient, of a given band and is given by [33]

$$x' = E\left[x|q', y\right] = \frac{\int_l^u x f_{X|y}(x|y)dx}{\int_l^u f_{X|y}(x|y)dx} \tag{8.3}$$

where x' is the reconstructed DCT coefficient, y is the corresponding DCT coefficient of $Y_{WZ}{}^{DCT}$, $E[.]$ is the expectation operator, and l and u represent the lower and the upper bounds of q', respectively. In (8.3), the conditional probability density function $f_{X|y}(.)$ models the residual statistics between corresponding coefficients in $X_{WZ}{}^{DCT}$ and $Y_{WZ}{}^{DCT}$. According to Section 8.3, $f_{X|y}(.)$ is typically assumed to be a Laplacian distribution. After some analytical manipulations in (8.3), the reconstructed DCT coefficient can be obtained from (8.4) where Δ corresponds to the quantization bin size [33]. In (8.4), α is the Laplacian distribution parameter estimated on-line at the decoder for each DCT coefficient (see Section 8.5.4).

$$x' = \begin{cases} l+b & , \quad y<l \\ y + \frac{(\gamma+\frac{1}{\alpha})e^{-\alpha\gamma} - (\delta+\frac{1}{\alpha})e^{-\alpha\delta}}{2-e^{-\alpha\gamma}-e^{-\alpha\delta}} & , \quad y \in [l, u[\quad \text{with} \quad b = \frac{1}{\alpha} + \frac{\Delta}{1-e^{\alpha\Delta}}, \gamma = y-l, \delta = u-y \\ u-b & , \quad y \geqslant u \end{cases} \tag{8.4}$$

As can be seen in (8.4), the reconstruction function shifts the reconstructed DCT coefficient value towards the center of the decoded quantization bin. The DCT coefficients bands to which no WZ bits are sent are replaced by the corresponding DCT bands of the side information.

8.6 THE DISCOVER CODEC PERFORMANCE

The target of this section is to perform a detailed evaluation of the DISCOVER WZ video codec. With this purpose, precise test conditions are first defined. Afterward, the most relevant performance metrics to evaluate a WZ video codec will be assessed, notably: (i) RD performance; (ii) feedback channel rate performance; and (iii) codec complexity. For a more detailed performance evaluation, please see [4].

8.6.1 Performance Evaluation Conditions

Considering the main objective of this section, it is essential to meaningfully define the performance evaluation conditions and the video codec control parameters in order to set a clear and useful benchmarking. As usual in the DVC literature, only the luminance component is coded, and thus all metrics in this section refer only to the luminance. However, there are no reasons preventing use of the DISCOVER WZ codec for the chrominance components with similar performance.

8.6.1.1 Test Material

The video test material used and some relevant test conditions are described in the following:

- **Sequences:** Foreman (with the Siemens logo), Hall Monitor, Coast Guard, and Soccer; these sequences represent different types of content, (see Figure 8.6).
- **Frames:** all frames (this means 299 for Foreman, 329 for Hall Monitor, 299 for Coast Guard, and 299 for Soccer at 30 Hz).
- **Spatial resolution:** QCIF.
- **Temporal resolution:** 15 Hz, which means 7.5 Hz for the WZ frames when GOP = 2 is used.
- **GOP length:** 2 (if not otherwise indicated), 4 and 8.

Although the most frequently used test conditions in the literature employ QCIF at 30 Hz, it was decided to use QCIF at 15 Hz since this combination is more realistic from a practical point of view. This decision penalizes the WZ video coding results reported in this chapter in comparison to standard video codecs since the side information tends to be poorer owing to the longer time gaps between the key frames.

8.6.1.2 Quantization

Different (decoded) quality can be achieved by changing the number of quantization levels, M_k, used for each DCT band b_k. In this section, eight rate-distortion (RD) points are considered, corresponding to the various 4×4 quantization matrices depicted in Figure 8.7. Within a 4×4 quantization matrix, the value at position k in Figure 8.7 (position numbering within the matrix is made in a zigzag scanning order) indicates the number of quantization levels associated with the DCT coefficients band b_k. The value 0 means that no Wyner–Ziv bits are transmitted for the corresponding band. In the following, the various matrices will be referred to as Q_i with $i = 1, \ldots, 8$; the higher is Q_i, the higher are the bit rate and the quality.

8.6.1.3 Side Information Creation Process

This section presents the most important control parameters related to the side information creation process described previously. In the MCI framework, the forward

(a)

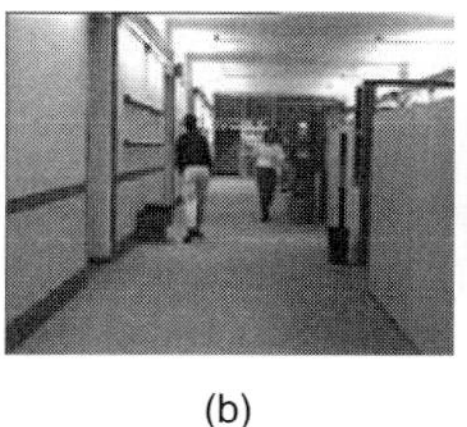
(b)

(c)

(d)

FIGURE 8.6

Sample frames for test sequences. (a) Foreman (frame 80); (b) Hall Monitor (frame 75); (c) Coast Guard (frame 60); (d) Soccer (frame 8); remind that only luminance is coded.

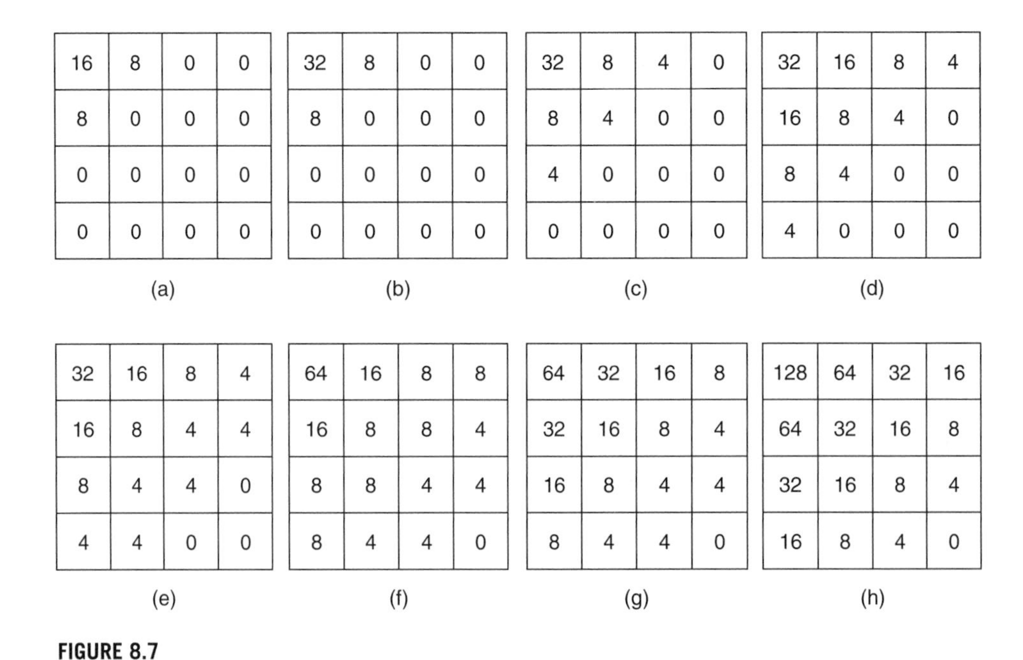

FIGURE 8.7

Eight quantization matrices associated with different RD points.

motion estimation works with 16×16 block sizes, and ± 32 pixels are used for the search range. In the second iteration of the bidirectional motion estimation, the motion vectors are refined using 8×8 block sizes, and an adaptive search range is used to restrict the motion vector trajectory to be close to the neighboring motion vectors while still leaving room to increase its accuracy and precision. The motion search is performed using the half-pixel precision, and the reference frames are first low-pass filtered with the mean filter using a 3×3 mask size.

8.6.1.4 *Key Frames Coding*

The key frames are always encoded with H.264/AVC Intra (Main profile) because this is the best performing Intra video coding standard available. The key frames are coded with constant quantization parameters (QP) as defined in Table 8.1. The key frames quantization parameters have been found using an iterative process that stops when the average quality (PSNR) of the WZ frames is similar to the quality of the Intra frames (H.264/AVC coded). The values in Table 8.1 have been obtained for GOP = 2, QCIF@15 Hz. The selection of these QP values for the key frames was made targeting to have almost constant (on average) decoded video quality for the full set of frames (key frames and WZ frames).

Investing the same total bit rate in a different way between WZ and key frames may lead to a better overall RD performance. For example, by investing more bits in the key frames, the overall RD performance may improve at the cost of a less stable (in time)

Table 8.1 Key Frames Quantization Parameters for the Various RD Points for QCIF@15 Hz

	Q_1	Q_2	Q_3	Q_4	Q_5	Q_6	Q_7	Q_8
Foreman	40	39	38	34	34	32	29	25
Hall Monitor	37	36	36	33	33	31	29	24
Coast Guard	38	37	37	34	33	31	30	26
Soccer	44	43	41	36	36	34	31	25

video quality. This is a choice to be made by the encoder knowing the consequences in terms of subjective impact. For comparison purposes, the QP values in Table 8.1 will be used all along this section under the assumption stated above.

8.6.2 RD Performance Evaluation

This section targets RD performance evaluation. This means the assessment of how the rate for the key frames (H.264/AVC Intra bits) and the rate for the WZ frames (LDPC syndrome bits + CRC bits) translates into quality. All metrics regard both the key frames and WZ frames bits since they correspond to the overall decoded quality. In this section, error-free channels are considered, but it is well acknowledged that DVC solutions are especially interesting when error resilience is a critical requirement. A detailed error resilience performance evaluation is available in [45], and a detailed performance evaluation as a function of the GOP size is available in [52].

8.6.2.1 Measuring the Overall Rate-distortion Performance

Defining the Metrics

Although many metrics are relevant to evaluate the compression efficiency performance, it is recognized that the most frequently used quality metric is the average PSNR (with all the well known limitations it implies) over all the frames of a video sequence coded for a certain quantization matrix (as defined in Section 8.6.1.2). When the (luminance) PSNR metric is represented as a function of the used bit rate—in this case, the overall bit rate for the luminance component—very important performance charts are obtained since they allow to easily compare the overall RD performance with other coding solutions, including standard coding solutions largely well known and used.

In this section, the RD performance of the DISCOVER WZ codec is compared with the corresponding performance of three standard coding solutions that share an important property in terms of encoder complexity: the complex and expensive motion estimation task is not performed by any of them. Section 8.6.3 will present the codec complexity evaluation that complements the RD evaluation proposed in this section.

The three state-of-the-art standard solutions used in this section for comparison purposes are:

- **H.263+ Intra** H.263+ video coding without exploiting temporal redundancy; this "Intra comparison" is still the one that appears most in the DVC literature but H.263+ Intra is clearly no longer the best standard Intra coding available and thus "beating" H.263+ Intra is much easier than "beating" H.264/AVC Intra (of course, their encoding complexity is also rather different).
- **H.264/AVC Intra** H.264/AVC video coding in Main profile without exploiting the temporal redundancy; this type of Intra coding is the most efficient Intra (video) coding standard solution, even more than JPEG2000 for many conditions. However, it is important to notice that H.264/AVC Intra (JM9.5 reference software) exploits quite efficiently the spatial correlation (at a higher encoding complexity cost when compared to H.263+ Intra) with several 4×4 and 16×16 Intra spatial prediction modes, a feature that is (still) missing in the DISCOVER WZ codec.
- **H.264/AVC Inter No Motion** Video coding with H.264/AVC in Main profile (JM9.5 reference software) exploiting temporal redundancy in a IB...IB... structure (depending on GOP size) but without performing any motion estimation, which is the most computationally expensive encoding task. The so-called No Motion coding mode typically achieves better performance than Intra coding because it can partly exploit the temporal redundancy by using a DPCM temporal scheme. Moreover, it requires far less complexity than full-motion compensated Inter coding because no motion search is performed. This type of comparison (excluding encoder motion estimation as in WZ coding) is not typically provided in most DVC published papers because it is still difficult to "beat" the RD performance of H.264/AVC Inter No Motion with DVC-based solutions.

Results and Analysis

Figures 8.8 and 8.9 show the RD performance, according to the test conditions, for the four test video sequences selected; also an RD performance comparison for GOP sizes 2, 4, and 8 is presented. For these results, the following conclusions can be drawn:

- **DISCOVER codec vs. H.263+ Intra Coding** For the Coast Guard and Hall Monitor sequences, coding gains exist for all RD points and for all GOP sizes with average gains up to 9 dB (for the Hall Monitor sequence at GOP = 4) when compared to H.263+ Intra. Therefore, it is possible to conclude that the DISCOVER codec can exploit (at least partly) the temporal correlation in the video content. On the other hand, for the Foreman sequence, the gains are not as impressive for GOP = 2 (3 dB) and decrease when the GOP size increases; for GOP = 8, a minor coding loss is observed. The major reason is that for content with high and medium motion, when the key frames are separated by a long gap in time, the side information quality decreases (and thus the overall RD performance) since

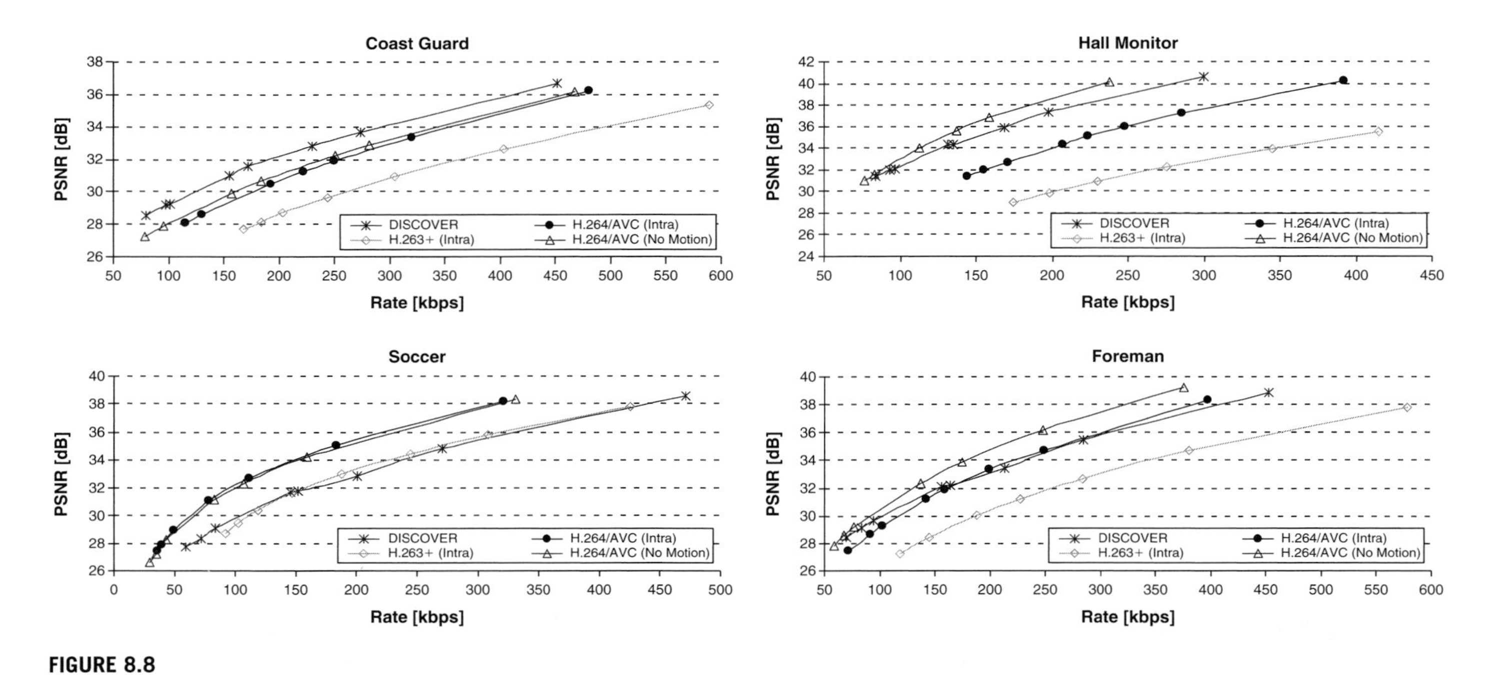

FIGURE 8.8

RD performance for GOP = 2: Coast Guard, Hall Monitor, Soccer, and Foreman (QCIF at 15 Hz).

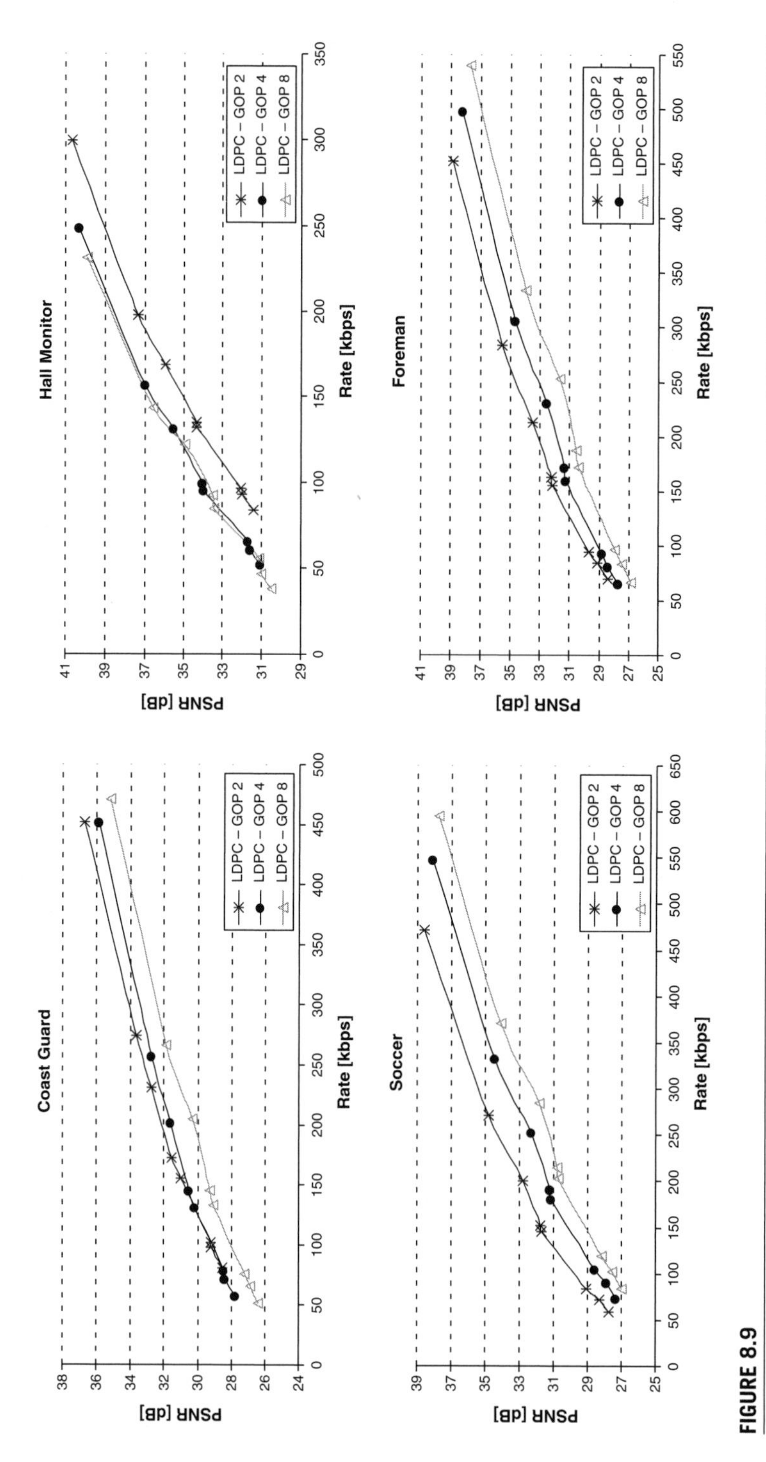

FIGURE 8.9

RD performance comparison for GOP = 2, 4, and 8: Coast Guard, Hall Monitor, Soccer, and Foreman (QCIF at 15 Hz).

it is more difficult to predict the frames in between. The worst performance is achieved for the Soccer sequence, for all GOP sizes, since it is always below H.263+ Intra. The major reason for this loss is the complex and erratic motion that normally leads to a significant amount of errors in the side information and causes a poorer RD performance, especially when the GOP size is quite large (8 frames at 15 fps correspond to more than half a second in a Soccer game). It is important to notice that better RD performance results would be obtained for 30 Hz, which is the most reported condition in the literature.

- **DISCOVER codec vs. H.264/AVC Intra Coding** In this case, the DISCOVER codec is compared to the best performing (video) Intra encoder available, the H.264/AVC Intra codec. The Hall Monitor sequence has the best performance, with consistent gains for all GOP sizes (up to 4 dB). The Coast Guard sequence at GOP = 2 has gains up to 1 dB, and some RD points for the Foreman sequence (GOP = 2) show minor gains. The rest of the sequence@GOP size combinations have a lower performance when compared to H.264/AVC Intra; in these cases, the difficulty is to obtain good side information for high motion sequences with large GOP sizes. However, the results are quite encouraging since for a significant amount of (sequence@GOP size) scenarios, the DISCOVER RD performance beats H.264/AVC Intra; notably, for GOP = 2, the DISCOVER codec only loses for the Soccer sequence, which is a rather difficult sequence to be coded at 15 Hz. It is also important to note that the H.264/AVC Intra codec includes several spatial prediction modes, RD optimization, and advanced entropy coding tools as CABAC in order to achieve a higher RD performance (at some encoding complexity cost). So, the conclusion to be confirmed in Section 8.6.3 is that the DISCOVER codec can, for most cases at GOP = 2, achieve better performance with lower encoding complexity.
- **DISCOVER codec vs. H.264/AVC Inter No Motion Coding** In this case, the DISCOVER codec is compared to a low complexity Inter codec (compared to full H.264/AVC Inter) that is still able to exploit some temporal correlation. An interesting observation is that the H.264/AVC No Motion curve is very close to the H.264/AVC Intra curve for the Soccer and Coast Guard sequences (for all GOP sizes) and for the Foreman sequence at GOP = 8. Since for these cases the temporal correlation is quite low, the H.264/AVC No Motion solution selects the Intra mode for the majority of the blocks (mode decision Intra/Inter is still allowed). The DISCOVER codec is quite competitive for the Coast Guard sequence where some gains (up to 1 dB) in comparison to H.264/AVC No Motion are observed. In such cases, the DISCOVER side information creation framework exploits very efficiently the temporal correlation. However, for the Hall Monitor sequence, an RD loss is observed, especially for higher bit rates, where the H.264/AVC SKIP macroblock mode (no information other than the mode is coded) brings additional benefits for this low-motion sequence.
- **DISCOVER codec RD performance for different GOP Sizes** For all sequences but Hall Monitor, GOP = 2 wins, showing the difficulty in getting good side information for longer GOP sizes due to the decreased performance

of the frame interpolation process. For Hall Monitor, which is a rather stable sequence, GOP = 4 and GOP = 8 are already more efficient than GOP = 2 since motion estimation and frame interpolation are more reliable.

8.6.2.2 Comparing the RD Performance of LDPC and Turbo Codes

During most of the DISCOVER project, the Slepian–Wolf codec part of the video codec was based on turbo codes. Later, turbo codes were substituted by LDPC codes [22], essentially owing to their better RD performance and low complexity, especially at the decoder side.

The turbo encoder enclosed a parallel concatenation of two identical constituent recursive systematic convolutional (RSC) encoders of rate ½, and a pseudo-random L-bit interleaver is employed to decorrelate the L-bit input sequence between the two RSC encoders. The pseudo-random interleaver length L corresponds to the DCT coefficients band size, that is, the ratio between the frame size and the number of different DCT coefficients bands. Each RSC encoder outputs a parity stream and a systematic stream. For the turbo encoder, each rate ½ RSC encoder is represented by the generator matrix $\left[1\ \frac{1+D+D^3+D^4}{1+D^3+D^4}\right]$ and the trellis of the second RSC encoder is not terminated. The puncturing period P is 48, which allows a fine-grained control of the bit rate. After turbo encoding a bit-plane of a given DCT coefficients band (starting with the most significant bit-plane), the systematic part (a copy of the turbo encoder input) is discarded, and the parity bits are stored in the buffer. Upon decoder request, the parity bits produced by the turbo encoder are transmitted according to a pseudo-random puncturing pattern. In each request, the WZ encoder sends one parity bit in each P parity bit from each RSC encoder parity stream; each parity bit is sent only once. The location of the parity bits to be sent in each request is pseudo-randomly generated (it is known at both the encoder and decoder).

The Slepian–Wolf decoder encloses an iterative turbo decoder constituted by two soft-input soft-output (SISO) decoders. Each SISO decoder is implemented using the Logarithmic *Maximum A Posteriori* (Log-MAP) algorithm. Since an iterative turbo decoding is used, the *extrinsic* information computed by one SISO decoder becomes the a priori information of the other SISO decoder. Through this information exchange procedure between the two SISO decoders, the a posteriori estimates of the data bits are updated until the maximum number of iterations allowed for iterative decoding is reached. A confidence measure based on the a posteriori probabilities ratio (check [38] for more details) is used as an error-detection criterion to determine the current bit-plane error probability P_e of a given DCT band. If P_e is higher than 10^{-3}, the decoder requests for more parity bits from the encoder via the feedback channel; otherwise, the bit-plane turbo decoding task is considered successful.

Defining the Metrics

This section compares the RD performances of the turbo codes-based and LDPC codes-based DISCOVER WZ codecs for equal conditions in terms of all other modules.

Results and Analysis

Figure 8.10 shows the RD performance comparison between the DISCOVER WZ video codecs using turbo codes and LDPC codes. Observing the RD performance curves shown in Figure 8.10, it can be noticed that the LDPC codes always have better performance than turbo codes for all GOP sizes and all sequences. The following conclusions can be drawn:

- For lower bit rates, the performance of the turbo and LDPC codes is quite similar because when the correlation between the side information and the WZ frames is high, the turbo codes achieve a good performance and similar results to the LDPC codes.
- For medium and high bit rates, the LDPC codes have a better performance when compared to the turbo codes, with coding gains up to 35 kbps for GOP = 8. The LDPC codes show better performance for the bands/bit-planes, which have a low correlation between the side information and the WZ frame (i.e., for side information with low quality).
- When the GOP size increases, the performance gap between the turbo codes and the LDPC codes increases, with a clear advantage for the LDPC codes (for GOP = 2, 4, 8 coding gains up to 10, 27, 35 kbps occur, respectively). One major reason for this effect is that the LDPCA codes (as other types of LDPC codes) always have a maximum rate of 1, that is, the maximum number of syndrome bits sent cannot exceed the number of bits that represent the original data. This property comes from the code linear properties, which allow recovering the original source information when the rate is equal to 1 by solving the parity-check matrix equations. This property does not exist for the iterative turbo codes (rate expansion is possible), and it is responsible for the turbo codes loss of efficiency, especially at larger GOP sizes, where the correlation between WZ and SI is lower and thus compression rates higher than 1 are necessary to guarantee successful decoding.

8.6.2.3 Measuring the Bit-plane Compression Factor

Defining the Metrics

The LDPCA syndrome encoder consists of an irregular LDPC encoder concatenated with an accumulator. The encoder buffers the accumulated syndrome and transmits it incrementally to the decoder, following the decoder requests. The total number of parity bits per bit-plane created by the LDPCA encoder is equal to the number of the input bit-plane bits. This way the compression factor can only be higher than 1 (and never lower than 1 as can happen for turbo codes); it allows measuring the capability of the DISCOVER WZ codec to compress the source data (bit-planes). The total average compression factor at frame, CF_Q, and bit-plane, CF_{Qij}, levels for a certain quality level, Q, is given by (8.5) and (8.6):

$$CF_Q = \frac{\sum_{i=1}^{B_Q} \sum_{j=1}^{M_i} \sum_{l=1}^{N} \frac{c_{ijl}}{w_{ijl}}}{N} \tag{8.5}$$

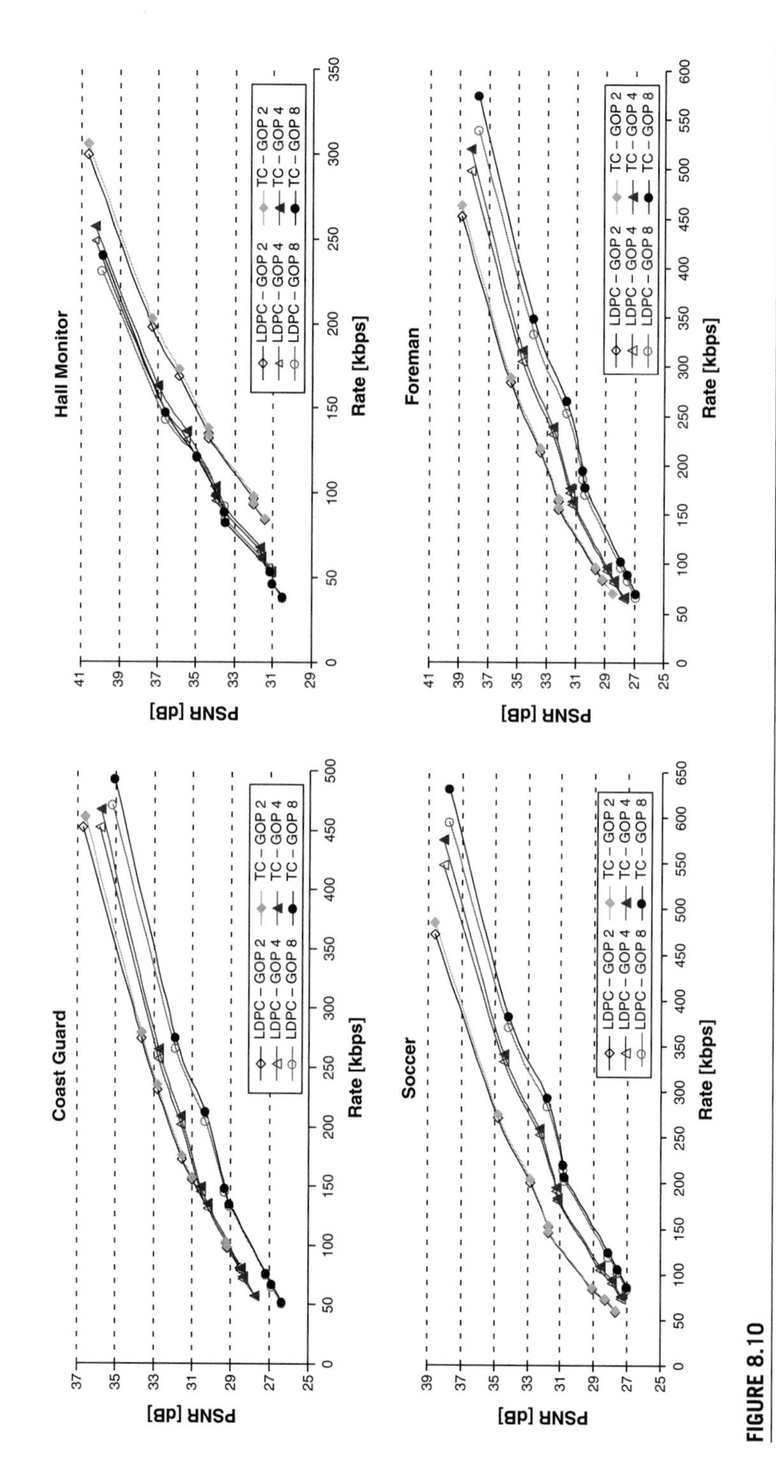

FIGURE 8.10

RD performance comparison using turbo codes and LDPC codes.

$$CF_{Qij} = \frac{\sum_{l=1}^{N} \frac{C_{ijl}}{w_{ijl}}}{N} \tag{8.6}$$

where M_i is the number of bit-planes of each band i, N the total number of WZ frames, C_{ijl} the number of bits in each original coefficient bit-plane j of each band i at frame l, and w_{ijl} is the amount of parity bits sent for each bit-plane j of band i at frame l, B_Q is the number of bands considering the quality level, Q. CF_{Qij} given by (8.6) represents the average compression factor at bit-plane j of band i for a certain quality level Q.

Performing a study of the bit-plane compression factors for the various DCT coefficients bands is extremely useful because it allows detecting where inefficiencies are located to develop more advanced solutions. For example, it may allow locating bands for which WZ coding is not efficient due to the lack of temporal correlation; this approach is followed by the PRISM codec [16].

Results and Analysis

Figure 8.11 shows the average bit-plane compression factor per band for $Q_i = 4$ (medium bit rates) and 8 (high bit rates) for GOP = 2 and for the four test video sequences. The following conclusions can be inferred:

- In a general way, the least significant bit-planes (LSBs) have lower compression factors when compared to the most significant bit-planes (MSB and MSB-1). This behavior is observed because the correlation between the side information and the original bit-planes decreases when the bit-plane number index increases, especially for bands for which syndrome bits are sent for a high number of bit-planes (e.g., DC and AC1-2). This suggests that more advanced quantization and/or correlation noise models can further improve the performance in such cases. Note that the compression factor is always greater than 1 and that bit rate expansion does not occur for LDPC codes.
- In a general way, for the AC bands, the highest compression factor is achieved for the MSB-1. In this case, the decoder makes use of the a priori information about the decoding of the MSB (obtained through the correlation model). Since the MSB-1 and MSB are quite correlated, it is possible to achieve higher compression factors. This correlation is lower for other bit-planes, and, therefore, lower compression ratios are achieved.
- The last AC bands usually have high compression ratios for both bit-planes because they only have four wide bins (2 bit-planes); therefore, the correlation between the side information and the WZ frame is high, and large compression ratios can be achieved. When more bit-planes are considered in each band, the number of bins increases (and thus they are smaller) and more errors occur; that is, the mismatch between the side information bin and the WZ frame bin increases (a low correlation is observed), decreasing the compression factor.

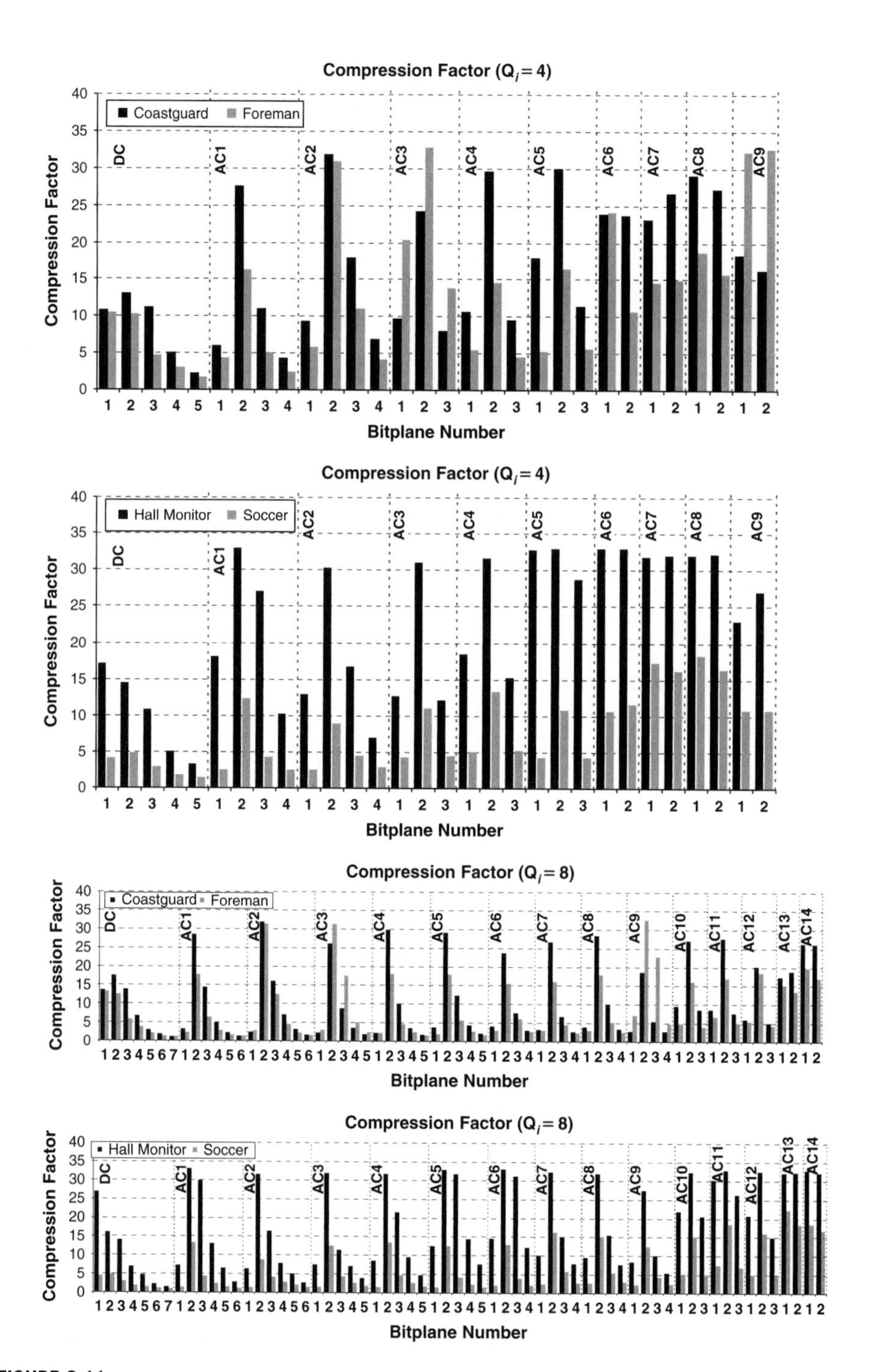

FIGURE 8.11

Bit-plane compression factor for $Q_i = 4$ and 8 (GOP = 2).

8.6.2.4 *Measuring the Feedback Channel Rate*

In the DISCOVER WZ video coding architecture, the feedback channel has the role to adapt the bit rate to the changing statistics between the side information (an estimation of the frame to be encoded) and the WZ frame to be encoded, that is, to the quality (or accuracy) of the frame interpolation process. Therefore, contrary to conventional codecs, it is the decoder's responsibility to perform rate control and, in this way, to guarantee that only a minimum number of parity bits are sent to correct the mismatches/errors present in each side information bit-plane. The DISCOVER WZ codec does not use a pure decoder rate control but rather a hybrid rate control-approach, which means that fewer requests are made, lowering the feedback channel rate. Still, it is important to be aware of the feedback channel usage and impact in order to design more efficient and lower complexity WZ video codecs.

Defining the Metrics

In order to measure the feedback channel rate for each DCT band and bit-plane, it is assumed that only one bit is required by the decoder to inform the encoder if more parity bits are needed to successfully decode the current bit-plane. If more parity bits are needed, the decoder sends the bit "1" via the feedback channel; otherwise, the bit "0" is transmitted, and the encoder, receiving such bit, sends parity bits for the next bit-plane to be decoded. This is clearly a simplistic solution since more bits are needed, depending on the transmission protocol used; however, this simple approach is enough to have an idea on the feedback channel rate.

Since only one bit is transmitted via the feedback channel for each decoder request, the total feedback channel rate at frame R_Q, and bit-plane R_{Qij} levels for a certain quality level Q, can be obtained from (8.7) and (8.8), respectively.

$$R_Q = \frac{\sum_{i=1}^{B_Q} \sum_{j=1}^{M_i} \sum_{l=1}^{N} n_{ijl}}{N} \times f \tag{8.7}$$

$$R_{Qij} = \frac{\sum_{l=1}^{N} n_{ijl}}{N} \times f \tag{8.8}$$

In (8.7) and (8.8), f is the WZ frame rate and n_{ijl} is the number of bits sent via the feedback channel for jth bit-plane of WZ frame l, DCT band i; N is the total number of WZ frames, M_i is the number of bit-planes for DCT band i, and B_Q is the number of bands considering a certain quality level, Q. R_{Qij} is a partial result of (8.7), representing the average feedback channel rate per frame for bit-plane j of DCT band i for a certain quality level Q.

Results and Analysis

Figure 8.12 shows the average feedback channel rate for $Q_i = 4$ and 8, for the four selected video sequences, and for GOP = 2. As can be observed, the number of bits sent through the feedback channel, for each band, increases with the bit-plane number

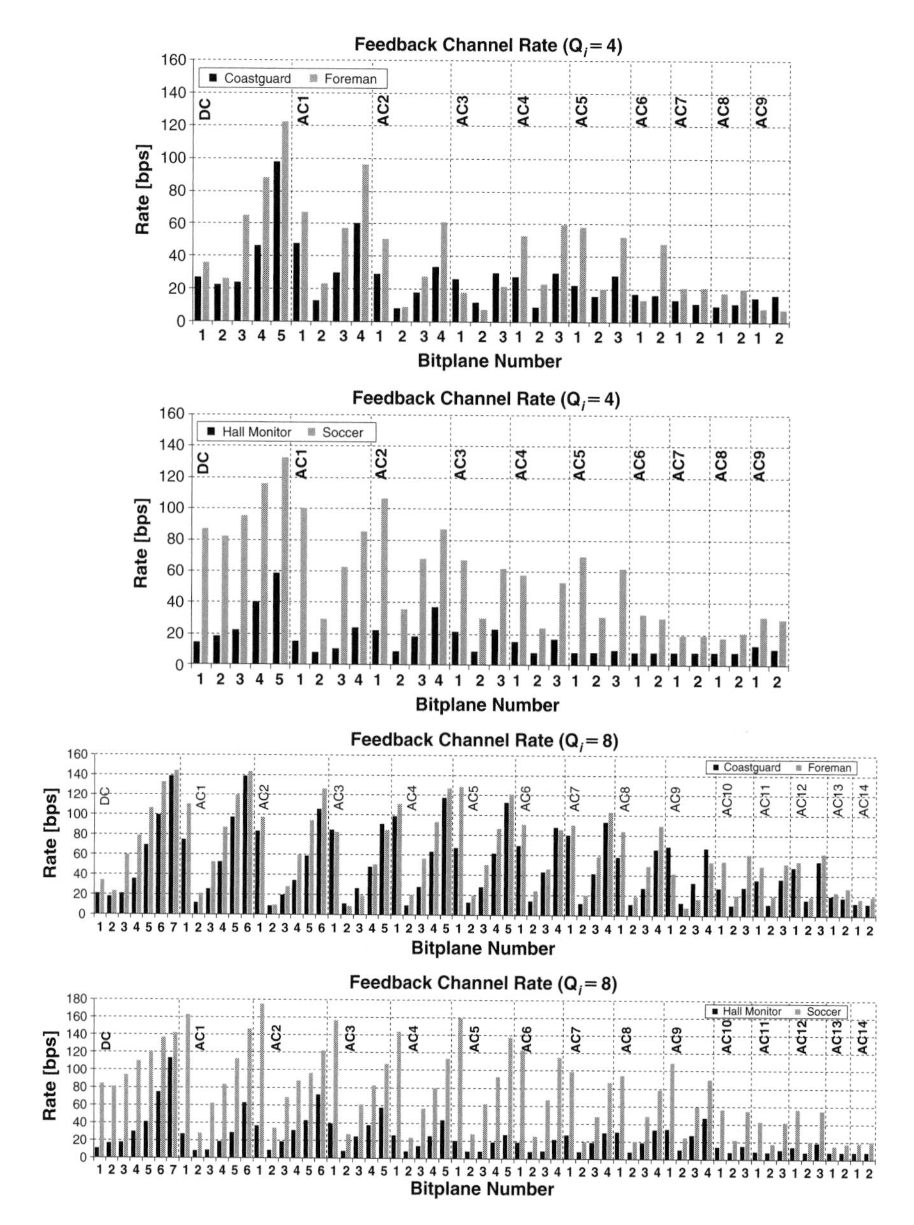

FIGURE 8.12

Feedback channel rate for $Q_i = 4$ and 8 (GOP = 2).

to be decoded; the lower correlation between the side information and the WZ frame for LSB bit-planes is the main reason for that behavior.

Table 8.2 to Table 8.5 show the total feedback rate for the selected test sequences for each Q_i for GOP = 2, 4, and 8, respectively. It can easily be seen that the feedback

Table 8.2 Feedback Channel Rate (bps) for the Coast Guard Sequence

Q_i	GOP 2	GOP 4	GOP 8
1	132.59	327.36	551.85
2	179.94	437.61	697.66
3	216.34	525.05	853.44
4	377.68	996.63	1634.79
5	403.73	1075.75	1757.23
6	593.47	1565.25	2510.35
7	829.51	2150.83	3323.16
8	1514.88	3917.22	5835.58

Table 8.3 Feedback Channel Rate (bps) for the Hall Monitor Sequence

Q_i	GOP 2	GOP 4	GOP 8
1	109.60	222.51	321.82
2	146.09	294.29	417.55
3	173.04	362.08	574.15
4	263.69	606.95	978.49
5	289.79	665.22	1104.34
6	410.73	942.56	1485.84
7	512.53	1181.35	1761.99
8	821.26	1937.46	2804.30

Table 8.4 Feedback Channel Rate (bps) for the Soccer Sequence

Q_i	GOP 2	GOP 4	GOP 8
1	365.68	875.29	1250.50
2	430.33	1031.28	1455.37
3	498.08	1207.75	1723.70
4	857.81	2094.51	2974.20
5	919.06	2238.27	3175.56
6	1211.20	2925.92	4137.77
7	1542.28	3720.81	5242.56
8	2372.39	5663.22	7951.66

Table 8.5 Feedback Channel Rate (bps) for the Foreman Sequence

Q_i	GOP 2	GOP 4	GOP 8
1	222.49	580.70	889.67
2	284.64	722.59	1087.97
3	333.48	850.91	1282.86
4	590.37	1555.05	2367.78
5	663.12	1730.02	2629.46
6	916.41	2366.74	3553.20
7	1181.90	3065.26	4609.26
8	1937.21	4935.68	7266.50

channel rate is rather negligible; the maximum feedback channel rate corresponds to less than 8 kbps for $Q_i = 8$, for GOP = 8, for the sequence Soccer (as expected since Soccer has the worse side information due to the high motion). Although it is not shown, the hybrid encoder/decoder rate control plays a central role, lowering the feedback channel rate when compared to a pure decoder rate control solution.

8.6.3 Complexity Performance Evaluation

Because evaluating the RD performance addresses only one side of the problem, this section intends to perform an evaluation of the complexity performance for the DISCOVER WZ video codec. Although it is commonly claimed that DVC "encoding complexity is low" and "decoding complexity is high," few complexity evaluation results are available in the literature.

While it is possible to measure the encoding and decoding complexities in many ways, some of them rather sophisticated, it is also possible to get a rather good estimation of relative complexities using rather simple complexity metrics. Here, the encoding and decoding complexities will be measured by means of the encoding and decoding times for the full sequence, in seconds, under controlled conditions. It is well known that the encoding (and decoding) times are highly dependent on the used hardware and software platforms. For the present results, the hardware used was an x86 machine with a dual core Pentium D processor at 3.4 GHz with 2048 MB of RAM. Regarding the software conditions, the results were obtained with a Windows XP operating system, with the C++ code written using version 8.0 of the Visual Studio C++ compiler, with optimizations parameters on, such as the release mode and speed optimizations. Besides the operating system, nothing was running in the machine when gathering the performance results to avoid influencing them. Under these conditions, the results have a relative and comparative value, in this case allowing comparing the DISCOVER WZ video codec with alternative solutions, for example, H.264/AVC JM9.5 reference software, running in the same hardware and software conditions. While the degree of optimization of the software has an impact on the running time, this is a dimension that was impossible to fully control in this case and thus will have to be kept in mind when dealing with the provided performance results.

8.6.3.1 Measuring the Encoding Complexity

This section targets the complexity evaluation of the encoding process. The encoding complexity includes two major components: the WZ frames encoding and the key frames encoding parts. The larger the GOP size, the smaller the number of key frames coded and, thus, the lower will be the share of the key frames in the overall complexity.

Defining the Metrics

Although it is possible to measure the encoding complexity in many ways, some of them rather sophisticated, it is also possible to get a rather good "feeling" of relative complexities using rather simple complexity metrics. In this section, the encoding complexity will be measured by the encoding time for the full sequence, in seconds, under the conditions stated above.

Results and Analysis

Figure 8.13 and Table 8.6 to Table 8.9 show the encoding complexity results for various GOP sizes measured in terms of encoding time, distinguishing between key

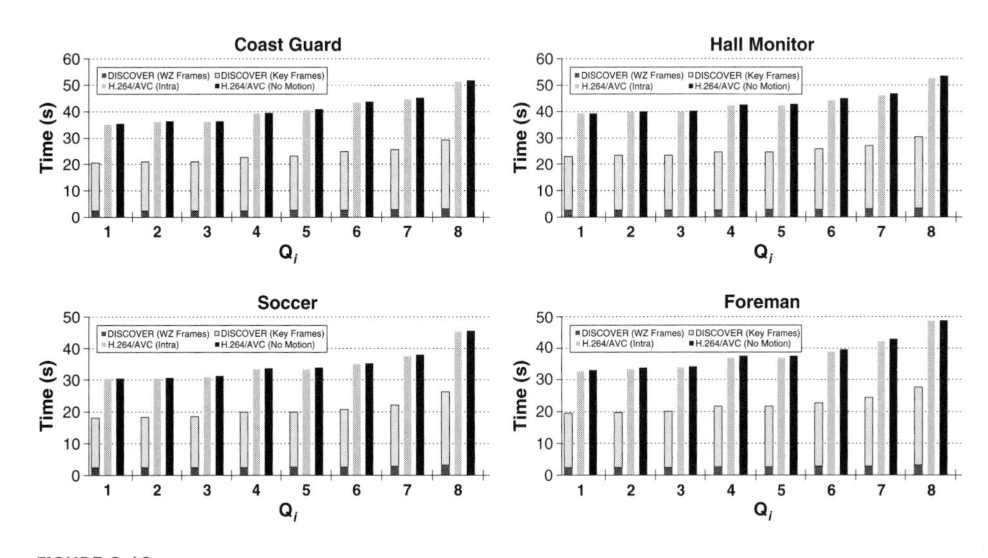

FIGURE 8.13

Encoding complexity measured in terms of encoding time (GOP = 2).

frames (blue) and WZ frames (red) encoding times. The results allow concluding that:

- For the DISCOVER WZ codec, the WZ frames encoding complexity is negligible when compared to the key frames encoding complexity, even for GOP = 2. The DISCOVER encoding complexity is always much lower than the H.264/AVC encoding complexity, for both the H.264/AVC Intra and H.264/AVC No Motion solutions.
- Although the H.264/AVC Intra encoding complexity does not vary with the GOP size and the H.264/AVC No Motion encoding complexity is also rather stable with a varying GOP size, the DISCOVER encoding complexity decreases with the GOP size. For longer GOP sizes, the overall encoding complexity decreases with the increase of the share of WZ frames regarding the key frames. In this case, the key frames share decreases, although their encoding complexity is still the dominating part.
- If encoding complexity is a critical requirement for an application, the results in this section together with the RD performance results previously shown indicate that the DISCOVER WZ video codec with GOP = 2 is a credible solution since it has a rather low encoding complexity and beats H.264/AVC Intra in terms of RD performance for most cases.
- Another important result is that the WZ encoding complexity does not increase significantly when the Q_i increases, that is, when the bit rate increases.
- The encoding complexity is rather similar for the LDPC and turbo coding alternatives for all GOP sizes. This shows that the adopted LDPC encoding solution does not significantly increase the encoding complexity when compared to the

Table 8.6 Encoding Time (s) Comparison for the Coast Guard Sequence

	H.264/AVC				DISCOVER						Ratios		
	Intra	No Motion			Using LDPC Codes			Using Turbo Codes			H.264/AVC Intra vs. DISCOVER LDPC		
QP		GOP 2	GOP 4	GOP 8	GOP 2	GOP 4	GOP 8	GOP 2	GOP 4	GOP 8	GOP 2	GOP 4	GOP 8
38	35.04	35.36	35.63	34.89	20.58	12.22	8.28	20.31	11.84	7.88	1.70	2.87	4.23
37	36.10	36.31	36.59	35.86	21.07	12.47	8.53	20.81	12.06	8.09	1.71	2.90	4.23
37	36.12	36.33	36.59	35.86	21.11	12.56	8.63	20.81	12.16	8.17	1.71	2.88	4.18
34	39.22	39.53	39.89	39.02	22.74	13.52	9.06	22.41	13.04	8.57	1.72	2.90	4.33
33	40.43	40.88	41.28	40.52	23.25	13.77	9.31	22.92	13.30	8.79	1.74	2.94	4.34
31	43.35	43.64	43.89	42.98	24.79	14.70	9.83	24.46	14.21	9.31	1.75	2.95	4.41
30	44.47	45.13	45.39	44.56	25.49	15.14	10.22	25.10	14.61	9.63	1.74	2.94	4.35
26	51.29	51.88	52.31	51.13	29.09	17.31	11.52	28.72	16.67	10.91	1.76	2.96	4.45

Table 8.7 Encoding Time (s) Comparison for the Hall Monitor Sequence

	H.264/AVC				DISCOVER						Ratios		
	Intra	No Motion			Using LDPC Codes			Using Turbo Codes			H.264/AVC Intra vs. DISCOVER LDPC		
QP		**GOP 2**	**GOP 4**	**GOP 8**	**GOP 2**	**GOP 4**	**GOP 8**	**GOP 2**	**GOP 4**	**GOP 8**	**GOP 2**	**GOP 4**	**GOP 8**
37	39.30	39.27	39.55	38.58	23.05	13.71	9.19	22.85	13.21	8.62	1.70	2.87	4.28
36	39.70	40.00	40.25	39.42	23.31	13.86	9.41	23.15	13.37	8.81	1.70	2.87	4.22
36	39.73	40.08	40.26	39.44	23.34	13.86	9.42	23.17	13.38	8.82	1.70	2.87	4.22
33	42.22	42.66	43.06	42.17	24.65	14.72	9.95	24.54	14.22	9.39	1.71	2.87	4.24
33	42.29	42.77	43.16	42.20	24.73	14.72	9.97	24.57	14.23	9.41	1.71	2.87	4.24
31	44.30	44.83	44.97	44.02	25.89	15.52	10.50	25.82	14.90	9.83	1.71	2.85	4.22
29	46.22	46.91	47.41	46.38	26.96	16.12	10.94	26.71	15.50	10.28	1.71	2.87	4.23
24	52.58	53.50	54.09	53.09	30.32	18.17	12.50	29.97	17.49	11.59	1.73	2.89	4.21

Table 8.8 Encoding Time (s) Comparison for the Soccer Sequence

	H.264/AVC				DISCOVER						Ratios		
	Intra	No Motion			Using LDPC Codes			Using Turbo Codes			H.264/AVC Intra vs. DISCOVER LDPC		
QP		GOP 2	GOP 4	GOP 8	GOP 2	GOP 4	GOP 8	GOP 2	GOP 4	GOP 8	GOP 2	GOP 4	GOP 8
44	30.19	30.38	30.70	30.13	18.17	11.06	7.64	17.84	10.67	7.31	1.66	2.73	3.95
43	30.53	30.77	31.06	30.45	18.30	11.16	7.86	17.98	10.77	7.42	1.67	2.74	3.88
41	30.94	31.31	31.80	31.22	18.62	11.28	7.89	18.31	10.88	7.44	1.66	2.74	3.92
36	33.27	33.71	34.44	33.75	19.83	12.05	8.40	19.48	11.62	7.91	1.68	2.76	3.96
36	33.36	33.88	34.58	33.77	19.88	12.08	8.42	19.61	11.62	7.93	1.68	2.76	3.96
34	34.94	35.31	35.69	35.05	20.69	12.67	8.80	20.34	12.09	8.28	1.69	2.76	3.97
31	37.55	37.97	38.47	37.69	22.17	13.42	9.38	21.78	12.87	8.79	1.69	2.80	4.00
25	45.33	45.70	46.09	45.09	26.23	15.89	10.80	25.83	15.27	10.22	1.73	2.85	4.20

Table 8.9 Encoding Time (s) Comparison for the Foreman Sequence

	H.264/AVC				DISCOVER						Ratios		
	Intra	No Motion			Using LDPC Codes			Using Turbo Codes			H.264/AVC Intra vs. DISCOVER LDPC		
QP		**GOP 2**	**GOP 4**	**GOP 8**	**GOP 2**	**GOP 4**	**GOP 8**	**GOP 2**	**GOP 4**	**GOP 8**	**GOP 2**	**GOP 4**	**GOP 8**
40	32.67	33.14	33.64	32.97	19.44	11.58	8.02	19.14	11.26	7.58	1.68	2.82	4.08
39	33.30	33.81	34.41	33.72	19.75	11.80	8.09	19.47	11.42	7.69	1.69	2.82	4.11
38	33.78	34.34	34.88	34.15	20.05	11.95	8.22	19.73	11.57	7.79	1.69	2.83	4.11
34	37.11	37.60	38.21	37.34	21.67	12.94	8.79	21.39	12.48	8.33	1.71	2.87	4.22
34	37.13	37.61	38.29	37.34	21.68	13.00	8.80	21.39	12.53	8.34	1.71	2.86	4.22
32	38.94	39.55	40.18	39.23	22.78	13.66	9.37	22.38	13.14	8.79	1.71	2.85	4.15
29	42.33	43.11	43.73	42.72	24.49	14.58	9.94	24.13	14.07	9.34	1.73	2.90	4.26
25	48.56	48.89	49.54	48.17	27.72	16.61	11.23	27.36	15.98	10.55	1.75	2.92	4.32

turbo encoding; the LDPC main benefits in terms of complexity only regard the decoding phase as it will be shown next.

8.6.3.2 Measuring the Decoding Complexity

This section targets the complexity evaluation of the decoding process. Again, the decoding complexity includes two major components, which are the WZ frames decoding and the key frames decoding parts. The larger is the GOP size, the smaller the number of key frames coded and, thus, the lower the complexity share of the key frames in the overall decoding complexity.

Defining the Metrics

Following the options taken for the encoding complexity evaluation, the decoding complexity evaluation will be measured using an equivalent metric for the decoder. This means the decoding time for the full sequence, in seconds, under the same conditions and software/hardware platform.

Results and Analysis

Tables 8.10 to 8.13 show the decoder complexity results for GOP = 2, 4, and 8, measured in terms of decoding time. No charts are presented because only the bars corresponding to the WZ frames decoding times would be visible since the key frames decoding times are negligible regarding the WZ frames decoding times. The results allow concluding that:

- For the DISCOVER codec, the key frames decoding complexity is negligible regarding the WZ frames decoding complexity, even for GOP = 2 (when there

Table 8.10 Decoding Time (s) Comparison for the Coast Guard Sequence

	H.264/AVC				DISCOVER					
	Intra	No Motion			Using LDPC Codes			Using Turbo Codes		
QP		GOP 2	GOP 4	GOP 8	GOP 2	GOP 4	GOP 8	GOP 2	GOP 4	GOP 8
38	1.55	1.56	1.56	1.53	430.27	709.75	986.66	440.89	691.36	890.94
37	1.58	1.61	1.56	1.56	485.89	776.86	1048.20	512.61	796.70	994.88
37	1.61	1.64	1.58	1.57	531.17	894.77	1226.23	579.48	908.95	1150.61
34	1.64	1.66	1.70	1.66	796.69	1525.11	2168.36	874.58	1463.97	1885.91
33	1.70	1.75	1.77	1.69	824.23	1628.22	2329.80	931.80	1564.95	2024.33
31	1.72	1.81	1.78	1.80	1144.45	2254.47	3193.66	1229.36	2093.36	2708.59
30	1.75	1.88	1.81	1.81	1497.25	2802.48	3856.28	1566.00	2668.41	3375.45
26	1.92	2.05	2.03	1.94	2461.73	4462.38	5872.11	2500.95	4309.33	5428.16

Table 8.11 Decoding Time (s) Comparison for the Hall Monitor Sequence

	H.264/AVC				DISCOVER					
	Intra	No Motion			Using LDPC Codes			Using Turbo Codes		
QP		GOP 2	GOP 4	GOP 8	GOP 2	GOP 4	GOP 8	GOP 2	GOP 4	GOP 8
37	1.72	1.63	1.61	1.55	360.52	524.11	641.33	417.14	607.66	685.36
36	1.75	1.64	1.61	1.56	391.14	554.22	692.58	469.14	662.31	738.91
36	1.77	1.65	1.62	1.57	414.50	599.67	739.19	532.13	750.00	849.83
33	1.78	1.69	1.63	1.58	545.75	886.86	1175.91	711.56	1052.36	1225.17
33	1.83	1.70	1.66	1.58	566.70	907.28	1195.64	766.23	1138.89	1319.63
31	1.84	1.75	1.66	1.60	732.00	1197.86	1612.94	946.53	1422.38	1639.22
29	1.89	1.75	1.69	1.63	859.97	1417.86	1863.14	1129.14	1699.98	1988.53
24	2.08	1.95	1.83	1.79	1210.47	2011.45	2526.56	1567.86	2407.45	2846.80

Table 8.12 Decoding Time (s) Comparison for the Soccer Sequence

	H.264/AVC				DISCOVER					
	Intra	No Motion			Using LDPC Codes			Using Turbo Codes		
QP		GOP 2	GOP 4	GOP 8	GOP 2	GOP 4	GOP 8	GOP 2	GOP 4	GOP 8
44	1.44	1.49	1.46	1.43	1072.96	1673.20	1955.53	824.28	1300.64	1527.80
43	1.44	1.50	1.48	1.44	1092.14	1733.53	2015.23	893.97	1411.63	1661.63
41	1.45	1.52	1.52	1.46	1288.43	2076.72	2436.09	1015.11	1622.06	1917.89
36	1.52	1.60	1.61	1.53	1981.15	3210.11	3775.03	1601.84	2577.00	3051.41
36	1.53	1.60	1.62	1.53	2140.70	3435.02	4029.86	1708.45	2748.72	3246.23
34	1.56	1.60	1.65	1.57	2636.76	4190.78	4882.00	2166.38	3473.70	4088.52
31	1.64	1.67	1.72	1.64	2969.98	4785.70	5559.48	2590.81	4165.30	4920.08
25	1.83	1.87	1.92	1.83	3874.17	6243.50	7407.88	3704.17	5997.52	7089.14

are more key frames). This confirms the well known WZ coding trade-off where the encoding complexity benefits are paid in terms of decoding complexity. Contrary to the encoding complexity, the longer is the GOP size, the higher is the overall decoding complexity since the higher is the number of WZ frames.

Table 8.13 Decoding Time (s) Comparison for the Foreman Sequence

	H.264/AVC				DISCOVER					
	Intra	No Motion			Using LDPC Codes			Using Turbo Codes		
QP		GOP 2	GOP 4	GOP 8	GOP 2	GOP 4	GOP 8	GOP 2	GOP 4	GOP 8
40	1.55	1.53	1.53	1.50	664.06	1150.11	1486.70	590.47	983.92	1237.19
39	1.55	1.55	1.53	1.52	729.45	1237.47	1605.34	680.80	1124.59	1402.70
38	1.58	1.55	1.58	1.53	848.45	1482.45	1904.23	768.25	1280.75	1606.08
34	1.64	1.66	1.66	1.64	1362.45	2536.06	3293.84	1219.88	2105.47	2663.94
34	1.66	1.67	1.69	1.64	1541.00	2824.58	3641.94	1346.77	2319.48	2930.91
32	1.72	1.73	1.77	1.70	2041.53	3640.48	4586.58	1765.06	3030.48	3808.64
29	1.83	1.83	1.84	1.81	2352.56	4273.28	5551.36	2148.23	3738.67	4728.45
25	1.92	1.97	1.98	1.94	3254.92	5901.63	7640.66	3207.05	5535.03	6974.56

- The DISCOVER decoding complexity is always much higher than the H.264/AVC decoding complexity, both for the H.264/AVC Intra and H264/AVC No Motion solutions. While the H.264/AVC Intra decoding complexity does not vary with the GOP size and the H.264/AVC No Motion decoding complexity is also rather stable with a varying GOP size, the DISCOVER decoding complexity increases with the GOP size.
- As can be observed, the WZ decoding complexity increases significantly when the Q_i increases (i.e., when the bit rate increases) since the number of bitplanes to LDPC decode is higher and the LDPC decoder (and the number of times that is invoked) is responsible for most of the decoding complexity share.
- Regarding the decoding complexity comparison between LDPC and turbo codes, the results seem to say that, while LDPC wins for more quiet sequences, for example, Hall Monitor and Coast Guard for GOP = 2, turbo codes win for sequences with more motion, for example, Soccer and Foreman.

The decoding complexity results also reveal that the most significant complexity burden is associated with the Slepian–Wolf decoding and the repetitive request-decode operation; this burden would be even higher if no encoder rate estimation was made. This fact highlights how important it is to reduce the number of decoder requests, not only by improving the side information quality, but also by adopting adequate rate control strategies such as efficient hybrid encoder-decoder rate control where the encoder estimates the rate and the feedback channel serves only to complement the estimated rate when the (under)estimation is not good enough. The side information creation time share typically decreases when Q_i increases since the number of bit-planes to LDPC decode is higher; on the contrary, the LDPC decoding time share increases with Q_i when compared to the remaining modules.

8.7 FINAL REMARKS

This chapter has reviewed the state-of-the-art on distributed video coding, going from the basics and the first WZ video codecs to the more recent developments represented by the DISCOVER WZ video codec. Although it is difficult to state, at this stage, if any video coding product will ever use distributed source coding (DSC) principles, and for what purpose, it is most interesting to study and research toward this possibility.

In terms of RD performance, the DISCOVER video codec already wins against the H.264/AVC Intra codec, for most test sequences, and for GOP = 2; for more quiet sequences, the DISCOVER codec may even win against the H.264/AVC No Motion codec. For longer GOP sizes, winning against H.264/AVC Intra is more difficult, highlighting the importance and difficulty of getting good side information, notably when key frames are farther away. The feedback channel rate is rather low since a reduced number of requests are made after the initial encoder rate estimation. However, the use of a feedback channel adds delay and requires a real-time setup. To address this limitation, allowing Stanford-based WZ video codecs to be applicable to applications that do not have a feedback channel, for example, storage applications, a WZ video codec with encoder rate control (without feedback channel) has also been developed in the DISCOVER project [37]. This first feedback channel-free WZ video codec pays a RD performance loss up to 1.2 dB, especially for the highest bit rates, to avoid usage of the feedback channel by using a pure encoder-based rate control.

The DISCOVER encoding complexity is always much lower than the H.264/AVC Intra encoding complexity, even for GOP = 2 where it performs better in terms of RD performance. Since the DISCOVER codec performs better than H.264/AVC Intra for GOP = 2, for most sequences, this highlights that Wyner–Ziv coding is already a credible solution when encoding complexity is a very critical requirement (even if at the cost of some additional decoding complexity). Good examples of these applications may be deep space video transmission, video surveillance, and video sensor networks.

Further WZ video coding research should address issues such as side information creation, progressive side information refinement, correlation noise modeling, novel channel codes, rate control, and spatial and temporal adaptive WZ coding. Recent developments have shown that WZ video coding is still at a rather immature stage compared with the level of optimization of the alternative solutions, and thus significant gains can still be obtained for most of the WZ video architecture modules.

It is presently more and more accepted that DSC principles are leading to a variety of tools that may help to solve different problems, not only coding but also, authentication, tampering detection [53], and secure biometrics [54]. The future will tell in which application domains distributed source coding principles will find success.

ACKNOWLEDGMENTS

Much of the work presented in this chapter was developed within DISCOVER, a European Project (http://www.discoverdvc.org), and VISNET II, a European Network of Excellence (http://www.visnet-noe.org), both funded under the European Commission IST FP6 program.

The authors would like to acknowledge that the DISCOVER performance results included in this chapter were the result of a collaborative effort developed by all the DISCOVER partners.

REFERENCES

[1] T. Wiegand, G. J. Sullivan, G. Bjontegaard, and A. Luthra, "Overview of the H.264/AVC Video Coding Standard," *IEEE Trans. on Circuits and Systems for Video Technology*, vol. 13, no. 7, pp. 560–576, July 2003.

[2] J. Slepian and J. Wolf, "Noiseless Coding of Correlated Information Sources," *IEEE Trans. on Information Theory*, vol. 19, no. 4, pp. 471–480, July 1973.

[3] A. Wyner and J. Ziv, "The Rate-Distortion Function for Source Coding with Side Information at the Decoder," *IEEE Trans. on Information Theory*, vol. 22, no. 1, pp. 1–10, January 1976.

[4] DISCOVER Page, http://www.img.lx.it.pt/~discover/home.html

[5] X. Artigas, J. Ascenso, M. Dalai, S. Klomp, D. Kubasov, and M. Ouaret, "The DISCOVER Codec: Architecture, Techniques and Evaluation," Picture Coding Symposium, Lisbon, Portugal, November 2007.

[6] X. Artigas, S. Malinowski, C. Guillemot, and L. Torres, "Overlapped Quasi-Arithmetic Codes for Distributed Video Coding," International Conference on Image Processing, San Antonio, TX, September 2007.

[7] A. Wyner, "Recent Results in the Shannon Theory," *IEEE Trans. on Information Theory*, vol. 20, no. 1, pp. 2–10, January 1974.

[8] R. Zamir, "The Rate Loss in the Wyner–Ziv Problem," *IEEE Trans. on Information Theory*, vol. 42, no. 6, pp. 2073–2084, November 1996.

[9] S. S. Pradhan, J. Chou, and K. Ramchandran, "Duality between Source Coding and Channel Coding and Its Extension to the Side Information Case," *IEEE Trans. on Information Theory*, vol. 49, no. 5, pp. 1181–1203, May 2003.

[10] F. Pereira et al., "Distributed Video Coding: Selecting the Most Promising Application Scenarios," *Signal Processing: Image Communication*, vol. 23, no. 5, pp. 339–352, June 2008.

[11] C. Guillemot, F. Pereira, L. Torres, T. Ebrahimi, R. Leonardi, and J. Ostermann, "Distributed Monoview and Multiview Video Coding," *IEEE Signal Processing Magazine*, vol. 24, no. 5, pp. 67–76, September 2007.

[12] A. Aaron, R. Zhang, and B. Girod, "Wyner–Ziv Coding of Motion Video," Asilomar Conference on Signals, Systems and Computers, Pacific Grove, CA, November 2002.

[13] A. Aaron, S. Rane, E. Setton, and B. Girod, "Transform-Domain Wyner–Ziv Codec for Video," Visual Communications and Image Processing, San Jose, CA, January 2004.

[14] B. Girod, A. Aaron, S. Rane, and D. Rebollo-Monedero, "Distributed Video Coding," *Proceedings of the IEEE*, vol. 93, no. 1, pp. 71–83, January 2005.

[15] R. Puri and K. Ramchandran, "PRISM: A New Robust Video Coding Architecture Based on Distributed Compression Principles," 40th Allerton Conference on Communication, Control, and Computing, Allerton, IL, October 2002.

[16] R. Puri, A. Majumdar, and K. Ramchandran, "PRISM: A Video Coding Paradigm with Motion Estimation at the Decoder," *IEEE Trans. on Image Processing*, vol. 16, no. 10, pp. 2436–2448, October 2007.

[17] S. S. Pradhan and K. Ramchandran, "Distributed Source Coding Using Syndromes (DISCUS): Design and Construction," IEEE Data Compression Conference, Snowbird, UT, pp. 158–167, March 1999.

[18] J. Bajcsy, and P. Mitran, "Coding for the Slepian-Wolf Problem with Turbo Codes," Global Communication Symposium, San Antonio, TX, pp. 1400–1404, November 2001.

[19] J. Garcia-Frias and Y. Zhao, "Compression of Correlated Binary Sources Using Turbo Codes," *IEEE Communication Letters*, vol. 5, pp. 417–419, October 2001.

[20] A. Aaron and B. Girod, "Compression with Side Information Using Turbo Codes," IEEE Data Compression Conference, Snowbird, UT, April 2002.

[21] A. D. Liveris, Z. Xiong, and C. N. Georghiades, "Compression of Binary Sources with Side Information at the Decoder Using LDPC Codes," *IEEE Communication Letters*, vol. 6, pp. 440–442, October 2002.

[22] D. Varodayan, A. Aaron, and B. Girod, "Rate-Adaptive Codes for Distributed Source Coding," *EURASIP Signal Processing Journal*, Special Issue on Distributed Source Coding, vol. 86, no. 11, pp. 3123–3130, November 2006.

[23] Q. Xu, V. Stankovic, and Z. Xiong, "Distributed Joint Source-Channel Coding of Video Using Raptor Codes," *IEEE Journal on Selected Areas in Communications*, vol. 25, no. 4, pp. 851–861, May 2007.

[24] D. J. C. MacKay, "Fountain Codes," *IEE Proceedings-Communications*, vol. 152, no. 6, pp. 1062–1068, December 2005.

[25] Q. Xu, V. Stankovic, and Z. Xiong, "Wyner–Ziv Video Compression and Fountain Codes for Receiver-Driven Layered Multicast," *IEEE Trans. on Circuits and Systems for Video Technology*, vol. 17, no. 7, pp. 901–906, July 2007.

[26] J. Ascenso, C. Brites, and F. Pereira, "Improving Frame Interpolation with Spatial Motion Smoothing for Pixel Domain Distributed Video Coding," EURASIP Conference on Speech and Image Processing, Multimedia Communications and Services, Smolenice, Slovak Republic, June 2005.

[27] J. Ascenso, C. Brites, and F. Pereira, "Content Adaptive Wyner–Ziv Video Coding Driven by Motion Activity," International Conference on Image Processing, Atlanta, GA, October 2006.

[28] L. Natário, C. Brites, J. Ascenso, and F. Pereira, "Extrapolating Side Information for Low-Delay Pixel-Domain Distributed Video Coding," International. Workshop on Very Low Bitrate Video Coding, Sardinia, Italy, September 2005.

[29] D. Varodayan, Y.-C. Lin, A. Mavlankar, M. Flierl, and B. Girod, "Wyner–Ziv Coding of Stereo Images with Unsupervised Learning of Disparity," Picture Coding Symposium, Lisbon, Portugal, November 2007.

[30] C. Brites, J. Ascenso, and F. Pereira, "Modeling Correlation Noise Statistics at Decoder for Pixel Based Wyner–Ziv Video Coding," Picture Coding Symposium, Beijing, China, April 2006.

[31] C. Brites, J. Ascenso, and F. Pereira, "Studying Temporal Correlation Noise Modeling for Pixel Based Wyner–Ziv Video Coding," International Conference on Image Processing, Atlanta, GA, October 2006.

[32] C. Brites and F. Pereira, "Correlation Noise Modeling for Efficient Pixel and Transform Domain Wyner–Ziv Video Coding," *IEEE Trans. on Circuits and Systems for Video Technology*, accepted for publication.

[33] D. Kubasov, J. Nayak, and C. Guillemot, "Optimal Reconstruction in Wyner-Ziv Video Coding with Multiple Side Information," IEEE Multimedia Signal Processing Workshop, Chania, Crete, Greece, October 2007.

[34] A. Trapanese, M. Tagliasacchi, S. Tubaro, J. Ascenso, C. Brites, and F. Pereira, "Intra Mode Decision Based on Spatio-Temporal Cues in Pixel Domain Wyner-Ziv Video Coding," International Conference on Acoustics, Speech, and Signal Processing, Toulouse, France, May 2006.

[35] A. Aaron, S. Rane, and B. Girod, "Wyner-Ziv Video Coding with Hash-based Motion Compensation at the Receiver," International Conference on Image Processing, Singapore, October 2004.

[36] J. Ascenso and F. Pereira, "Adaptive Hash-based Side Information Exploitation for Efficient Wyner-Ziv Video Coding," International Conference on Image Processing, San Antonio, TX, September 2007.

[37] C. Brites and F. Pereira, "Encoder Rate Control for Transform Domain Wyner-Ziv Video Coding," International Conference on Image Processing, San Antonio, TX, September 2007.

[38] D. Kubasov, K. Lajnef, and C. Guillemot, "A Hybrid Encoder/Decoder Rate Control for a Wyner-Ziv Video Codec with a Feedback Channel," IEEE Multimedia Signal Processing Workshop, Chania, Crete, Greece, October 2007.

[39] J. D. Areia, J. Ascenso, C. Brites, and F. Pereira, "Low Complexity Hybrid Rate Control for Lower Complexity Wyner-Ziv Video Decoding," European Signal Processing Conference, Lausanne, Switzerland, August 2008.

[40] M. Tagliasacchi, J. Pedro, F. Pereira, and S. Tubaro, "An Efficient Request Stopping Method at the Turbo Decoder in Distributed Video Coding," European Signal Processing Conference, Poznan, Poland, September 2007.

[41] S. Shamai, S. Verdú, and R. Zamir, "Systematic Lossy Source/Channel Coding," *IEEE Trans. on Information Theory*, vol. 44, no. 2, pp. 564-579, March 1998.

[42] A. Sehgal, A. Jagmohan, and N. Ahuja, "Wyner-Ziv Coding of Video: An Error-Resilient Compression Framework," *IEEE Transactions on Multimedia*, vol. 6, no. 2, pp. 249-258, April 2004.

[43] S. Rane, P. Baccichet, and B. Girod, "Modeling and Optimization of a Systematic Lossy Error Protection System Based on H.264/AVC Redundant Slices," Picture Coding Symposium, Beijing, China, April 2006.

[44] J. Wang, A. Majumdar, K. Ramchandran, and H. Garudadri, "Robust Video Transmission over a Lossy Network Using a Distributed Source Coded Auxiliary Channel," Picture Coding Symposium, San Francisco, CA, December 2004.

[45] J. Pedro et al., "Studying Error Resilience Performance for a Feedback Channel Based Transform Domain Wyner-Ziv Video Codec," Picture Coding Symposium, Lisbon, Portugal, November 2007.

[46] A. Sehgal, A. Jagmohan, and N. Ahuja, "Scalable Video Coding Using Wyner-Ziv Code," Picture Coding Symposium, San Francisco, CA, December 2004.

[47] M. Tagliasacchi, A. Majumdar, and K. Ramchandran, "A Distributed Source Coding Based Spatio-Temporal Scalable Video Codec," Picture Coding Symposium, San Francisco, CA, December 2004.

[48] H. Wang, N. Cheung, and A. Ortega, "A Framework for Adaptive Scalable Video Coding Using Wyner-Ziv Techniques," *EURASIP Journal on App. Signal Processing*, vol. 2006, Article ID 60971.

[49] Q. Xu and Z. Xiong, "Layered Wyner-Ziv Video Coding," *IEEE Trans. on Image Processing*, vol. 15, no. 12, pp. 3791-3803, December 2006.

[50] M. Ouaret, F. Dufaux, and T. Ebrahimi, "Codec-Independent Scalable Distributed Video Coding," International Conference on Image Processing, San Antonio, TX, September 2007.

[51] J. Ascenso, C. Brites, and F. Pereira, "Design and Performance of a Novel Low-Density Parity-Check Code for Distributed Video Coding," IEEE International Conference on Image Processing, San Diego, CA, October 2008.

[52] F. Pereira, J. Ascenso, and C. Brites, "Studying the GOP Size Impact on the Performance of a Feedback Channel-based Wyner–Ziv Video Codec," IEEE Pacific Rim Symposium on Image Video and Technology, Santiago, Chile, December 2007.

[53] Y.-C. Lin, D. Varodayan, and B. Girod, "Image Authentication and Tampering Localization Using Distributed Source Coding," IEEE Multimedia Signal Processing Workshop, Chania, Crete, Greece, October 2007.

[54] S. C. Draper, A. Khisti, E. Martinian, A. Vetro, and J. S. Yedidia, "Using Distributed Source Coding to Secure Fingerprint Biometrics," International Conference on Acoustics, Speech, and Signal Processing, Honolulu, HI, April 2007.

CHAPTER

Model-based Multiview Video Compression Using Distributed Source Coding Principles

Jayanth Nayak

Mayachitra, Inc., Santa Barbara, CA

Bi Song, Ertem Tuncel, and Amit K. Roy-Chowdhury

Department of Electrical Engineering, University of California, Riverside, CA

CHAPTER CONTENTS

9.1 INTRODUCTION

Transmission of video data from multiple sensors over a wireless network requires an enormous amount of bandwidth and could easily overwhelm the system. However, by exploiting the redundancy *between* the video data collected by different cameras, in addition to the inherent temporal and spatial redundancy *within* each video sequence, the required bandwidth can be significantly reduced. Well-established video compression standards, such as MPEG-1, MPEG-2, MPEG-4, H.261, H.263, and H.264, all rely on efficient transform coding of motion-compensated frames, using the discrete cosine transform (DCT) or computationally efficient approximations to it. However, they can only be used in a protocol that encodes the data of each sensor independently. Such methods would exploit spatial and temporal redundancy within each video sequence but would completely ignore the redundancy between the sequences.

In this chapter, we develop novel multiterminal, model-based video coding algorithms combining distributed source coding (DSC) and computer vision techniques. In broad terms, our schemes rely on model-based tracking of individual video sequences captured by cameras (which could be located arbitrarily in space), leading to removal of spatial, temporal, and *inter-camera* redundancies. The presence of the three-dimensional (3D) model provides correspondence between overlapping feature points in the different views, provided that the tracking of the individual sequences is accurate. The performance of our algorithm depends, most crucially, on the quality of tracking and the coding efficiency of the distributed quantization scheme. The tracking must result in correspondences between pixels that are maximally correlated, and the distributed coding must optimally exploit this correlation.

Although DSC was introduced more than three decades ago by Slepian and Wolf [17], and underlying ideas as to how it should be practically implemented were outlined by Wyner [20], the papers by Zamir and Shamai [25] and Pradhan and Ramchandran [13] arguably demonstrated the feasibility of implementing distributed source codes for the first time. Following the publication of these papers, considerable effort has been devoted to adapting DSC ideas to application scenarios such as distributed image and video coding (e.g., [5, 6, 14, 19, 24, 26]), but the results have been mixed. The reason is that while distributed processing appears to be a natural choice in many scenarios, the conditions necessary for known DSC schemes to perform well are rarely satisfied, and the more mature non-DSC techniques either easily outperform the DSC schemes or the gains of DSC schemes are relatively small. We make similar observations in this work. More specifically, the gains we achieve over separate coding are diminished because it seems that the temporal redundancies are much larger than the intercamera ones. The exception is at very low rates where even small gains become significant. In this chapter, we therefore focus on very low bit rates.

Although early work on distributed video coding focused on the single-view case, some attention has been given to applying distributed coding for multiview video as well [1, 7, 12, 18]. Most of this work has been on block-based coding, and, as in our work, a key issue is construction of side information to optimally exploit intrasequence memory and intersequence correlation.

The rest of the chapter is organized as follows. Section 9.2 outlines the model-based tracking algorithm; Section 9.3 presents an overview of our approach to distributed video coding; Section 9.4 gives some experimental results; and finally, Section 9.5 presents the conclusion and discusses some avenues for future research.

9.2 MODEL TRACKING

Numerous methods exist for estimating the motion and shape of an object from video sequences. Many of these methods can handle significant changes in the illumination conditions by *compensating* for the variations [4, 8, 9]. However, few methods can *recover* the 3D motion *and* time-varying global illumination conditions from video sequences of moving objects. We achieve this goal by building on a recently proposed framework for combining the effects of motion, illumination, 3D shape, and camera parameters in a sequence of images obtained by a perspective camera [21, 22]. This theory allows us to develop a simple method for estimating the rigid 3D motion, as presented in [21]. However, this algorithm involves the computation of a bilinear basis (see Section 9.2.1) in each iteration, which is a huge computational burden. In this chapter, we show that it is possible to efficiently and accurately reconstruct the 3D motion and global lighting parameters of a rigid and nonrigid object from a video sequence within the framework of the inverse compositional (IC) algorithm [2]. Details of this approach are available in [23].

A well-known approach for 2D motion estimation and registration in monocular sequences is Lucas-Kanade tracking, which attempts to match a target image with a reference image or template. Building on this framework, a very efficient tracking algorithm was proposed in [8] by inverting the role of the target image and the template. However, this algorithm can only be applied to restricted class of warps between the target and template (for details, see [2]). A forward compositional algorithm was proposed in [16] by estimating an incremental warp for image alignment. Baker and Matthews [2] proposed an IC algorithm for efficient implementation of the Lucas-Kanade algorithm to save computational cost in reevaluation of the derivatives in each iteration. The IC algorithm was then used for efficiently fitting active appearance models [11] and the well-known 3D morphable model (3DMM) [15] to face images under large pose variations. However, none of these schemes estimates the lighting conditions in the images. A version of 3DMM fitting [3] used a Phong illumination model, estimation of whose parameters in the presence of extended light sources can be difficult.

Our lighting estimation can account for extended lighting sources and attached shadows. Furthermore, our goal is to estimate 3D motion, unlike in [4, 8, 16], works that perform 2D motion estimation. The warping function in the present work is different from [2, 15], as we explain in Section 9.2.2. Since our IC approach estimates 3D motion, it allows us to perform the expensive computations only once every few frames (unlike once for every frame as in the image alignment approaches of Baker and Matthews [2]). Specifically, these computations are done only when there is a significant change of pose.

9.2.1 Image Appearance Model of a Rigid Object

Our goal is to describe the appearance of an image in terms of the 3D rigid motion and the overall lighting in the scene. For this purpose, we derive a generative image appearance model that is a function of these parameters. Given a sequence of images, we can estimate the parameters that lead to the best fit with this model. Details of the generative model, as well as the estimation framework, are available in [21, 22]. A brief overview is provided here.

In [21], it was proved that if the motion of the object (defined as the translation of the object centroid $\Delta\mathbf{T} \in \mathbb{R}^{3\times 1}$ and the rotation vector $\Delta\Omega \in \mathbb{R}^{3\times 1}$ about the centroid in the camera frame) from time t_1 to new time instance $t_2 = t_1 + \delta t$ is small, then up to a first-order approximation, the reflectance image $I(x, y)$ at t_2 can be expressed as

$$I_{t_2}(\mathbf{v}) = \sum_{i=1}^{9} l_i^{t_2} b_i^{t_2}(\mathbf{v}) \tag{9.1}$$

where

$$b_i^{t_2}(\mathbf{v}) = b_i^{t_1}(\mathbf{v}) + \mathbf{A}(\mathbf{v}, \mathbf{n})\Delta\mathbf{T} + \mathbf{B}(\mathbf{v}, \mathbf{n})\Delta\Omega.$$

In these equations, $\mathbf{v}$ represents the image point projected from the 3D surface with surface normal $\mathbf{n}$, and $b_i^{t_1}(\mathbf{v})$ are the original basis images before motion (precise format of $b_i^{t_1}(\mathbf{v})$ is defined in [21]). $\mathbf{l}_t = \left[l_1^t \ldots l_{N_l}^t\right]^T$ is the vector of illumination parameters. $\mathbf{A}$ and $\mathbf{B}$ contain the structure and camera intrinsic parameters, and are functions of $\mathbf{v}$ and the 3D surface normal $\mathbf{n}$. For each pixel $\mathbf{v}$, both $\mathbf{A}$ and $\mathbf{B}$ are $N_l \times 3$ matrices, where $N_l \approx 9$ for Lambertian objects with attached shadows. A derivation of (9.1) and explicit expressions for $\mathbf{A}$ and $\mathbf{B}$ are presented in [21]. For the purposes of this chapter, we only need to know the form of the equations.

The left side of Equation (9.1) is the image at time t_2, which is expressed in terms of the basis images and the lighting coefficients on the right-hand side. The basis images, in turn, depend on the motion of the object between two consecutive time instants. Thus, Equation (9.1) expresses the image appearance in terms of the object's 3D pose and the scene lighting.

We can express the result in (9.1) succinctly using tensor notation as

$$\mathcal{I}_{t_2} = \left(\mathcal{B}_{t_1} + \mathcal{C}_{t_1} \times_2 \begin{bmatrix} \Delta\mathbf{T} \\ \Delta\Omega \end{bmatrix}\right) \times_1 \mathbf{l}_{t_2}, \tag{9.2}$$

where $\times_n$ is called the *mode-n product* [10] and $\mathbf{l} \in \mathbb{R}^{N_l}$ is the N_l-dimensional vector of l_i components. More specifically, the mode-n product of a tensor

$\mathcal{A} \in \mathbb{R}^{I_1 \times I_2 \times \ldots \times I_n \times \ldots \times I_N}$ by a vector $\mathbf{V} \in \mathbb{R}^{I_n \times 1}$, denoted by $\mathcal{A} \times_n \mathbf{V}$, is the $I_1 \times I_2 \times \ldots \times 1 \times \ldots \times I_N$ tensor

$$(\mathcal{A} \times_n \mathbf{V})_{i_1 \ldots i_{n-1} 1 i_{n+1} \ldots i_N} = \sum_{i_n} a_{i_1 \ldots i_{n-1} i_n i_{n+1} \ldots i_N} v_{i_n}.$$

Thus, the image at t_2 can be represented using the parameters computed at t_1. For each pixel (p, q) in the image, $\mathcal{C}_{pq} = \begin{bmatrix} \mathbf{A} & \mathbf{B} \end{bmatrix}$ of size $N_l \times 6$. Thus for an image of size $M \times N$, $\mathcal{C}$ is $N_l \times 6 \times M \times N$, $\mathcal{B}_{t_1}$ is a subtensor of dimension $N_l \times 1 \times M \times N$, comprising the basis images $b_i^{t_1}(\mathbf{u})$, and $\mathcal{I}_{t_2}$ is a subtensor of dimension $1 \times 1 \times M \times N$, representing the image.

9.2.2 Inverse Compositional Estimation of 3D Motion and Illumination

We now present the estimation algorithm for the bilinear model in Equation (9.2). The detailed proof of convergence and extension to video for the rigid motion case can be found in [23].

We begin by estimating the 3D motion assuming that illumination is constant across two consecutive frames. We will then estimate variations in illumination. Let $\mathbf{p} \in \mathbb{R}^{6 \times 1}$ denote the pose of the object. Then the image synthesis process can be considered as a rendering function of the object at pose $\mathbf{p}$ in the camera frame to the pixel coordinates $\mathbf{v}$ in the image plane as $f(\mathbf{v}, \mathbf{p})$. Using the bilinear model described above, it can be implemented with Equation (9.2). Given an input image $I(\mathbf{v})$, we want to align the synthesized image with it so as to obtain

$$\hat{\mathbf{p}} = \arg\min_{\mathbf{p}} \sum_{\mathbf{v}} \left(f(\mathbf{v}, \mathbf{p}) - I(\mathbf{v}) \right)^2, \tag{9.3}$$

where $\hat{\mathbf{p}}$ denotes the estimated pose for this input image $I(\mathbf{v})$. This is the cost function of Lucas–Kanade tracking in [2] modified for 3D motion estimation.

We will consider the problem of estimating the pose change, $\mathbf{m}_t \triangleq \Delta \mathbf{p}_t$, between two consecutive frames, $I_t(\mathbf{v})$ and $I_{t-1}(\mathbf{v})$. Let us introduce a warp operator $\mathbf{W}: \mathbb{R}^2 \to \mathbb{R}^2$ such that, if we denote the pose of $I_t(\mathbf{v})$ as $\mathbf{p}$, the pose of $I_t(\mathbf{W}_{\mathbf{p}}(\mathbf{v}, \Delta \mathbf{p}))$ is $\mathbf{p} + \Delta \mathbf{p}$ (see Figure 9.1). Specifically, a 2D point on the image plane is projected onto the 3D object surface. Then we transform the pose of the object surface by $\Delta \mathbf{p}$ and back-project the point from the 3D surface onto the image plane. Thus, $\mathbf{W}_{\mathbf{p}}$ represents the displacement in the image plane due to a pose transformation of the 3D model. Note that this warping involves a 3D pose transformation (unlike [2]). In [15], the warping was from a point on the 3D surface to the image plane and was used for fitting a 3D model to an image. Our new warping function can be used for the IC estimation of 3D rigid motion and illumination in video sequence, which [2] and [15] do not address.

A key property of $\{\mathbf{W}_{\mathbf{p}}\}$ is that these warps form a group with respect to function composition (see [23] for a detailed proof of this and other properties of the set of warps), which is necessary for applying the IC algorithm. Here we will only show how the inverse of a warp is another warp. The inverse of the warp $\mathbf{W}$ is defined

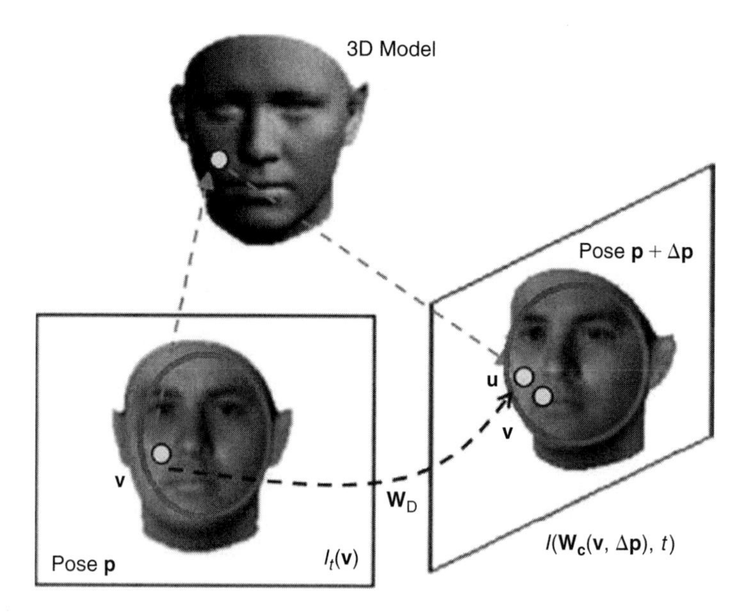

FIGURE 9.1

Illustration of the warping function **W**. A point **v** in image plane is projected onto the surface of the 3D object model. After the pose transformation with $\triangle\mathbf{p}$, the point on the surface is backprojected onto the image plane at a new point **u**. The warping function maps from $\mathbf{v} \in \mathbb{R}^2$ to $\mathbf{u} \in \mathbb{R}^2$. The red ellipses show the common part in both frames upon which the warping function **W** is defined.

to be the $\mathbb{R}^2 \rightarrow \mathbb{R}^2$ mapping such that if we denote the pose of $I_t(\mathbf{v})$ as $\mathbf{p}$, the pose of $I_t(\mathbf{W_p}(\mathbf{W_p}(\mathbf{v}, \triangle\mathbf{p}), \triangle\mathbf{p})^{-1})$ is $\mathbf{p}$ itself. As the warp $\mathbf{W_p}(\mathbf{v}, \triangle\mathbf{p})$ transforms the pose from $\mathbf{p}$ to $\mathbf{p} + \triangle\mathbf{p}$, the inverse $\mathbf{W_p}(\mathbf{v}, \triangle\mathbf{p})^{-1}$ should transform the pose from $\mathbf{p} + \triangle\mathbf{p}$ to $\mathbf{p}$, that is, $\mathbf{W_p}(\mathbf{v}, \triangle\mathbf{p})^{-1} = \mathbf{W}_{\mathbf{p}+\triangle\mathbf{p}}(\mathbf{v}, -\triangle\mathbf{p})$.

Using this warp operator, for any frame $I_t(\mathbf{v})$, we can write the cost function as

$$\hat{\mathbf{m}}_t = \arg\min_{\mathbf{m}} \sum_{\mathbf{v}} \left(f(\mathbf{v}, \hat{\mathbf{p}}_{t-1}) - I_t(\mathbf{W}_{\hat{\mathbf{p}}_{t-1}}(\mathbf{v}, -\mathbf{m}))\right)^2. \tag{9.4}$$

Rewriting the cost function (9.4) in the IC framework [2], we consider minimizing

$$\arg\min_{\triangle\mathbf{m}} \sum_{\mathbf{v}} \left(f(\mathbf{W}_{\hat{\mathbf{p}}_{t-1}}(\mathbf{v}, \triangle\mathbf{m}), \hat{\mathbf{p}}_{t-1}) - I_t(\mathbf{W}_{\hat{\mathbf{p}}_{t-1}}(\mathbf{v}, -\mathbf{m}))\right)^2 \tag{9.5}$$

with the update rule

$$\mathbf{W}_{\hat{\mathbf{p}}_{t-1}}(\mathbf{v}, -\mathbf{m}) \leftarrow \mathbf{W}_{\hat{\mathbf{p}}_{t-1}}(\mathbf{v}, -\mathbf{m}) \circ \mathbf{W}_{\hat{\mathbf{p}}_{t-1}}(\mathbf{v}, \triangle\mathbf{m})^{-1}. \tag{9.6}$$

The compositional operator $\circ$ in Equation (9.6) means the second warp is composed into the first warp, that is, $\mathbf{W}_{\hat{\mathbf{p}}_{t-1}}(\mathbf{v}, -\mathbf{m}) \equiv \mathbf{W}_{\hat{\mathbf{p}}_{t-1}}(\mathbf{W}_{\hat{\mathbf{p}}_{t-1}}(\mathbf{v}, \triangle\mathbf{m})^{-1}, -\mathbf{m})$. According to the definition of the warp $\mathbf{W}$, we can replace $f(\mathbf{W}_{\hat{\mathbf{p}}_{t-1}}(\mathbf{v}, \triangle\mathbf{m}), \hat{\mathbf{p}}_{t-1})$

in (9.5) with $f(\mathbf{v}, \hat{\mathbf{p}}_{t-1} + \triangle\mathbf{m})$. This is because $f(\mathbf{v}, \hat{\mathbf{p}}_{t-1} + \triangle\mathbf{m})$ is the image synthesized at $\hat{\mathbf{p}}_{t-1} + \triangle\mathbf{m}$, while $f(\mathbf{W}_{\hat{\mathbf{p}}_{t-1}}(\mathbf{v}, \triangle\mathbf{m}), \hat{p}_{t-1})$ is the image synthesized at $\hat{\mathbf{p}}_{t-1}$ followed with the warp of the pose increments $\triangle\mathbf{m}$. Applying the first-order Taylor expansion on it, we have

$$\sum_{\mathbf{v}} \left(f(\mathbf{v}, \hat{\mathbf{p}}_{t-1}) + \frac{\partial f(\mathbf{v}, \mathbf{p})}{\partial \mathbf{p}} |_{\mathbf{p}=\hat{\mathbf{p}}_{t-1}} \triangle\mathbf{m} - I_t(\mathbf{W}_{\hat{\mathbf{p}}_{t-1}}(\mathbf{v}, -\mathbf{m})) \right)^2 .$$

Taking the derivative of the above expression with respect to $\triangle\mathbf{m}$ and setting it to be zero, we have

$$\sum_{\mathbf{v}} \left(f(\mathbf{v}, \hat{\mathbf{p}}_{t-1}) + \mathcal{G}^{\mathbf{T}}_{\mathbf{v}|\hat{\mathbf{p}}_{t-1}} \triangle\mathbf{m} - I_t(\mathbf{W}_{\hat{\mathbf{p}}_{t-1}}(\mathbf{v}, -\mathbf{m})) \right) \mathcal{G}_{\mathbf{v}|\hat{\mathbf{p}}_{t-1}} = 0,$$

where $\mathcal{G}_{\mathbf{v}|\hat{\mathbf{p}}_{t-1}}$ is the derivative $\frac{\partial f(\mathbf{v},\mathbf{p})}{\partial \mathbf{p}}|_{\mathbf{p}=\hat{\mathbf{p}}_{t-1}}$. Solving for $\triangle\mathbf{m}$, we get:

$$\triangle\mathbf{m} = \mathbf{H}_{\mathbf{IC}} \sum_{\mathbf{v}} \mathcal{G}_{\mathbf{v}|\hat{\mathbf{p}}_{t-1}} \left(I_t \left(\mathbf{W}_{\hat{\mathbf{p}}_{t-1}} (\mathbf{v}, -\mathbf{m}) \right) - f \left(\mathbf{v}, \hat{\mathbf{p}}_{t-1} \right) \right) \tag{9.7}$$

where

$$\mathbf{H}_{\mathbf{IC}} = \left[\sum_{\mathbf{v}} \mathcal{G}_{\mathbf{v}|\hat{\mathbf{p}}_{t-1}} \mathcal{G}^{\mathbf{T}}_{\mathbf{v}|\hat{\mathbf{p}}_{t-1}} \right]^{-1} .$$

Note that the derivative $\mathcal{G}_{\mathbf{v}|\hat{\mathbf{p}}_{t-1}}$ and Hessian $\mathbf{H}_{\mathbf{IC}}$ in (9.7) do not depend on the updating variable $\mathbf{m}$, which is moved into the warp operator $\mathbf{W}$. The computational complexity of $\mathbf{W}_{\hat{\mathbf{p}}_{t-1}}(\mathbf{v}, -\mathbf{m})$ will be significantly lower than that of recomputing $\mathcal{G}_{\mathbf{v}|\hat{\mathbf{p}}_{t-1}+\mathbf{m}}$ and the corresponding Hessian $\mathbf{H}$ in every iteration.

Reintroducing the illumination variation, the lighting parameter $\mathbf{l}$ can be estimated using

$$\hat{\mathbf{l}} = (\mathcal{B}_{l(l)} \mathcal{B}^{\mathbf{T}}_{l(l)})^{-1} \mathcal{B}_{l(l)} \mathcal{I}^{\mathbf{T}}_{(l)},$$

where the subscript (l) indicates the unfolding operation [10] along the illumination dimension. That is, assuming an Nth-order tensor $\mathcal{A} \in \mathbb{R}^{I_1 \times I_2 \times \ldots \times I_N}$, the matrix unfolding $\mathbf{A}_{(n)} \in \mathbb{R}^{I_n \times (I_{n+1} I_{n+2} \ldots I_N I_1 I_2 \ldots I_{n-1})}$ contains the element $a_{i_1 i_2 \ldots i_N}$ at the position with row number i_n and column number equal to $(i_{n+1} - 1) I_{n+2} I_{n+3} \ldots I_N I_1 I_2 \ldots I_{n-1} + (i_{n+2} - 1) I_{n+3} I_{n+4} \ldots I_N I_1 I_2 \ldots I_{n-1} + \cdots + (i_N - 1) I_1 I_2 \ldots I_{n-1} + (i_1 - 1) I_2 I_3 \ldots I_{n-1} + \cdots + i_{n-1}$.

Following the same derivation as (9.7), we have

$$\triangle\mathbf{m} = \mathbf{H}_{\mathbf{IC}} \sum_{\mathbf{v}} (\mathcal{C}_{\mathbf{v}|\hat{\mathbf{p}}_{t-1}} \times_1 \hat{\mathbf{l}}) \left(I_t(\mathbf{W}_{\hat{\mathbf{p}}_{t-1}}(\mathbf{v}, -\mathbf{m})) - \mathcal{B}_{\mathbf{v}|\hat{\mathbf{p}}_{t-1}} \times_1 \hat{\mathbf{l}} \right) \tag{9.8}$$

where

$$\mathbf{H}_{\mathbf{IC}} = \left[\sum_{\mathbf{v}} (\mathcal{C}_{\mathbf{v}|\hat{\mathbf{p}}_{t-1}} \times_1 \hat{\mathbf{l}})(\mathcal{C}_{\mathbf{v}|\hat{\mathbf{p}}_{t-1}} \times_1 \hat{\mathbf{l}})^{\mathbf{T}} \right]^{-1} .$$

9.3 DISTRIBUTED COMPRESSION SCHEMES

The distributed compression schemes discussed in this section can all be represented by the block diagram of Figure 9.2. The choices we make for the following blocks result in the various schemes as described later in this section.

- Feature extraction
- Mode decision
- Side information extraction

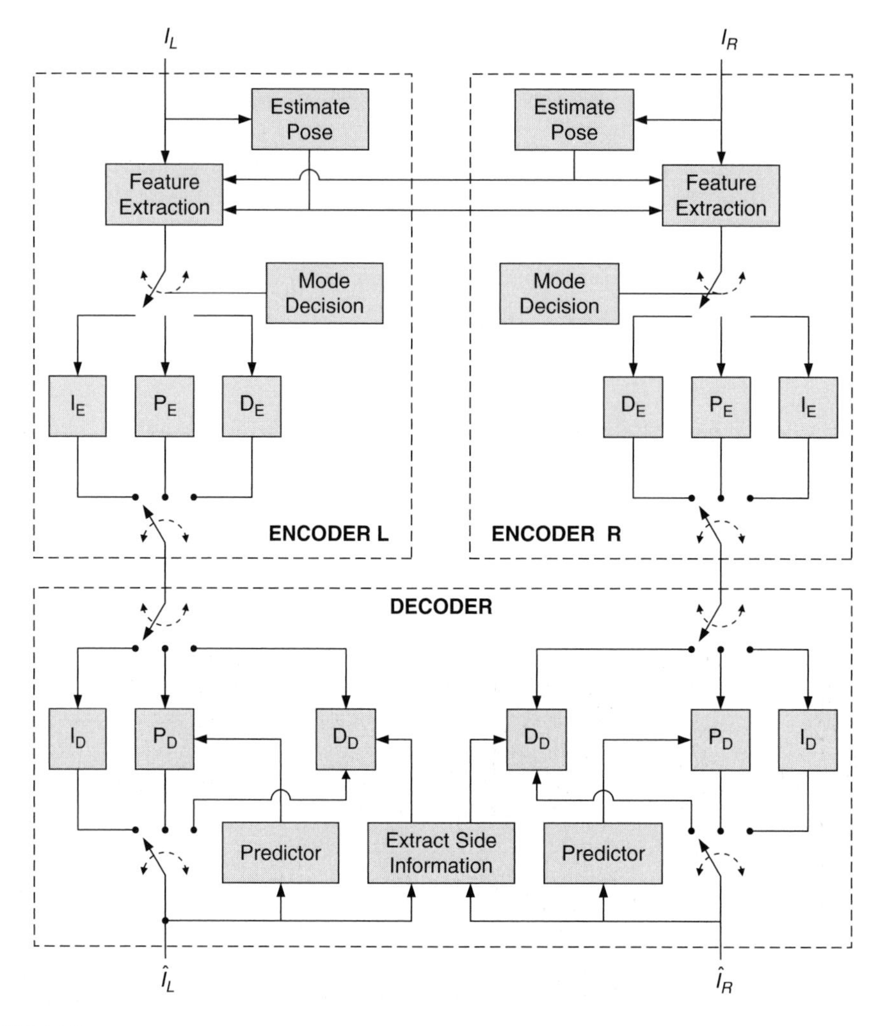

FIGURE 9.2

Block diagram of coding schemes.

9.3.1 Feature Extraction and Coding

In order to extract the image features, we need to detect them in the first frame and track them in subsequent frames. Since we estimate a 3D pose, a 3D mesh model is registered to the first frame and then tracked using the method in Section 9.2. The mesh model consists of a set of mesh points $\mathcal{V} = \{V_i = (x_i, y_i, z_i), i = 1, \ldots, N\}$ and a set of triangles $\mathcal{T} = \{T_{abc} = (V_a, V_b, V_b)\} \subset \mathcal{V}^3$ formed by mesh points. Figure 9.3 depicts the 2D projection of a triangular mesh model on a pair of frames. The model pose parameters, namely, the translation and rotation parameters, are computed independently at each view and transmitted to the decoder at high fidelity. Each encoder also receives the initial pose parameters of the other encoder.

The visibility of a mesh point V in both views is computed at each encoder by computing its position relative to each triangle $T = (V_1, V_2, V_3)$ of which it is not a member. Given the estimated pose of a certain view, let (x, y, z) and (x_i, y_i, z_i) be the coordinates of V and V_i, respectively. The triangle, which is the convex hull of the three points, occludes V if the segment connecting the origin to V passes through the triangle. To test occlusion, we project the point onto the basis formed by the vertices of T:

$$\begin{bmatrix} c_x \\ c_y \\ c_z \end{bmatrix} = \begin{bmatrix} x_1 & x_2 & x_3 \\ y_1 & y_2 & y_3 \\ z_1 & z_2 & z_3 \end{bmatrix}^{-1} \begin{bmatrix} x \\ y \\ z \end{bmatrix} \tag{9.9}$$

The triangle occludes the point if and only if c_x, c_y, and c_z are all positive and $c_x + c_y + c_z > 1$. Observe that if all coefficients are positive and sum to 1, the point lies in the convex hull of the three points forming T, that is, it lies on the triangle. If the mesh point is visible with respect to every triangle, we declare the point visible.

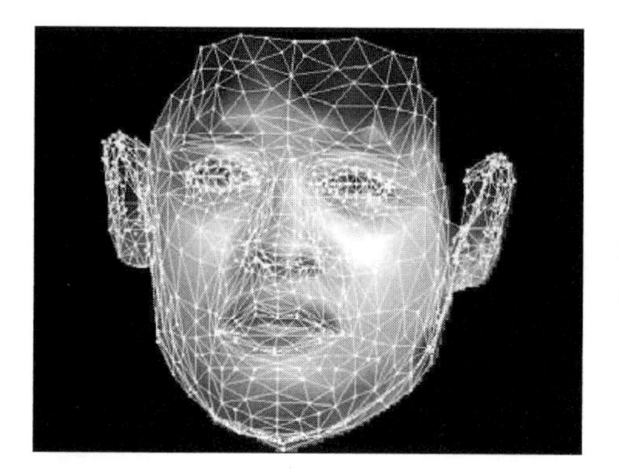

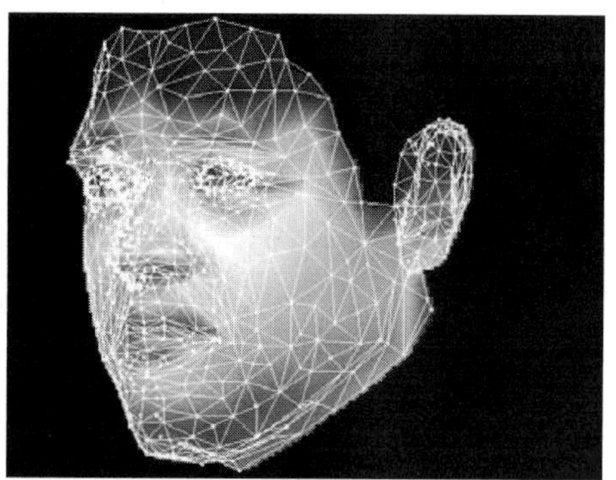

FIGURE 9.3

Mesh points $\mathcal{V}$ and triangles $\mathcal{T}$ are overlaid on the two views.

We consider three feature extraction techniques.

- Point Sampled Intensity (PSI): The feature set for an image is the set of intensity values at the visible mesh points projected onto the image plane.
- Triangles with Constant Intensity (TCI): For all triangles containing at least one visible vertex, this feature is the average of the intensity values over the visible pixels in the triangle.
- Triangles with Linearly Varying Intensity (TLI): For triangles with all vertices visible, we compute a linear least squares fit to the intensity profile. The intensity values at the vertices according to the estimated planar intensity profile form a part of the feature set. For those triangles where not all vertices are visible, we use the average intensity feature as in TCI.

For distributed encoding of the extracted features, we utilized a very simple binning technique whereby each feature is independently encoded using

$$\mathcal{C}(i) = i \bmod W$$

where $\mathcal{C}(i)$ is the codeword corresponding to the quantization index i of the coded feature and W is a parameter controlling the rate of the code. Even though the above binning scheme can be improved by more sophisticated DSC techniques (e.g., LDPC- or turbo-based codes) in general, the improvement would be marginal at very low bit rates, which we particularly focus on.

9.3.2 Types of Frames

In the compression schemes that we consider, each frame of each view can belong to one of three classes:

- Intraframe or I-frame: Each feature is encoded and decoded independently of both past frames in the same view and all frames in the other view.
- Predictive frame or P-frame: If any feature is common to the current frame and the previous frame in the same view, we only encode the difference between the two features. Features that are unique to the current frame are coded as in an I-frame.
- Distributed frame or D-frame: The features for which side information is available at the decoder are distributively coded, while the rest of the features are coded as in an I-frame.

We considered two schemes for coding sequences:

- Scheme 1: We begin with an I-frame in both views. For the subsequent DRefreshPeriod-1 sampling instants, both views are coded as P-frames. At the following instant, the left view is transmitted as an I-frame and the other view is transmitted as a D-frame. The next DRefreshPeriod-1 are transmitted as P-frames. For the next instant, the right view is an I-frame while the left is a D-frame.

Table 9.1 Frame Types of Right and Left Views in the Two Schemes and Separate Coding. (DRefreshPeriod = 3 and IRefreshPeriod = 9)

Frame Number		1	2	3	4	5	6	7	8	9	10
Scheme 1	Left	I	P	P	I	P	P	D	P	P	I
	Right	I	P	P	D	P	P	I	P	P	I
Scheme 2	Left	I	P	P	D	D	D	I	P	P	I
	Right	I	D	D	I	P	P	D	D	D	I
Separate	Left	I	P	P	I	P	P	I	P	P	I
	Right	I	P	P	I	P	P	I	P	P	I

This process then repeats. Every IRefreshPeriod instants, both views are coded as I-frames.

- Scheme 2: As in Scheme 1, we begin with I-frames in both views. For the subsequent DRefreshPeriod-1 sampling instants, the left view is coded as a P-frame, while the other view is coded as a D-frame. At the next instant, the right view is coded as an I-frame, and the other view is coded as a D-frame. This process is repeated in every block of DRefreshPeriod frame pairs with the roles of the right and left views reversing from block to block. Again, as in Scheme 1, every IRefreshPeriod instants, both views are coded as I-frames.

Table 9.1 shows some representative code sequences.

9.3.3 Types of Side Information

The side information in a D-frame can be one of three types:

- Other view (OD): The corresponding features between the frame being encoded and the one that is observed at the current instant at the other view form the side information. We can only encode features that are common to both views.
- Previous frame (PD): We use the corresponding features in the previous frame from the same view as side information. We can only encode features that are present in the previous and current views.
- Previous frame and Other view (POD): The side information is the optimal linear estimate of the source given the previous frame in the same view and the current frame in the other view. Only features that are common to the three frames involved can be distributively encoded by this method.

  The estimation coefficients are to be computed at the decoder before reconstructing a given view. So, if for example, frame R_n is distributively coded, we use the correlations from decoder reconstructions of R_{n-1}, R_{n-2}, and L_{n-1} to obtain approximations to the optimal estimation coefficients.

9.4 EXPERIMENTAL RESULTS

In this section, we evaluate the performance of the various schemes in compressing a sequence of a face viewed from two cameras under varying illumination. The face mesh model is assumed to be known to both encoders as well as the

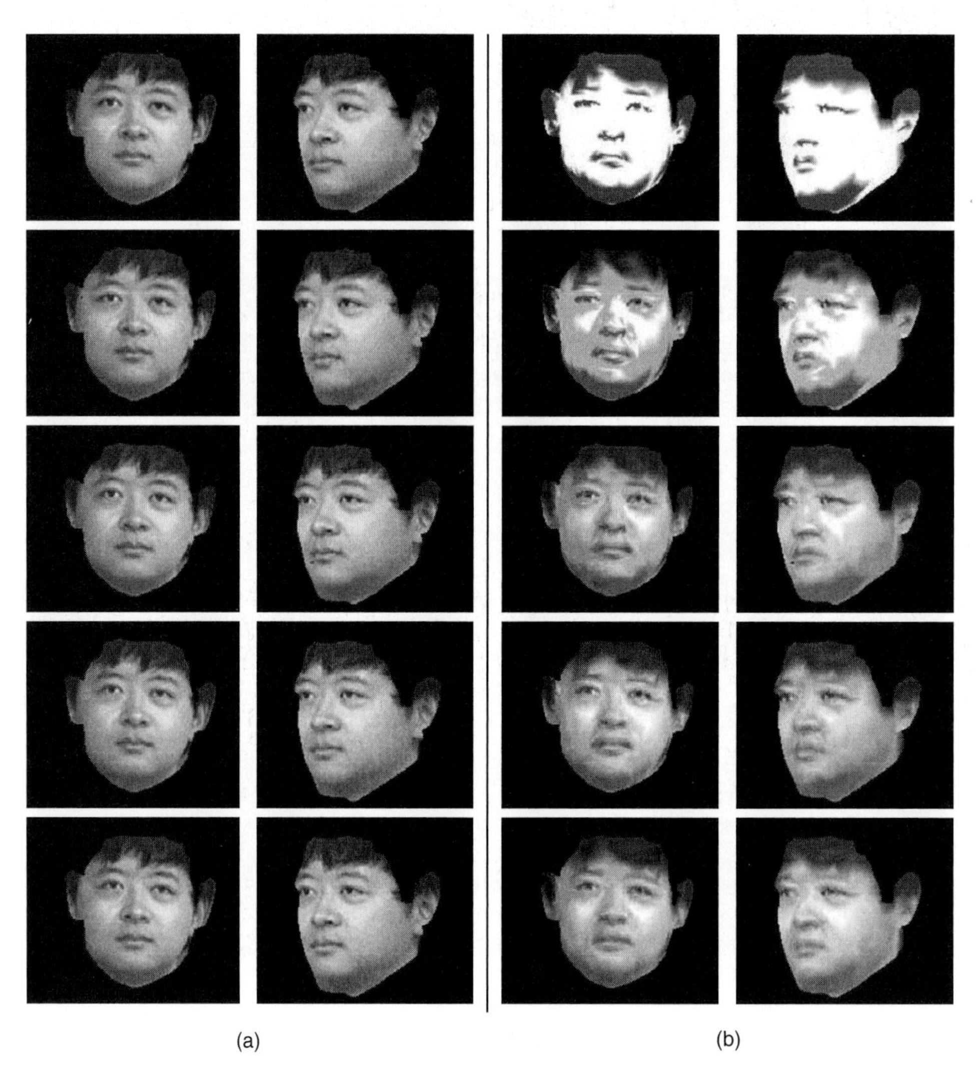

FIGURE 9.4

(a) Five consecutive frames from the original sequence. (b) Separate coding of the two views using the PSI technique and Scheme 1. The bit rate is ≈20 kbps and the PSNR is ≈16.5 dB.

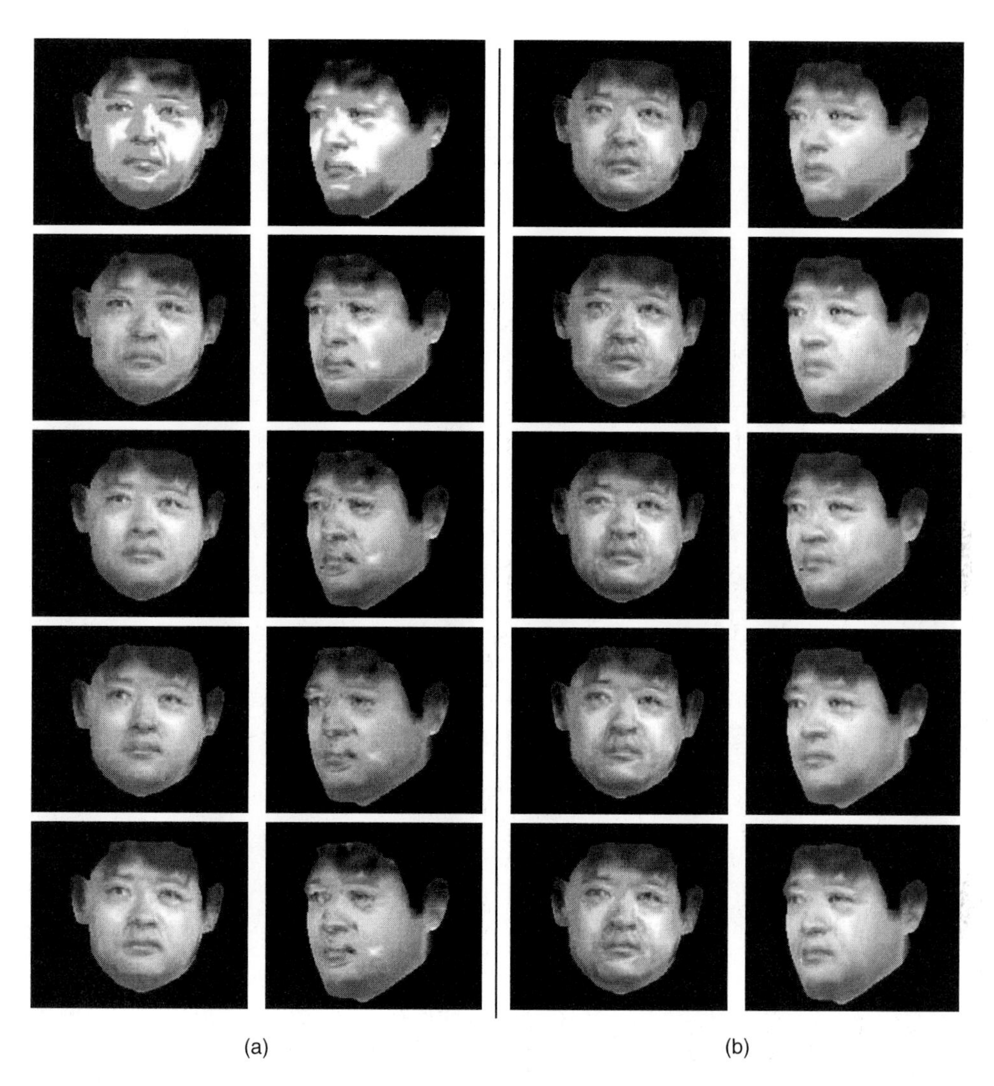

FIGURE 9.5

Distributed coding of the sequence using the PSI technique for feature extraction and POD-type side information. (a) Reconstruction using Scheme 1 with bit rate ≈22 kbps and PSNR ≈21 dB. (b) Reconstruction using Scheme 2 with bit rate ≈18 kbps and PSNR ≈22.8 dB.

decoder. Transmission of the initial model pose parameters forms the sole interencoder communication.

For both schemes and all feature and side information types, we tested the performance on a 15-frame sequence from each view. Five consecutive original frames are shown in Figure 9.4(a). We fixed DRefreshPeriod = 5 and IRefreshPeriod = 15.

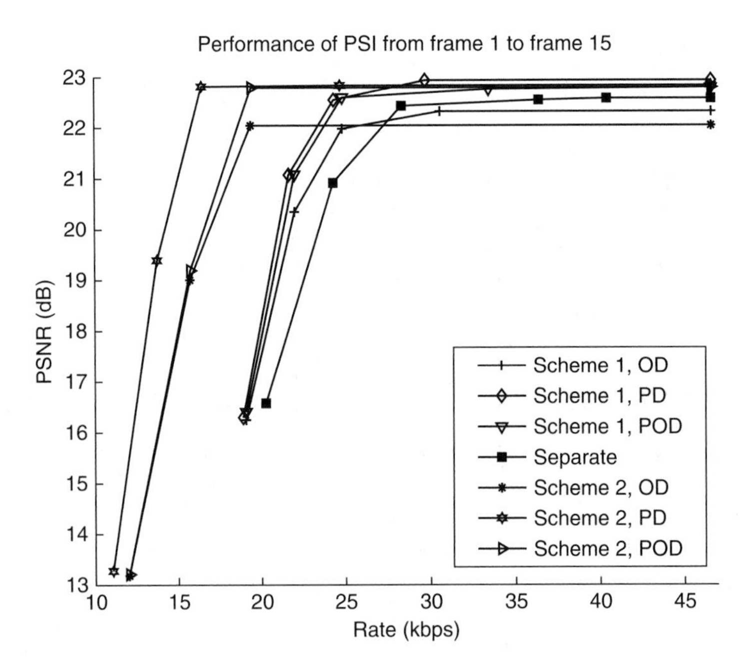

FIGURE 9.6

The rate-distortion trade-off using the PSI feature extraction technique.

In Figures 9.4(b), 9.5(a), and 9.5(b), examples of reconstructed sequences are shown, respectively, for separate coding, Scheme 1, and Scheme 2, all using the PSI technique for feature extraction, and the latter two using POD-type side information. The bit rates are fixed at around 20 kbps. The complete rate-distortion trade-off obtained by running all methods with various parameters is shown in Figure 9.6, where it is clearly seen that the gains are more significant at lower bit rates.

We then increase the target bit rate to around 41 kbps by adopting the TCI technique for feature extraction. The reason for the rate increase even with the same quantization parameters is that the number of triangles is about twice as large as the number of mesh points. As apparently seen from the reconstructed sequences shown in Figures 9.7 and 9.8, this results in a quality increase. That is because the average of intensity values within a triangle is a much better representative of the triangle than a reconstruction based on only the three corner values. However, as can be seen from the complete rate-distortion trade-off shown in Figure 9.9, the increase in the reconstruction quality comes at the expense of reduced distributed coding gains.

Using the TLI technique for feature extraction, we further increase the bit rate approximately by a factor of 3, since most triangles are now represented by three parameters. The reconstructed sequences are shown in Figures 9.10 and 9.11, and the

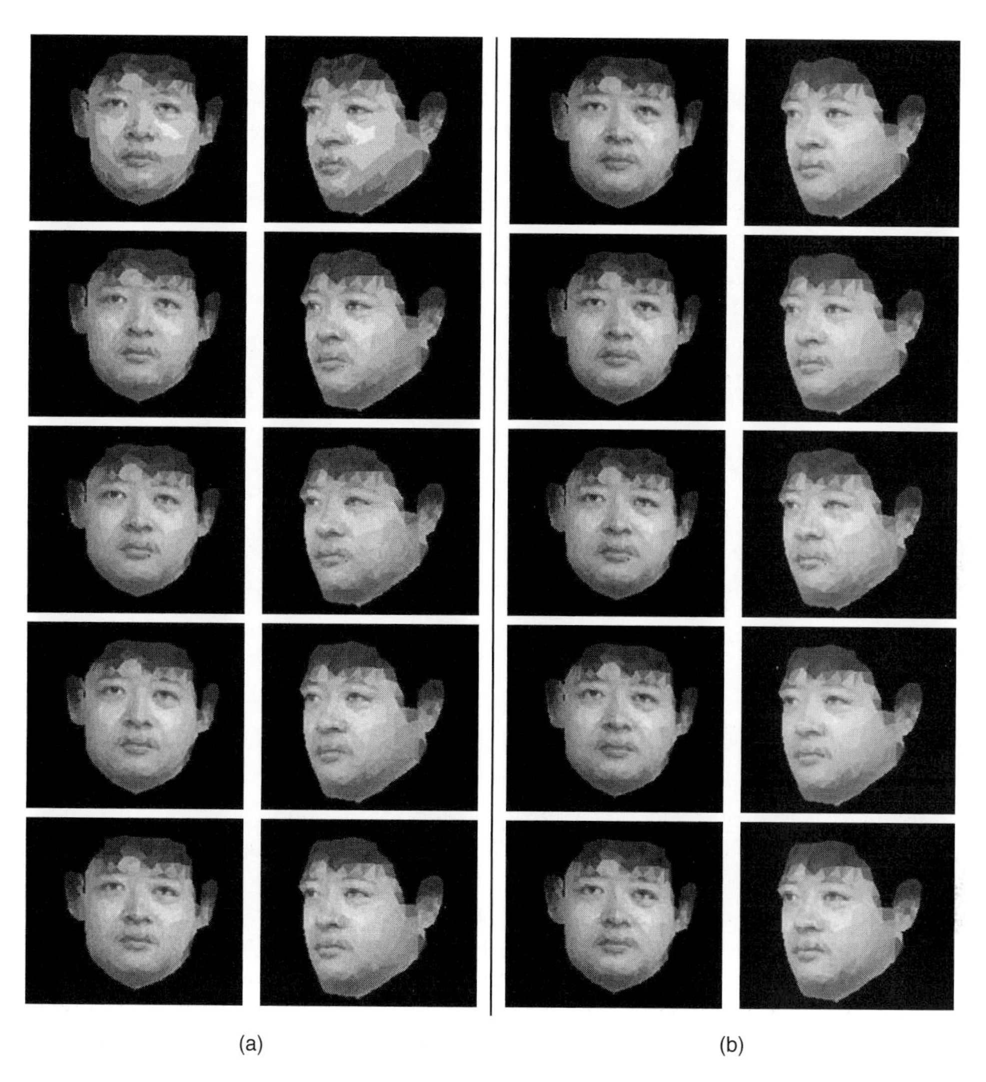

FIGURE 9.7

Coding using the TCI technique for feature extraction. (a) Separate coding of the two views using Scheme 1 with bit rate ≈41 kbps and PSNR ≈23.9 dB. (b) Distributed coding using Scheme 2 and PD-type side information with the same bit rate PSNR ≈25.5 dB.

complete rate-distortion trade-off is depicted in Figure 9.12. The distributed coding gains are further diminished in this regime.

Even though PD-type side information yields the best performance at low bit rates as seen from Figures 9.6, 9.9, and 9.12, reconstruction based on POD-type side

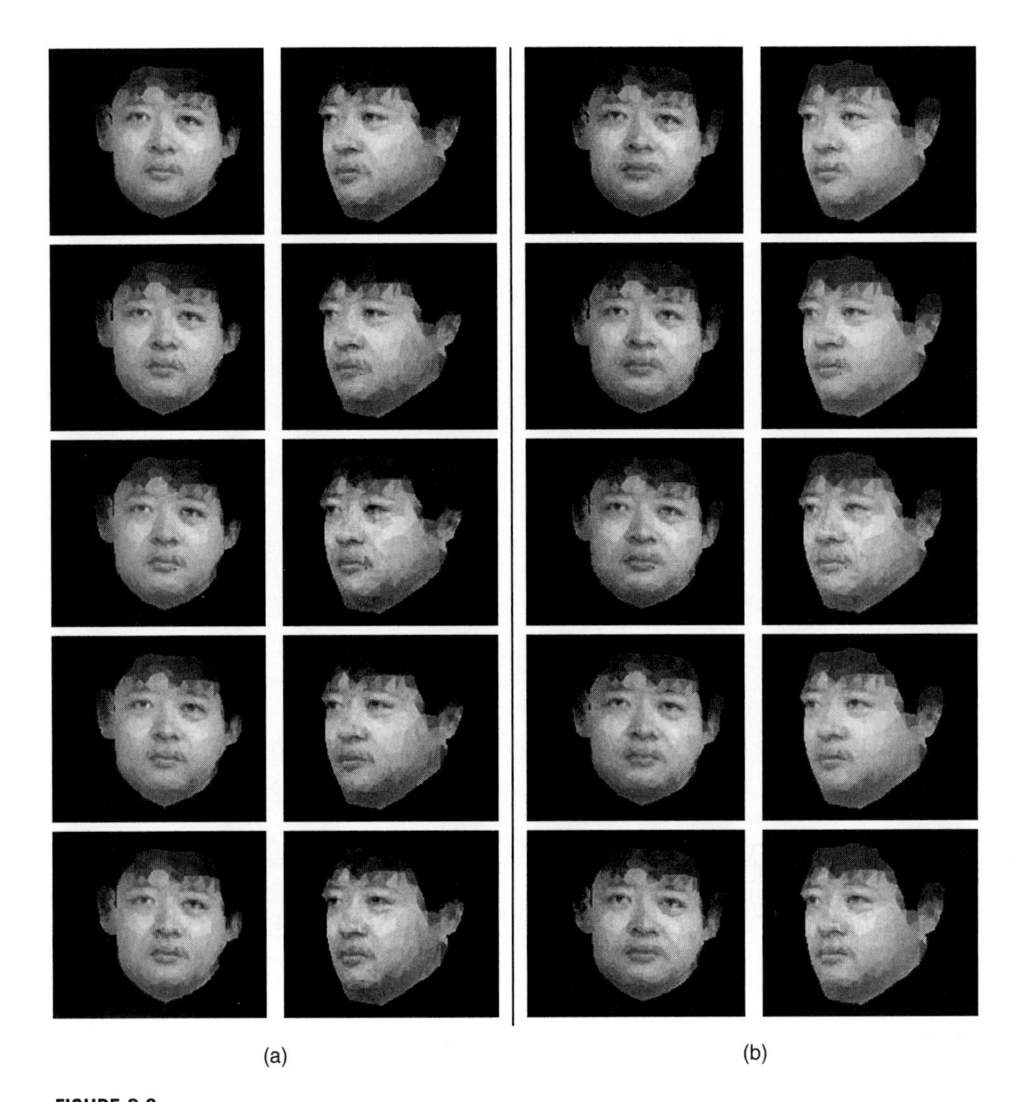

FIGURE 9.8

Coding using the TCI technique for feature extraction and POD-type side information. (a) Distributed coding using Scheme 1 with bit rate $\approx$40 kbps and PSNR $\approx$24.4 dB. (b) Distributed coding using Scheme 2 with bit rate $\approx$43 kbps and the same PSNR.

information is more beneficial in scenarios in which the channel carrying the D-frame is less reliable and can result in a very noisy reference for the future D-frames. On the other hand, since the temporal correlation is much higher than the correlation between the views, relying solely on the other view always yields worse performance.

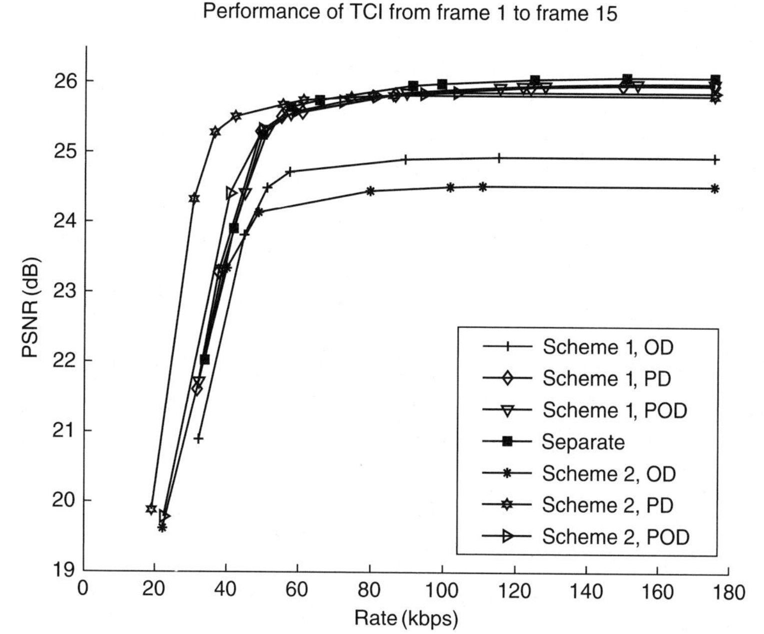

FIGURE 9.9

The rate-distortion trade-off using the TCI feature extraction technique.

9.5 CONCLUSIONS

In this article, we have presented a method for distributed compression of two video sequences by using a combination of 3D model-based motion estimation and distributed source coding. For the motion estimation, we propose to use a newly developed inverse compositional estimation technique that is computationally efficient and robust. For the coding method, a binning scheme was used, with each feature being coded independently. Different methods for rendering the decoded scene were considered. Detailed experimental results were shown. Analyzing the experiments, we found that the distributed video compression was more efficient than separate motion-estimation-based coding at very low bit rates. At higher bit rates, the ability of the motion estimation methods to remove most of the redundancy in each video sequence left very little to be exploited by considering the overlap between the views. We believe that this should be taken into account in the future while designing distributed coding schemes in video.

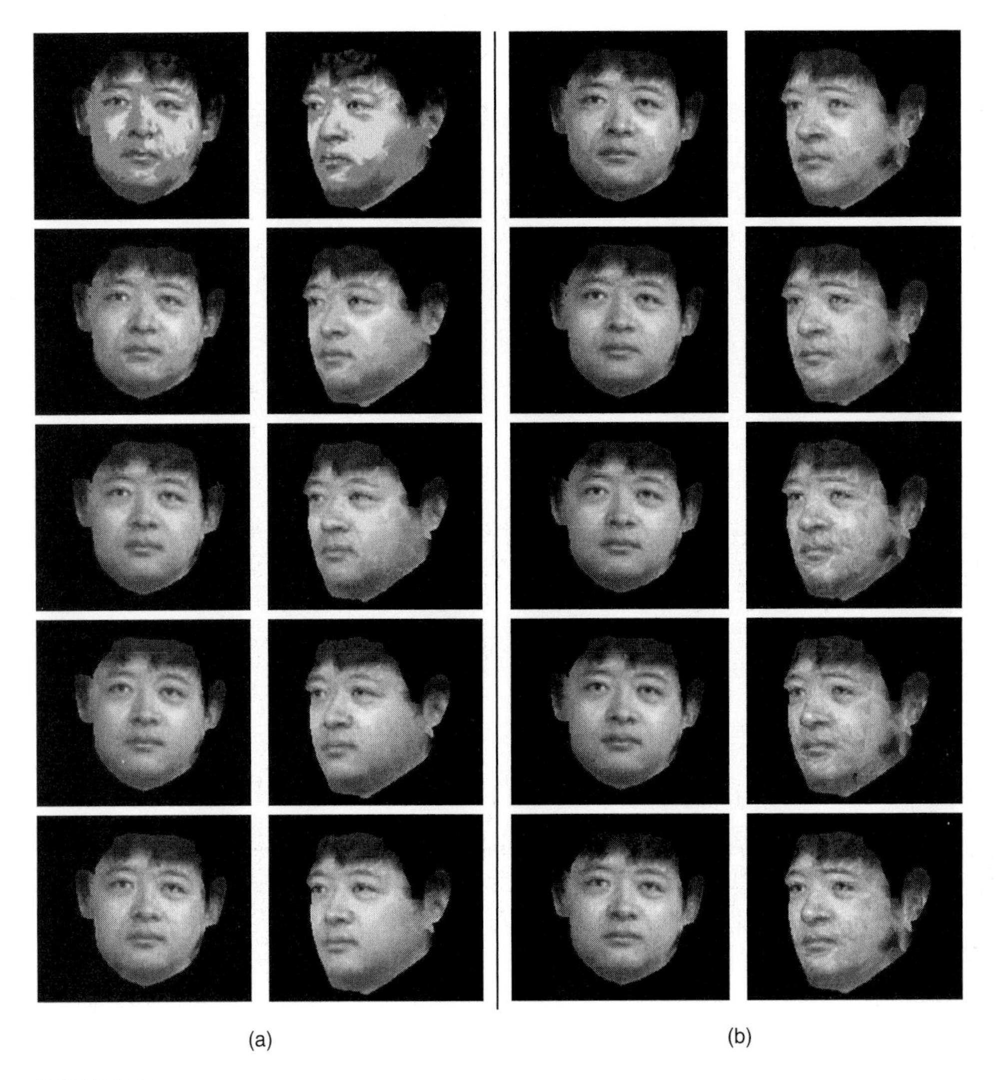

FIGURE 9.10

Coding using the TLI technique for feature extraction. (a) Separate coding of the two views using Scheme 1 with bit rate $\approx$ 120 kbps and PSNR $\approx$28 dB. (b) Distributed coding using Scheme 1 and OD-type side information with bit rate $\approx$128 kbps and PSNR $\approx$27 dB.

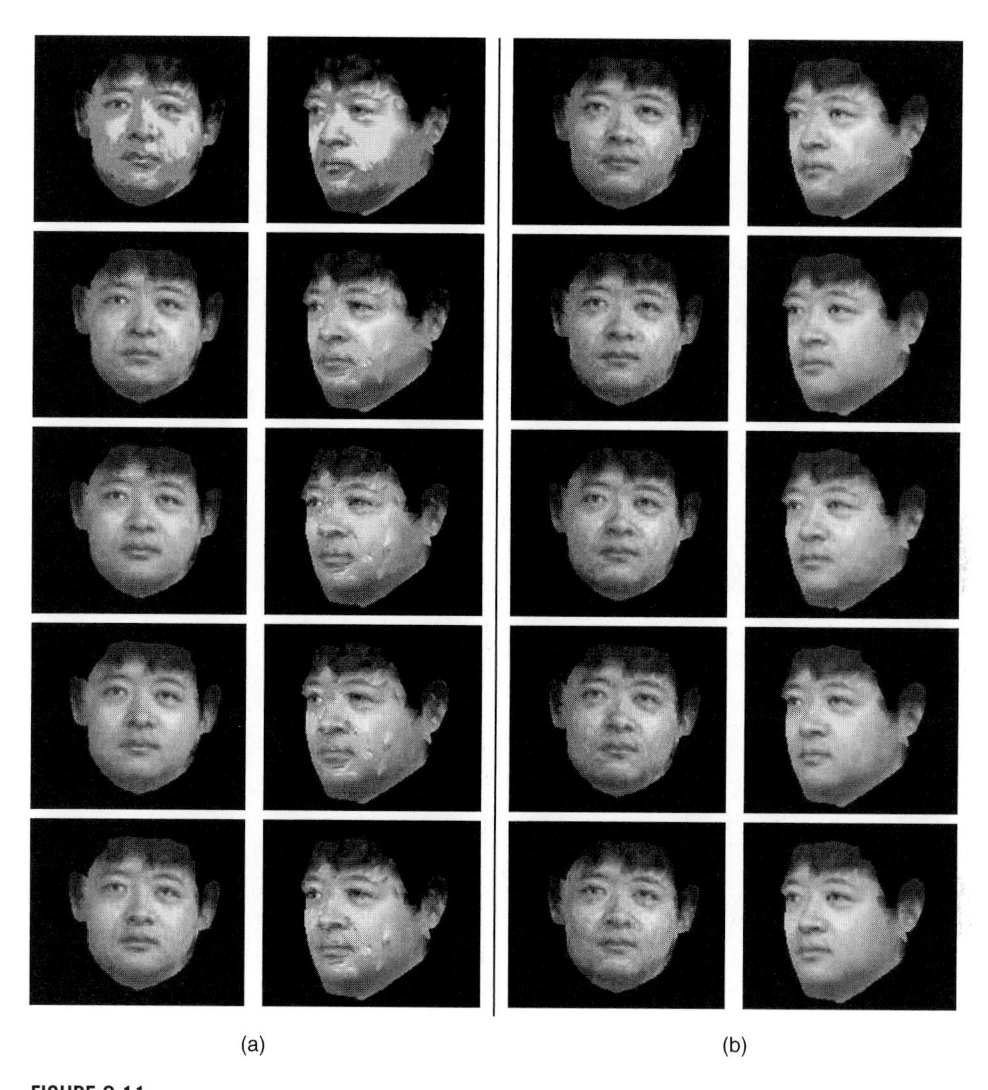

(a) (b)

FIGURE 9.11

Coding using the TLI technique for feature extraction and POD-type side information. (a) Distributed coding using Scheme 1 with bit rate ≈ 128 kbps and PSNR ≈ 28.7 dB. (b) Distributed coding using Scheme 2 with bit rate ≈ 115 kbps and PSNR ≈ 26.5 dB.

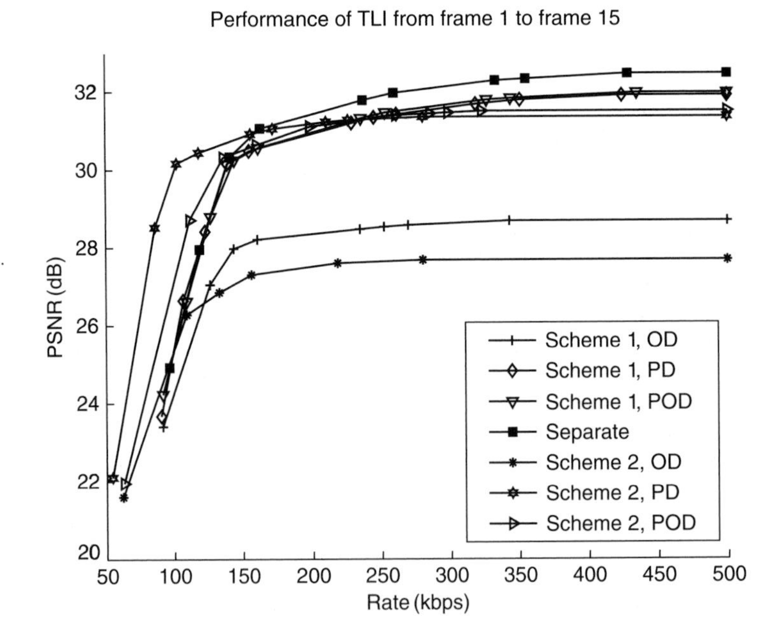

FIGURE 9.12

The rate-distortion trade-off using the TLI feature extraction technique.

REFERENCES

[1] Artigas, X., Angeli, E., and Torres, L.: 2006, Side information generation for multiview distributed video coding using a fusion approach, *7th Nordic Signal Processing Symposium*, 250–253.

[2] Baker, S., and Matthews, I.: 2004, Lucas-Kanade 20 years on: A unifying framework, *International Journal of Computer Vision* **56**(3), 221–255.

[3] Blanz, V., and Vetter, T.: 2003, Face recognition based on fitting a 3D morphable model, *IEEE Trans. on Pattern Analysis and Machine Intelligence* **25**(9), 1063–1074.

[4] Freedman, D., and Turek, M.: 2005, Illumination-invariant tracking via graph cuts, *Proc. of IEEE Conference on Computer Vision and Pattern Recognition*.

[5] Gehrig, N., and Dragotti, P. L.: 2005, DIFFERENT: Distributed and fully flexible image encoders for camera sensor networks, *International Conference on Image Processing*.

[6] Girod, B., Margot, A., Rane, S., and Rebollo-Monedero, D.: 2005, Distributed video coding, *Proceedings of the IEEE* **93**(1), 71–83.

[7] Guo, X., Lu, Y., Wu, F., Gao, W., and Li, S.: 2006, Distributed multi-view video coding, *Visual Communications and Image Processing 2006* **6077**(1), 60770T.1–60770T.8.

[8] Hager, G. D., and Belhumeur, P.: 1998, Efficient region tracking with parametric models of geometry and illumination, *IEEE Trans. on Pattern Analysis and Machine Intelligence* **20**(10), 1025–1039.

[9] Jin, H., Favaro, P., and Soatto, S.: 2001, Real-time feature tracking and outlier rejection with changes in illumination, *IEEE Intl. Conf. on Computer Vision*.

[10] Lathauwer, L. D., Moor, B. D., and Vandewalle, J.: 2000, A multilinear singular value decomposition, *SIAM J. Matrix Anal. Appl.* **21**(4), 1253–1278.

[11] Matthews, I., and Baker, S.: 2004, Active appearance models revisited, *International Journal of Computer Vision* **60**(2), 135–164.

[12] Ouaret, M., Dufaux, F., and Ebrahimi, T.: 2006, Fusion-based multiview distributed video coding, *4th ACM International Workshop on Video Surveillance and Sensor Networks*, 139–144.

[13] Pradhan, S., and Ramchandran, K.: 1999, Distributed source coding using syndromes (DISCUS): Design and construction, *Data Compression Conference*, 158–167.

[14] Puri, R., and Ramchandran, K.: 2007, PRISM: A video coding architecture based on distributed compression principles, *IEEE Transactions on Image Processing* **16**(10), 2436–2448.

[15] Romdhani, S., and Vetter, T.: 2003, Efficient, robust and accurate fitting of a 3D morphable model, *IEEE International Conference on Computer Vision 2003*.

[16] Shum, H.-Y., and Szeliski, R.: 2000, Construction of panoramic image mosaics with global and local alignment, *International Journal of Computer Vision* **16**(1), 63–84.

[17] Slepian, D., and Wolf, J. K.: 1973, Noiseless coding of correlated information sources, *IEEE Transactions on Information Theory* **19**(4), 471–480.

[18] Song, B., Bursalioglu, O., Roy-Chowdhury, A., and Tuncel, E.: 2006, Towards a multi-terminal video compression algorithm using epipolar geometry, *IEEE Intl. Conf. on Acoustics, Speech and Signal Processing*.

[19] Wagner, R., Nowak, R., and Baranuik, R.: 2003, Distributed image compression for sensor networks using correspondence analysis and superresolution, *ICIP*, vol. 1: pp. 597–600.

[20] Wyner, A.: 1974, Recent results in the Shannon theory, *IEEE Transactions on Information Theory* **20**(1), 2–10.

[21] Xu, Y., and Roy-Chowdhury, A.: 2007, Integrating motion, illumination and structure in video sequences, with applications in illumination-invariant tracking, *IEEE Trans. on Pattern Analysis and Machine Intelligence* **29**(5), 793–807.

[22] Xu, Y., and Roy-Chowdhury, A.: 2008, A Theoretical analysis of linear and multi-linear models of image appearance, *IEEE International Conference on Computer Vision and Pattern Recognition*.

[23] Xu, Y., and Roy-Chowdhury, A.: 2008, Inverse compositional estimation of 3d motion and lighting in dynamic scenes, *IEEE Trans. on Pattern Analysis and Machine Intelligence*, vol. 30, no. 7, pp. 1300–1307, July 2008.

[24] Yang, Y., Stankovic, V., Zhao, W., and Xiong, Z.: 2007, Multiterminal video coding, *IEEE International Conference on Image Processing*, pp. III 28–28.

[25] Zamir, R., and Shamai, S.: 1998, Nested linear/lattice codes for Wyner-Ziv encoding, *Information Theory Workshop*, pp. 92–93.

[26] Zhu, X., Aaron, A., and Girod, B.: 2003, Distributed compression for large camera arrays, *IEEE Workshop on Statistical Signal Processing*.

CHAPTER

Distributed Compression of Hyperspectral Imagery

10

Ngai-Man Cheung and Antonio Ortega

Signal and Image Processing Institute, Department of Electrical Engineering, University of Southern California, Los Angeles, CA

CHAPTER CONTENTS

10.1 INTRODUCTION

Hyperspectral images are volumetric image cubes that consist of hundreds of spatial images. Each spatial image, or *spectral band*, captures the responses of ground objects at a particular wavelength (Figure 10.1). For example, the NASA AVIRIS (Airborne Visible/Infrared Imaging Spectrometer) measures the spectral responses in

Distributed Source Coding: Theory, Algorithms, and Applications

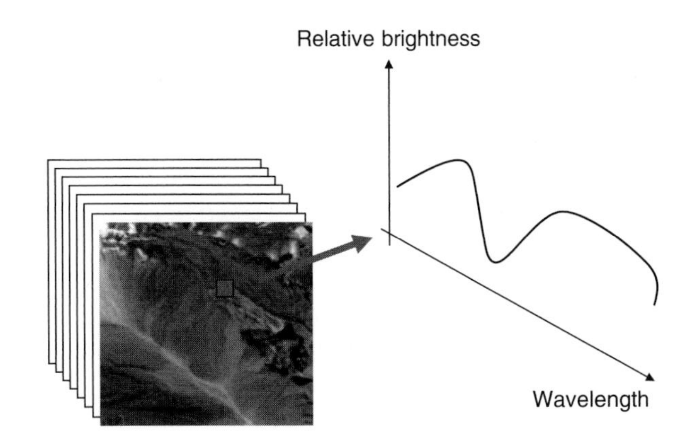

FIGURE 10.1

Hyperspectral image.

Table 10.1 Examples of Research and Commercial Imaging Spectrometers [45]

Sensor	Organization	Country	Number of Bands	Wavelength Range (μm)
AVIRIS	NASA	United States	224	0.4–2.5
AISA	Spectral Imaging Ltd.	Finland	286	0.45–0.9
CASI	Itres Research	Canada	288	0.43–0.87
PROBE-1	Earth Search Sciences Inc.	United States	128	0.4–2.45

224 contiguous spectral bands acquired at the visible and near-infrared regions [24] (see Table 10.1 for some examples of imaging spectrometers). Effectively, pixel values of spectral bands along the spectral direction represent the spectra of the captured objects, which can provide useful information to identify the ground objects. For example, scientists may use hyperspectral data to determine if a pixel may correspond to vegetation, soil, or water, which have quite distinct spectral responses [4]. Based on the frequency locations of specific *absorption bands*, the wavelength ranges where the materials selectively absorb the incident energy, it may be possible to infer the molecular compositions of the objects [4, 45]. Thanks to its rich spatial and spectral information contents, hyperspectral imagery has become an important tool for a variety of remote sensing and scanning applications; it is widely used in the sensing and discovering of ground minerals [4], in monitoring of the Earth's resources [51], and in military surveillance [28, 43].

The raw data size of hyperspectral images is very large. For example, a single hyperspectral image captured by NASA AVIRIS could contain up to 140M bytes of raw data [25]. Therefore, efficient compression is necessary for practical hyperspectral imagery applications. In addition, hyperspectral images are usually captured by satellites or spacecrafts that use embedded processors with limited resources, so encoding complexity is an important issue in hyperspectral image compression. These are thus clearly asymmetric applications where reductions in encoding time and cost are desirable even if they come at the cost of increases in decoding complexity. Our focus in this chapter will be on exploring how distributed source coding (DSC) techniques [14, 16, 34, 44, 59] can facilitate reductions in encoding cost. Encoding cost can be quantified in different ways depending on the specific implementation platform chosen. For example, in previous works, methods have been proposed to reduce encoder power consumption in field programmable gate array (FPGA) implementations using new closed-loop prediction techniques [31], or to speed up encoding using a parallel implementation [2, 8].

10.1.1 Hyperspectral Imagery Compression: State of the Art

In a hyperspectral dataset many spectral bands tend to be correlated. For example, Figure 10.2 shows the scenario in which pixel values of the current spectral band are predicted by the co-located pixel values of the adjacent spectral band after linear filtering, and the mean-square residuals after prediction are plotted against spectral band indexes. The figure suggests that neighboring bands tend to be correlated, and the degree of correlation varies relatively slowly over a broad range of spectral regions. Therefore, a key element of state-of-the-art lossy-to-lossless hyperspectral compression has been to develop techniques to exploit interband correlation (see Section 10.2 for

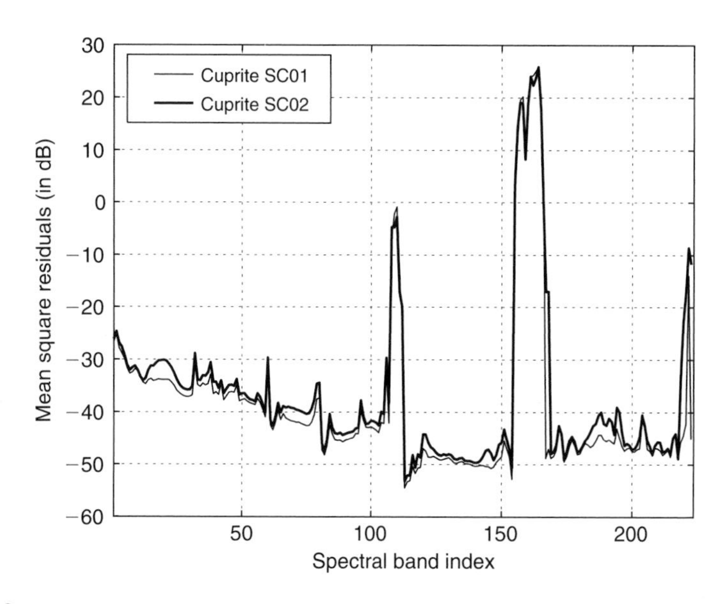

FIGURE 10.2

Mean-square residuals after linear prediction.

a more detailed discussion of the correlation characteristics that motivate the design of these algorithms):

- In *interband predictive approaches* [31, 37, 48], a band is predicted using previously encoded bands. The resulting prediction residuals can then be encoded using standard image coding techniques (e.g., transformation followed by quantization and entropy coding). Since, typically, the prediction residual has a much lower energy than the original band, encoding it usually requires fewer bits than encoding the original band. Interband prediction approaches are analogous to predictive video coding (e.g., the standard MPEG/H.26x compression algorithms), but a key difference is that motion estimation/compensation are not necessary since co-located pixels in different spectral bands represent the responses of the same ground object at different wavelengths. However, simple linear prediction to align the values of co-located pixels of consecutive spectral bands has been found to be useful to improve the prediction gain [31].
- *3D wavelet methods*, including 3D SPIHT [15, 21], 3D SPECK [47], and 3D ICER [20, 22], provide alternatives to predictive coding. 3D wavelet methods exploit interband correlation by performing filtering across spectral bands, with the expectation that most of the signal energy will be concentrated in low-pass subbands (corresponding to low spatial and "cross-band" frequencies). Similar to typical 2D wavelet image coding, bit-plane coding is usually applied to the transform coefficients obtained from filtering. The resulting bits can be compressed with entropy coding schemes such as zero-tree coding [41] or those based on context modeling [20].

As an example to illustrate 3D wavelet approaches, in 3D ICER [20, 22], a modified 3D Mallat decomposition is first applied to the image cube. Then mean values are subtracted from the spatial planes of the spatially low-pass subbands to account for the "systematic difference" in different image bands [22, 23].[1] Bit-plane coding is then applied to the transform coefficients, with each coefficient bit adaptively entropy-coded based on its estimated probability-of-zero statistics. The probability is estimated by classifying the current bit to be encoded into one of several contexts according to the significance status of the current coefficient and its spectral neighbors. A different probability model is associated with each context.

In addition, a lossy compression scheme using the Karhunen–Loève transform (KLT) and the discrete cosine transform (DCT) was proposed in [40]; several lossless compression schemes have also been proposed for hyperspectral data, for example, those based on vector quantization [32, 39], and spatial, spectral, or hybrid prediction [38, 52, 58].

[1] Note that mean subtraction in these 3D wavelet approaches is conceptually similar to using linear prediction to align the values of co-located pixels in order to account for the global offset in each spatial image specific to hyperspectral data [22].

10.1.2 Outline of This Chapter

Section 10.2 of this chapter discusses the general issues of hyperspectral data compression. Datasets and correlation characteristics of the commonly used NASA AVIRIS data will be discussed, followed by discussions on some potential problems in applying interband prediction approaches and 3D wavelet approaches for hyperspectral data. Motivated by these potential problems in conventional approaches, application of distributed coding principles for hyperspectral data compression will then be discussed in Section 10.3, with a focus on the potential advantages and associated challenges. Section 10.4 then summarizes several examples of DSC-based hyperspectral compression algorithms. Section 10.5 concludes the chapter.

10.2 HYPERSPECTRAL IMAGE COMPRESSION

10.2.1 Dataset Characteristics

As illustrated in Figure 10.1, hyperspectral images are three-dimensional image cubes, with two spatial dimensions and one spectral dimension. A well-known example is the NASA AVIRIS data, which consists of 224 spatial images [2, 24]. Each spatial line of AVIRIS data has 614 pixels, and datasets with 512 spatial lines are commonly available, leading to about 140M bytes of information. AVIRIS captures wavelengths from 400 to 2500 nanometers, at a spectral resolution of about 10 nanometers. Spatial resolution would depend on how high above the surface the scanning operations are taking place. For example, at an altitude of 20 kilometers, AVIRIS has a spatial resolution of 20 meters. In typical operations, spatial resolutions of 2 to 20 meters can be achieved [2, 45].

Pixel values of hyperspectral images along the spectral dimension provide important information regarding the extent of energy absorption and scattering at different wavelengths of the captured objects. Figure 10.3 shows some examples of spectral curves of hyperspectral image data. Different materials tend to exhibit different spectral responses [33]. Therefore, these spectral curves can be very useful for classifying the ground objects. It should be noted that, depending on the spatial resolution and hence the size of the ground resolution cell, the values of a pixel may include contributions from more than one material; that is, the hyperspectral data may represent *composite* or *mixed* spectrum. The contribution from each material may depend on its area within a ground resolution cell [45].

Alternatively, spectral information of each pixel may also be represented as a spectral vector in the n-dimensional space, where n is the number of spectral bands, and the response at each wavelength corresponding to the projection of the vector onto the respective coordinate axis. This vector representation provides a useful mean for automatic classification of the captured materials. For example, the similarity of two spectra can be measured by the angle between the corresponding spectral vectors, such as in the well-known spectral angle mapper (SAM) algorithm [13].

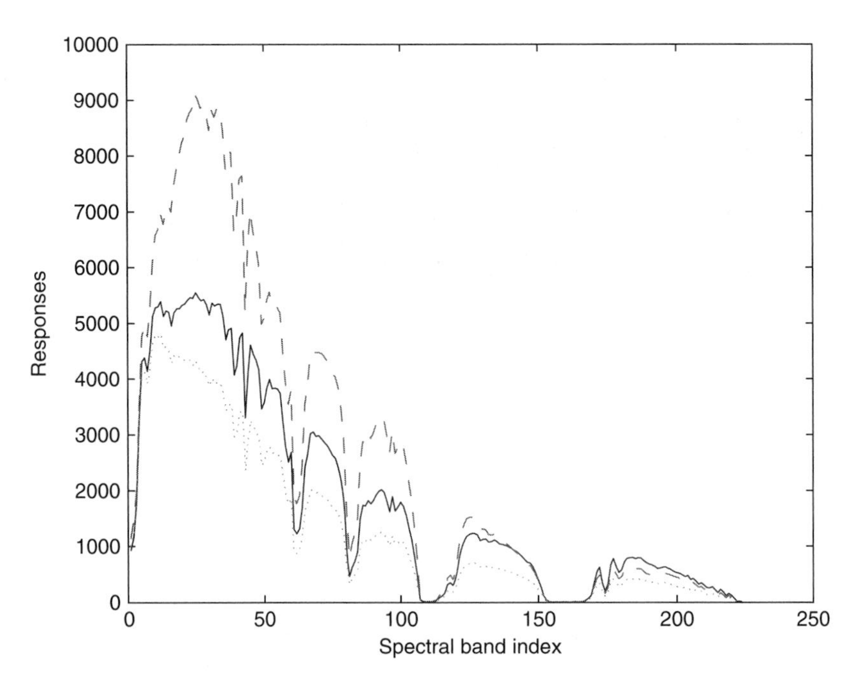

FIGURE 10.3

Some examples of spectral responses in hyperspectral image dataset: NASA AVIRIS Cuprite.

From the perspective of lossy compression, it is therefore important to preserve the spectral characteristics of hyperspectral images, in addition to the average reconstruction fidelity (measured by peak signal-to-noise ratio (PSNR), for example) of the datasets. Lossy hyperspectral compression algorithms may be compared by the classification performance of the reconstructed images, in addition to rate-distortion performance [8, 47].

10.2.2 Intraband Redundancy and Cross-band Correlation

Hyperspectral applications may exploit the spatial and/or the spectral redundancy to achieve compression. The degree of spatial redundancy within each spectral band would obviously depend on the captured scene. Moreover, while the spatial images of a hyperspectral dataset represent captured data of the same scene and therefore the spatial contents of most images would be similar, it is possible that some images may have different spatial contents and hence different amounts of intraband redundancy, depending on the response characteristics of the ground objects. For example, Figure 10.4 shows several spatial images in a hyperspectral dataset. Notice that in this example the image data of the first spectral band is rather noisy. Therefore, correlation between pixels may not be high, impacting the performance of using intraband coding schemes to compress the data.

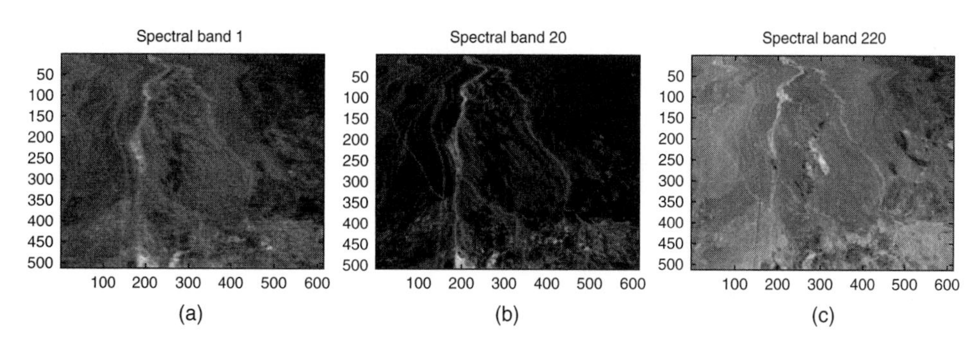

FIGURE 10.4

Some spatial images from the hyperspectral dataset Cuprite: (a) band 1; (b) band 20; (c) band 220. Notice that the image data in (a) is rather blurred and noisy, which may impact compression performance even when each image is encoded independently of the others.

Interband correlation in hyperspectral datasets tends to be significant. For example, Figure 10.2 in Section 10.1.1 illustrates some examples of cross-band correlation, and suggests that spectral bands are in general highly correlated. In fact, some 3D wavelet approaches use context-adaptive entropy coding where the context definitions rely on only the spectral correlation [20]. However, Figure 10.2 also suggests that significant changes in the degree of correlation may occur occasionally. Furthermore, interband correlation may not be spatially uniform, which leads to the observation that in a 3D Mallat decomposition of hyperspectral data, spatial low-pass and spectral high-pass subbands tend to retain a significant amount of structural information, which resembles that of the original spatial images [22]. Previous work has demonstrated that modified transforms that further decompose these spatial low-pass and spectral high-pass subbands may lead to coding gain in hyperspectral compression [12, 20, 22, 55].

Based on the previous discussions, we can conclude that efficient hyperspectral compression should aim to exploit *both* intraband and cross-band redundancy, and should do so in an *adaptive* switching and/or combining tools that exploit these different types of redundancies. For example, a (nondistributed) lossless compression scheme, 3D CALIC, uses local correlation information to switch between intraband and interband pixel predictors, leading to state-of-the-art coding performance [58]. Note, however, that correlation modeling, estimation, and adaptive encoding in the context of distributed coding remain challenging, as will be further discussed in what follows.

10.2.3 Limitations of Existing Hyperspectral Compression Techniques

As discussed in Section 10.1.1 interband prediction approaches and 3D wavelet approaches have been proposed for hyperspectral image compression. Although these algorithms can achieve state-of-the-art coding efficiency, they may face potential problems when applying them to on-board hyperspectral image compression.

These potential problems serve as motivation to study new hyperspectral compression techniques based on DSC.

10.2.3.1 Interband Prediction Approaches

In interband prediction approaches, previously encoded bands are used to predict the current band and the resulting prediction residuals are then further compressed. Interband prediction approaches can in general achieve a high compression ratio with moderate memory requirements. There are several drawbacks, however.

First, interband prediction methods need to generate exact copies of the decoded bands at the encoder, so encoders need to perform both encoding and decoding,[2] and decoding complexity could be significant, for example, comparable to encoding complexity. Previous work has proposed to reduce the complexity of this decoding loop. For example, [31] has proposed using only full bit-planes to form the predictors at the encoder and decoder when using the Set Partitioning in Hierarchial Trees (SPIHT) algorithm [41] to compress the residue. This avoids bit-plane decoding at the encoder. However, since this approach does not utilize fractional bit-plane information in reconstruction, the predictors in general have worse qualities compared to that of conventional interband prediction methods, leading to degradation in coding performance.

Second, interband predictive methods are inherently serial, since each band is encoded based on a predictor obtained from previously encoded bands. Therefore, it is difficult to parallelize processing in order to improve the processing speed of an interband predictive encoder to handle the high data rate generated by hyperspectral imaging instruments.

Third, it is difficult to achieve efficient rate scalability. This is because, while interband prediction approaches would stipulate identical predictors to be used in encoding and decoding, bit-rate scaling by on-the-fly truncation of the embedded bitstream may, however, lead to different reconstructions at the decoder, depending on the truncation points. Therefore, rate scaling could result in mismatches between the predictors at the encoder and the decoder, leading to drift and degradation of the reconstruction quality.

10.2.3.2 3D Wavelet Approaches

3D wavelet methods exploit interband correlation by performing filtering across spectral bands, which in general can lead to energy concentration in low-pass subbands. While 3D wavelet methods can achieve good compression efficiency with excellent scalability, a main disadvantage is that they lead to complex memory management issues. A naive implementation would consist of loading several spectral bands in memory so as to perform cross-band filtering, leading to expensive memory requirements.

[2] The exception is the case when compression is lossless since the original images will be available for decoding and thus can be used for prediction.

More sophisticated approaches are possible (e.g., loading simultaneously only subbands corresponding to a given spatial frequency in various spectral bands), but these approaches have the drawback of requiring numerous iterations of memory access.

10.3 DSC-BASED HYPERSPECTRAL IMAGE COMPRESSION

Motivated by these potential drawbacks in existing hyperspectral image compression algorithms, several techniques based on DSC have been proposed recently [3, 5, 8, 26, 30, 46]. In this section, we first discuss the potential advantages of applying DSC for hyperspectral image compression, in terms of reduced encoding complexity, ease of parallel encoding, and scalability. The challenges of applying DSC for hyperspectral compression will also be discussed.

Figure 10.5 shows an example of how DSC may be applied to hyperspectral compression. To decorrelate the input pixel data, the current spectral band B_i may undergo some transformations. Outputs from the first stage (i.e., transform coefficients, depicted as X in Figure 10.5) could then be compressed by DSC techniques to exploit interband correlation, using the corresponding data Y from the neighboring *reconstructed* band as side information in joint decoding. The key distinct feature of DSC systems is that only the correlation information between X and Y (depicted as $P(X, Y)$ in Figure 10.5) is needed for encoding X. In particular, exact values of Y would not be required. Therefore, DSC-based encoders could in principle operate in "open loop," that is, without requiring that the encoders include decoding loops to replicate data reconstructed at the decoders, as included in typical (lossy) interband prediction approaches. Operating in open loop can lead to some system-level advantages, as will be discussed. If the correlation between X and Y is weak, encoders might switch to use intracoding to compress X. Also intraprediction might be applied before DSC encoding to exploit spatial redundancy. In practical applications, encoders may need to estimate the correlation information, which could be nontrivial in a distributed computing environment.

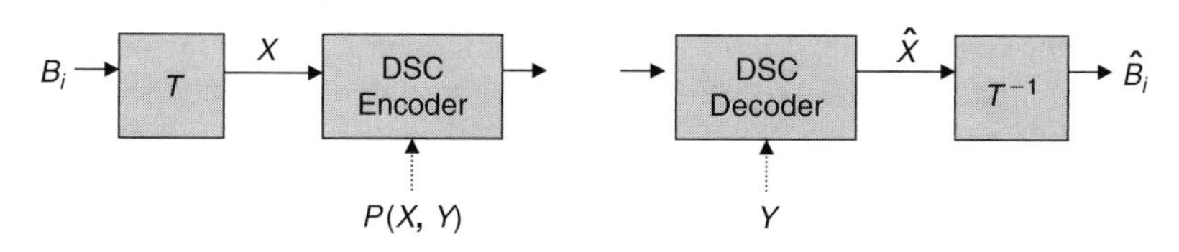

FIGURE 10.5

An example of DSC-based hyperspectral image compression. "T" and "T^{-1}" denote transformation and the inverse, respectively.

10.3.1 Potential Advantages of DSC-based Hyperspectral Compression

10.3.1.1 Asymmetric Compression Framework with a Reduced Complexity Encoder

Because of the limited available resources in satellites and spacecrafts, on-board encoding complexity is an important issue in hyperspectral compression. DSC may lead to an asymmetric framework where encoding complexity could be reduced at the expense of some increases in decoding complexity at the ground station. This is possible because, with DSC, encoding of the current spectral band would require only the correlation information. Therefore, computation and buffering storage for replicating the predictor would not be required in encoding, leading to complexity reduction. We remark that the complexity for "closing" the prediction loop could be nontrivial in lossy hyperspectral compression [31]. Note that a similar idea has also been applied to scalable video coding to reduce the encoding complexity due to closing multiple prediction loops of different quality layers [53, 54].

In addition, the memory requirements of a DSC-based encoder tend to be moderate because correlation information may be estimated using data from consecutive spectral bands [6]. This compares favorably to 3D wavelets methods, where pixel data of multiple spectral bands is needed in cross-band filtering.

10.3.1.2 Parallel Encoding of Datasets

The fact that DSC-based encoders can operate in open loop without being required to replicate the exact predictors may also lead to encoding algorithms that facilitate parallel implementation. For example, in encoding systems with multiple processors, compression of each entire spatial image may be assigned to each processor (see Figure 10.6 in Section 10.4.2 for an example). As long as the cross-band correlation information between adjacent spectral bands is available, each processor can proceed independently without further interaction during the course of encoding, leading to highly parallel encoding systems. Moreover, information-theoretic results [44] suggest such DSC-based parallel encoding can in principle achieve competitive coding efficiency, for example, comparable to that of nondistributed schemes, where all the spectral bands reside at the same processor for centralized encoding.

10.3.1.3 Scalability

DSC can also lead to hyperspectral compression systems that support more efficient rate scalability. Precisely, bit-rate scaling by on-the-fly truncation of an embedded bitstream may result in several different possible *predictor candidates* available at the decoder, each corresponding to the reconstruction at a truncation point. The actual predictor present at the decoder would depend on the specific truncation point chosen by the system. This *uncertainty* poses a challenge to interband prediction approaches, where precise status of the available predictor at the decoder needs to be maintained during encoding. In contrast, there is no need for a DSC encoder to know about the precise decoder predictor status. In particular, a DSC system may truncate an embedded bitstream, and as long as the side information matching (or exceeding) the degree of correlation used in encoding can be reconstructed at the decoder, successful Slepian-Wolf decoding can be achieved, leading to drift-free outputs [60]. In practice, the truncation points may be restricted by the block sizes used in Slepian-Wolf coding. Note that applying DSC to address multiple predictor candidates in video applications has been proposed in [7, 10] and to address scalable video streaming with feedback in [42].

10.3.2 Challenges in Applying DSC for Hyperspectral Imaging

Information-theoretic results state that distributed encoding can be as efficient as joint encoding provided that different bitstreams are jointly decoded [14, 44, 59]. However, achieving coding performance comparable to that of nondistributed approaches remains challenging for DSC hyperspectral image applications. This could be due to the difficulties involved in *modeling*, *estimating*, and *exploiting* the correlation in hyperspectral data, in a distributed setting (see Section 10.2.2 for a discussion of correlation characteristics).

First, while the coding efficiency of DSC depends strongly on the accuracy of the correlation models, real-world hyperspectral image data may exhibit nonstationary and complex correlation structures. For example, local variations of spatial or spectral correlation may occur in hyperspectral datasets. Also, considerable dependency between individual (cross-band) correlation noise symbols may exist in hyperspectral data [22].[3] Therefore, i.i.d. noise models commonly used in DSC applications could be inadequate to capture the statistics of cross-band correlation noise in hyperspectral image compression. While utilizing more sophisticated models that capture these complicated correlation structures may lead to better coding performance, it may be challenging to design the coding algorithms and the correlation estimation techniques for such advanced models.

Second, correlation information in hyperspectral data may need to be estimated under complexity constraints, so that the potential advantage of applying DSC to reduce encoder power consumption could be realized. For example, some relatively computationally intensive estimation techniques (e.g., the expectation-maximization (EM) algorithm [29]) may not be suitable for on-board correlation estimation in hyperspectral applications. Also, in distributed encoding scenarios, only a small number of samples may be exchanged in order to reduce the communication overheads, but using low sampling rates may fail to capture the local changes in correlation. These complexity and rate constraints therefore make on-board correlation estimation a difficult task [9]. While some DSC applications may employ feedback channels to acquire the correlation information from the decoder [1, 18], in hyperspectral applications, this information may need to be estimated on-board during encoding, as feedback from the ground station may incur considerable delay.

[3] Previous work has reported that spectral high-pass coefficients can exhibit spatial structure resembling that of the original spatial images [22].

Third, to efficiently exploit the correlation in hyperspectral images the compression algorithms may need to be highly adaptive to the local variations of spatial or spectral correlation (such as in the case of 3D CALIC as briefly discussed in Section 10.2.2). On the other hand, achieving such a level of adaptivity using DSC could be challenging. For example, existing Slepian–Wolf coders usually employ capacity approaching error-correcting codes to perform data compression, and these codes tend to require large block sizes, for example, of the order of 10^5 [27,60]; otherwise significant performance degradation may result. However, large encoding block sizes may not be able to exploit the changes in correlation in small blocks of hyperspectral data. Recent work has proposed addressing distributed compression based on arithmetic coding, which may achieve good coding performance with much smaller block sizes, for example, of the order of 10^3 [17], and this code could be more suitable for hyperspectral applications.

In addition, since one of the main applications of hyperspectral data is to identify the captured objects, it is important for DSC-based lossy compression algorithms to preserve the spectral characteristics (see Section 10.2.1).

10.4 EXAMPLE DESIGNS

In this section we discuss several DSC-based hyperspectral image compression algorithms proposed in the literature. Specifically, we discuss in Section 10.4.1 the two lossless compression algorithms proposed by Magli et al. [30], in Section 10.4.2 the wavelet-based lossy-to-lossless algorithm proposed by Tang et al. [46], Cheung et al. [8], and Cheung and Ortega [5], and in Section 10.4.3 the multispectral compression algorithm proposed by Li [26]. Both the first lossless algorithm in Section 10.4.1 and the wavelet-based algorithm in Section 10.4.2 exploit intra- and interband redundancy. However, the specific mechanisms are rather different: the first algorithm in Section 10.4.1 applies intraprediction to pixel data, and then the prediction residue would be further compressed by DSC exploiting cross-band correlation. The algorithm in Section 10.4.2, however, uses wavelet transform and bit-plane extraction similar to SPIHT to generate different types of source bits, each with different marginal and joint statistics (w.r.t. side information), and based on these statistics the bits are then adaptively compressed by either intra (zero-tree coding) or inter (DSC) approaches. In contrast to these algorithms that use capacity-approaching large-block-size error-correcting codes to compress the binary bits, the second algorithm in Section 10.4.1 uses a simple multilevel coset code to compress individual symbols so that better adaptation to local correlation statistics can be achieved. The algorithm in Section 10.4.3 uses different constraints to iteratively refine coarsely encoded spectral band images.

10.4.1 DSC Techniques for Lossless Compression of Hyperspectral Images

Lossless compression of hyperspectral imagery based on DSC has been proposed by Magli et al. [30]. In this work, DSC is used to exploit the correlation between different spectral bands. This is combined with other techniques, such as the prediction stage of 2D CALIC [57], to exploit intraband correlation. The main purpose is to shift complexity, in particular computational operations, from the encoder to the decoder, by relying on the decoder to exploit spatial correlation, similar in spirit to low-complexity video encoding [1, 16, 35, 36]. Specifically, the work proposed two algorithms for data compression, one based on powerful binary error-correcting codes and the other based on simpler multilevel coset codes (to be discussed in more detail in the subsequent sections).

10.4.1.1 Lossless Hyperspectral Compression Based on Error Correcting Codes

The first proposed algorithm is based on capacity approaching binary error-correcting codes such as Slepian–Wolf (SW) codes.

The algorithm operates on pixel data when compressing the current spectral band and exploits intraband redundancy by applying to each pixel a 2D CALIC predictor formed by spatially adjacent pixels. Denote the spatial prediction error array by E_X, and the spatial prediction error array of the previous band after linear filtering by E'_Y. E_X are losslessly compressed based on DSC techniques using E'_Y as side information. Specifically, E_X are decomposed into a raw bit-plane representation. Denote the extracted raw bit-planes by $E_{X_i}^{(b)}$, where $0 \leq i < N_{E_X}$ is the significance level of the bit-plane and N_{E_X} is the number of extracted bit-planes. Denote also the corresponding bit-planes in E'_Y by $E'^{(b)}_{Y_i}$. Binary sources $E_{X_i}^{(b)}$ are compressed by a LDPC-based SW coder, using $E'^{(b)}_{Y_i}$ in joint decoding at the decoder. The SW encoding rate of each bit-plane is derived from the conditional entropy of the input bits given the side information bits, $H(E_{X_i}^{(b)}|E'^{(b)}_{Y_i})$, assuming the bits are i.i.d. binary sources and the correlation can be modeled by a binary channel. Depending on the conditional entropy and the efficiency of the SW coder, some bit-planes would instead be compressed by an arithmetic coder.

The encoding rate or the conditional entropy $H(E_{X_i}^{(b)}|E'^{(b)}_{Y_i})$ would depend on the correlation between bits $E_{X_i}^{(b)}$ and $E'^{(b)}_{Y_i}$. In [30], this correlation information is assumed to be known, and there is no discussion related to efficient estimation of this information. It should be noted that this correlation estimation problem could be nontrivial and remains to be a challenging issue in practical application of DSC [18]. A straightforward approach that exchanges all the bit data for direct comparison would inevitably incur significant communication overhead and may upset the complexity saving due to DSC techniques (see more discussion in [11]).

The proposed algorithm was compared with several 2D and 3D lossless compression schemes using NASA AVIRIS datasets, and experimental results reported in [30] suggest the proposed algorithm can outperform 2D coding techniques (specifically, JPEG-LS) by about 1 bit per pixel (bpp) on average, while there is still a performance gap of about 1 bpp compared to 3D CALIC. According to [30], this performance gap is partly due to the sophisticated context modeling techniques in 3D CALIC that capture local spatial and spectral correlation (see Section 10.2.2 for correlation characteristics), and partly caused by the inefficiency in assuming a binary correlation model that ignores the correlation between bit-planes of different significances [56].

10.4.1.2 Lossless Hyperspectral Compression Based on Multilevel Coset Codes

In [30] another lossless hyperspectral compression algorithm based on simpler multilevel coset codes was proposed.

The algorithm adopts a block-based approach, where the spatial images are split into nonoverlapping small blocks, and correlation estimation is performed on each block independently to estimate the coding rate based on direct comparisons of a subset of pixels of the corresponding block in the adjacent spectral band. Using a block-based approach allows fine adaption in the estimation of interband dependencies, which tends to be nonstationary. This, however, complicates application of capacity approaching error-correcting codes in SW coding, which would in general require block lengths of several thousands of symbols.

To address this issue, Magli et al. [30] proposed using simple multilevel coset codes that encode each symbol (corresponding to a pixel in this case), independently using as side information the corresponding symbol in the neighboring spectral band. Specifically, for each pixel $x(i,j)$ in the current spectral band, Magli et al. [30] propose a correlation model[4]

$$x(i,j) = \mathsf{P}(y(i,j)) + n(i,j), \tag{10.1}$$

where $n(i,j)$ is the correlation noise, and $y(i,j)$ is the value of the corresponding pixel in neighboring spectral band. P is a prediction model to estimate $x(i,j)$ from $y(i,j)$. Depending on the statistics of $n(i,j)$, only the k least significant bits (LSB) of $x(i,j)$ would be communicated to the decoder, and the decoder would use the side information $y(i,j)$ and P to infer the discarded most significant bits (MSB) of $x(i,j)$. With accurate estimation of $n(i,j)$, the value of k that leads to exact recovery of $x(i,j)$ can be derived. In [30], k is determined for each block based on the maximum value of $n(i,j)$ over the block, estimated based on exchanging some subsets of pixels $x(i,j)$ and $y(i,j)$.

Experimental results reported in [30] suggest that, in this particular application using NASA AVIRIS datasets, the proposed block-based adaptive coding algorithm using simple multilevel codes can achieve similar performance to its counterpart based on powerful error-correcting codes, that is, on average outperforming JPEG-LS by about 1 bpp, while compared to 3D CALIC there is still a 1-bpp performance gap. Regarding execution time, for simulations running on a workstation, where correlation estimation complexity has been taken into account but the spectral bands were available in the same location, [30] reported the proposed algorithm is about 10 times faster than 3D CALIC.

[4] Note that here cross-band correlation is exploited between pixels data rather than the intraprediction residues as in the first algorithm.

10.4.2 Wavelet-based Slepian–Wolf Coding for Lossy-to-lossless Compression of Hyperspectral Images

Wavelet-based algorithms for lossy-to-lossless compression of hyperspectral images have been proposed [5, 8, 46]. The proposed algorithm combines set partitioning of wavelet coefficients such as that introduced in the popular SPIHT algorithm [41] with DSC techniques, so that sign bits would be compressed by SW coding, while magnitude bits would be encoded using adaptive combinations of DSC and zero-tree coding. Using DSC tools allows the encoder to operate in open loop without requiring access to decoded versions of neighboring spectral bands, and this facilitates parallel encoding of spectral bands in multiprocessor architectures. The memory requirement of the proposed algorithm is modest and similar to that of interband prediction approaches. Note that wavelet-based DSC algorithms have also been subsequently proposed for distributed video and sensors applications [19, 50].

10.4.2.1 System Overview

Figure 10.6 shows an overview of the proposed encoding system in [5, 8, 46]. The proposed encoder consists of multiple processors, and each processor compresses one entire spatial image at a time, using the algorithm to be discussed in detail in the next section. With a DSC approach, encoding needs only the correlation information in order to exploit interband redundancy. In particular, as long as the correlation information is available, each encoding thread compressing one spatial image would be able to proceed in parallel, and parallel encoding of a hyperspectral dataset can be achieved.

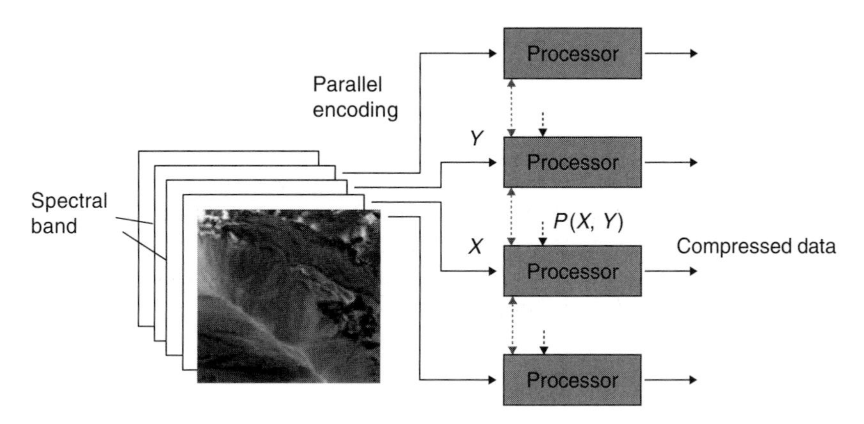

FIGURE 10.6

Parallel encoding system based on DSC as proposed in [5, 8, 46]. Each processor compresses one spatial image at a time.

One key issue for the proposed system is how to estimate the correlation information efficiently during encoding. However, correlation estimation may be nontrivial in this setting, for the following reasons:

- The spatial images of different spectral bands reside in different processors, and the communication bandwidth between the processors could be limited.
- Data exchanges between the processors may impact parallel encoding, as the processors may have to remain idle to wait for the data.

To address these constraints, sampling-based techniques have been proposed to perform correlation estimation, with small amounts of interprocessor communication overhead and minimal dependencies between different encoding threads [8].

10.4.2.2 Encoding Algorithm

Figure 10.7 depicts the encoding algorithm. To compress the current spectral band B_i, the proposed algorithm applies the wavelet transform and then bit-plane extraction similar to SPIHT to iteratively extract *sign, refinement, significance information* or, alternatively, using *raw* bit-planes from the wavelet coefficients [41,49]. The extracted bit-planes are then compressed by SW coding to exploit interband correlation, or zero-tree coding to exploit intraband redundancy, depending on the correlation characteristics and the coefficients distributions [5]. Bit-planes are progressively transmitted to the decoder starting from the most significant bit (MSB), and the number of bit-planes communicated would depend on the data rate or the quality requirements of the particular application. Lossless compression can be achieved by using integer wavelets and transmitting all the bit-planes.

In SW coding of bit-planes, the proposed algorithm uses as side information the corresponding bit-planes of the adjacent spectral band available at another processor. For example, the refinement bits of the current spectral band would be compressed using the same magnitude bits extracted from the adjacent spectral band (after linear prediction) as side information. The bits are assumed to be i.i.d. sources, and the correlation w.r.t. the side information bits is modeled as a binary channel parameterized by the *crossover probability* between the bits data,[5] which needs to be estimated during encoding to determine the SW encoding rate.

10.4.2.3 Correlation Estimation

A straightforward approach to estimate the crossover probability could be to explicitly generate and extract the side information bits from the adjacent spectral band at another processor, and exchange and compare the source and side information bits, to determine the crossover probability. However, information exchange in the middle

[5] Crossover probability is basically the probability that input bits are different from the corresponding bits in the side information. Small crossover probability therefore implies high correlation among the bits data.

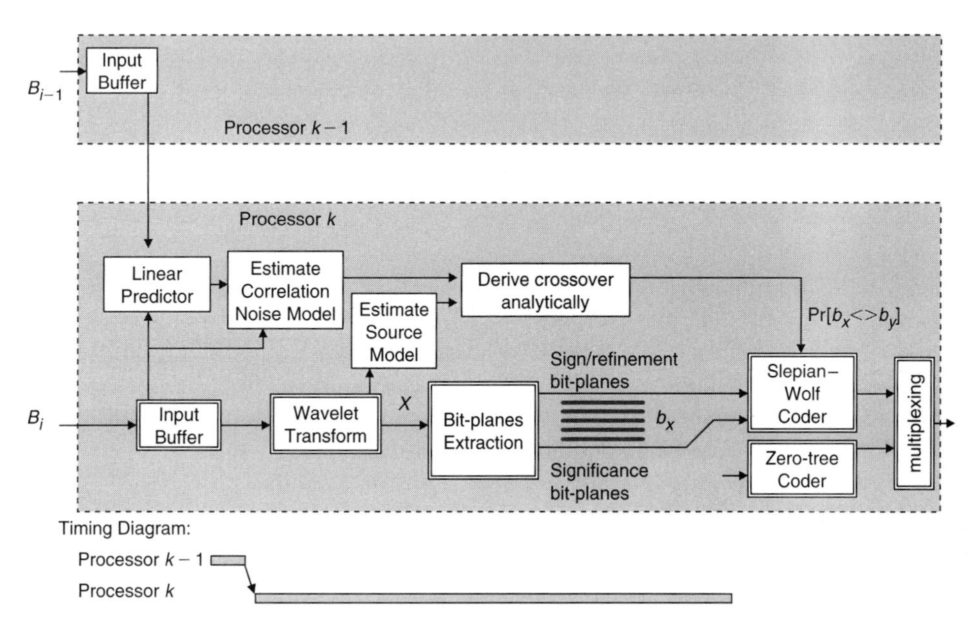

FIGURE 10.7

Wavelet-based Slepian–Wolf coding of hyperspectral images: encoding algorithm. The current spectral band B_i is compressed using the adjacent spectral band B_{i-1} as side information.

of processing is required in this direct approach in order to generate the side information bits. This information exchange could make parallel encoding inefficient, as the processors may have to remain idle to wait for the data.

Therefore, [11] proposed a *model-based estimation* that could be applicable to this scenario. The model-based estimation of crossover probability is a two-step process. In the first step, we estimate the joint p.d.f. $f_{XY}(x, y)$, between the transform coefficients of the current spectral band X and those of the adjacent spectral band Y. Assuming that the correlation between X and Y can be modeled as $Y = X + Z$, where Z is the correlation noise independent of X. Then, it can be shown that the joint p.d.f. can be derived from estimates of the source model $f_X(x)$ and the correlation noise model $f_Z(z)$. In particular, [11] proposed estimating the cross-band correlation noise model (between corresponding wavelet coefficients), $f_Z(z)$, by exchanging a small number of samples of the *original pixel-domain data*, so that communication overheads and dependencies between encoding threads can be significantly reduced, thus facilitating parallel encoding. Some results illustrating how the number of samples may affect the accuracy of the model-based estimation can also be found in [11].

After the first step and with the estimates of $f_{XY}(x, y)$ available, [11] proposed estimating the crossover probability analytically by integrating the joint p.d.f. over some regions of the sample space of X and Y. For example, the crossover probability of the raw bit-plane data at significance level l (when l is equal to zero this corresponds to the LSB) can be estimated by integrating $f_{XY}(x, y)$ over all the shaded regions A_i

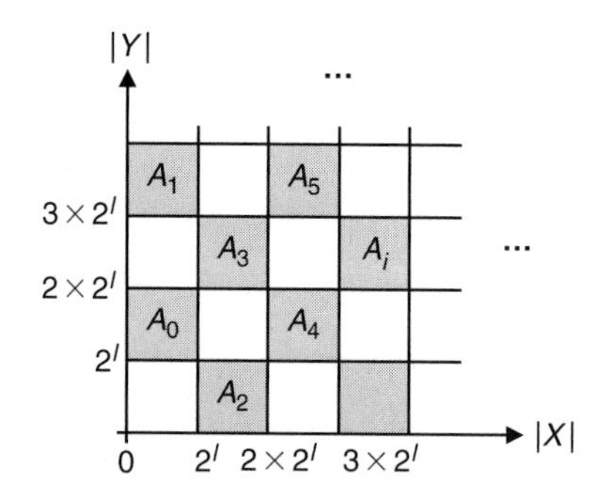

FIGURE 10.8

Model-based approach for crossover probability estimation. A_i are the events that lead to occurrence of raw bit-plane crossovers between X and Y at significance level l.

as shown in Figure 10.8. In practice, only a few regions that lead to nonzero integrals need to be considered. The crossover estimates for sign and refinement bit-planes can be derived following the same procedure, as discussed in [11].

10.4.2.4 *Adaptive Combination of DSC and Zero-tree Coding*

To improve the coding performance, [5] proposed an algorithm to adaptively encode the magnitude bits using DSC or zero-tree coding. Specifically, magnitude bits may be extracted as raw bit-planes and compressed using SW coding. Alternatively, magnitude bits may also be partitioned into refinement bit-planes and significance information bit-planes, and compressed using SW coding and zero-tree coding, respectively. To determine the switching strategy between the two schemes, modeling techniques were proposed in [5] to estimate the relative coding efficiencies of the SW coding and zero-tree coding approaches, using the marginal distribution $f_X(x)$ and the joint p.d.f. $f_{XY}(x, y)$. Estimates of the number of coded bits using DSC and zero-tree coding are derived from these p.d.f. at each significance level, and also separately for each wavelet subband.

Figure 10.9 shows some modeling results of two wavelet subbands, which illustrate how the number of coded bits (estimated by modeling) changes with the significance levels. The modeling results suggest that, at high significance levels, it tends to be more efficient to partition the magnitude bits into refinement bit-planes and significance information bit-planes, and compress them using SW coding and zero-tree coding, respectively (labeled "DSC + ZTC" in Figure 10.9). On the other hand, in the middle significance levels, it may be possible to achieve better coding performance by switching to another scheme that extracts the magnitude bits as raw bit-planes and encodes them using SW coding only (labeled "DSC" in Figure 10.9). The figures show that these switchings should occur at certain significance levels in order to obtain the optimal compression results. Different wavelet subbands would have different switching levels, and the optimal switching levels can be determined using the modeling techniques in [5].

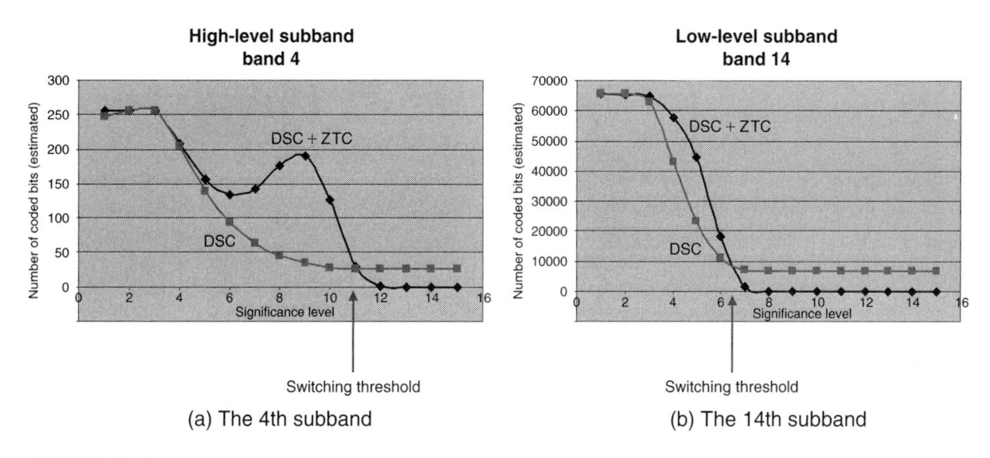

(a) The 4th subband (b) The 14th subband

FIGURE 10.9

The amounts of coded bits (estimated by modeling) vs. the significance levels of two wavelet subbands.

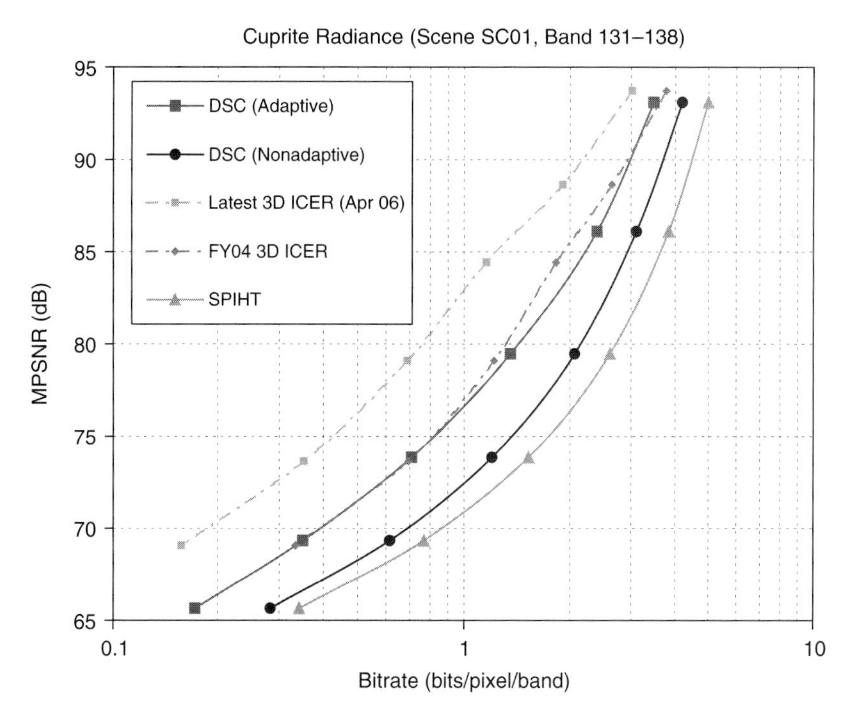

FIGURE 10.10

Coding performance of the DSC algorithm proposed in [5, 8, 46]: NASA AVIRIS Cuprite datasets. Reconstruction qualities are measured by MPSNR, the average PSNR of multiple spectral bands.

10.4.2.5 *Coding Performance*

Figure 10.10 shows the performance of the wavelet-based DSC algorithm along with several 3D wavelet algorithms (3D ICER) developed by NASA-JPL [20] for encoding NASA AVIRIS datasets. As shown in the figure, the DSC algorithm with adaptive coding is comparable to a simple 3D wavelet system (FY04 3D ICER) in terms of coding efficiency. The simple 3D wavelet system uses the standard dyadic wavelet decomposition and a context-adaptive entropy coding scheme to compress transform coefficients bits. However, there still is a performance gap when comparing the DSC algorithm to a more recent and sophisticated version of 3D wavelet (latest 3D ICER). The more recent 3D wavelet developed in NASA-JPL exploits the spatial correlation that still exists in the correlation noise [20]. The results indicate further improvement may be possible in the DSC algorithm, which currently uses a simple i.i.d. model for correlation noise and ignores the dependency between correlation noise symbols. Figure 10.10 also shows that when compared with 2D SPIHT the DSC algorithm can achieve 8 dB gains at some bit rates.

10.4.3 Distributed Compression of Multispectral Images Using a Set Theoretic Approach

DSC-based lossy compression of multispectral images was proposed in [26]. The main purpose of the proposed work is to compress image data corresponding to different spectral channels of the same scene, using several spatially separated encoders. A set-theoretic framework for distributed compression was proposed. Specifically, the cross-band correlation and observation constraint are assumed to be characterized by convex sets. Following from the convex set assumptions, joint decoding of multispectral images can be implemented via iterative projection-onto-convex-set (POCS) operations at the centralized decoder, and the convexity of constraint sets guarantees the convergence of the decoding algorithm.

In one of the scenarios considered in [26], spatial images $X_1, X_2, \ldots, X_K$ are compressed using an asymmetric protocol, where X_1 shall be intracoded at a relatively high quality, and the decoded version of X_1 would serve as side information for decoding the DSC compressed images $X_2, \ldots, X_K$. Li [26] proposed that the spatial images X_k, $2 \leq k \leq K$ could be wavelet transformed and coarsely quantized. The reconstruction of X_k can be iteratively refined, at the decoder, using the following constraints:

- Cross-band correlation structure: co-located wavelet coefficients of X_1 and X_k are assumed to be less than some threshold.
- Observation constraint of image source: each of the wavelet coefficients of X_k can only assume values in its own quantization bin.

Li [26] argues that these constraint sets are convex, and the process of refining X_k can be cast as the problem of identifying a point at the intersection of these convex sets, which can be implemented by alternatively projecting onto the defined convex sets.

Preliminary results in [26], however, indicate that the proposed algorithm is 2 dB less efficient compared to JPEG2000 intraband coding when applying it to RGB color images. This may be partly due to the assumed cross-band correlation model incapable of exploiting local correlation variations between co-located wavelet coefficients.

10.5 CONCLUSIONS

In this chapter we have discussed distributed compression of hyperspectral images. Motivated by the potential problems in conventional hyperspectral compression approaches, applications of DSC for hyperspectral data compression have been proposed in the literature. Potentially, DSC can lead to hyperspectral systems with reduced complexity encoders that facilitate parallel encoding and scalability, suitable for the specific operating conditions in hyperspectral data acquisition. Several examples of DSC-based hyperspectral compression algorithms proposed in the literature were summarized. While some of these DSC systems have demonstrated competitive rate-distortion efficiency, there are still performance gaps w.r.t. sophisticated nondistributed approaches. Future research direction could be to investigate improved algorithms with better adaptation to local variations in correlation.

REFERENCES

[1] A. Aaron, R. Zhang, and B. Girod. Wyner–Ziv coding of motion video. In *Proc. Asilomar Conf. Signals, Systems, and Computers*, November 2002.

[2] N. Aranki. *Parallel implementations of the discrete wavelet transform and hyperspectral data compression on reconfigurable platforms: Approach, methodology and practical considerations*. Ph.D. thesis, University of Southern California, August 2007.

[3] M. Barni, D. Papini, A. Abrardo, and E. Magli. Distributed source coding of hyperspectral images. In *Proc. Int'l Symp. Geoscience and Remote Sensing (IGARSS)*, 2005.

[4] C. H. Chen. *Information Processing for Remote Sensing*. World Scientific Publishing Company, 2000.

[5] N.-M. Cheung and A. Ortega. An efficient and highly parallel hyperspectral imagery compression scheme based on distributed source coding. In *Proc. Asilomar Conf. Signals, Systems, and Computers*, 2006.

[6] N.-M. Cheung and A. Ortega. A model-based approach to correlation estimation in wavelet-based distributed source coding with application to hyperspectral imagery. In *Proc. Int'l Conf. Image Processing (ICIP)*, 2006.

[7] N.-M. Cheung and A. Ortega. Distributed source coding application to low-delay free viewpoint switching in multiview video compression. In *Proc. Picture Coding Symposium (PCS)*, 2007.

[8] N.-M. Cheung, C. Tang, A. Ortega, and C. S. Raghavendra. Efficient wavelet-based predictive Slepian-Wolf coding for hyperspectral imagery. *EURASIP Journal on Signal Processing—Special Issue on Distributed Source Coding*, 86(11):3180–3195, November 2006.

[9] N.-M. Cheung, H. Wang, and A. Ortega. Correlation estimation for distributed source coding under information exchange constraints. In *Proc. Int'l Conf. Image Processing (ICIP)*, 2005.

[10] N.-M. Cheung, H. Wang, and A. Ortega. Video compression with flexible playback order based on distributed source coding. In *Proc. Visual Communications and Image Processing (VCIP)*, 2006.

[11] N.-M. Cheung, H. Wang, and A. Ortega. Sampling-based correlation estimation for distributed source coding under rate and complexity constraints. *IEEE Trans. Image Processing*, 2008.

[12] E. Christophe and W. A. Pearlman. Three-dimensional SPIHT coding of volume images with random access and resolution scalability. *EURASIP Journal on Image and Video Processing*. Submitted, 2008.

[13] R. N. Clark, G. Swayze, J. Boardman, and F. Kruse. Comparison of three methods for materials identification and mapping with imaging spectroscopy. In *Proc. JPL AVIRIS Airborne Geoscience Workshop*, 1993.

[14] T. M. Cover and J. A. Thomas. *Elements of Information Theory*. NY: Wiley, 1991.

[15] P. L. Dragotti, G. Poggi, and A.R.P. Ragozini. Compression of multispectral images by three-dimensional SPIHT algorithm. *IEEE Trans. Geoscience and Remote Sensing*, 38(1):416–428, January 2000.

[16] B. Girod, A. Aaron, S. Rane, and D. Rebollo-Monedero. Distributed video coding. *Proceedings of the IEEE, Special Issue on Advances in Video Coding and Delivery*, 93(1):71–83, January 2005.

[17] M. Grangetto, E. Magli, and G. Olmo. Distributed arithmetic coding. *IEEE Communications Letters*, 11(11):883–885, November 2007.

[18] C. Guillemot, F. Pereira, L. Torres, T. Ebrahimi, R. Leonardi, and J. Ostermann. Distributed monoview and multiview video coding. *IEEE Signal Processing Magazine*, 24(5):67–76, September 2007.

[19] X. Guo, Y. Lu, F. Wu, W. Gao, and S. Li. Wyner–Ziv video coding based on set partitioning in hierarchical tree. In *Proc. Int'l Conf. Image Processing (ICIP)*, 2006.

[20] A. Kiely, M. Klimesh, H. Xie, and N. Aranki. ICER-3D: A progressive wavelet-based compressor for hyperspectral images. Technical report, NASA-JPL IPN Progress Report, 2006. http://ipnpr.jpl.nasa.gov/progress_report/42-164/164A.pdf.

[21] B. Kim and W. A. Pearlman. An embedded wavelet video coder using three-dimensional set partitioning in hierarchical tree. In *Proc. Data Compression Conference (DCC)*, 1997.

[22] M. Klimesh, A. Kiely, H. Xie, and N. Aranki. Spectral ringing artifacts in hyperspectral image data compression. Technical report, NASA-JPL IPN Progress Report, 2005. http://ipnpr.jpl.nasa.gov/progress_report/42-160/160C.pdf.

[23] M. Klimesh, A. Kiely, H. Xie, and N. Aranki. Spectral Ringing Artifacts in Hyperspectral Image Data Compression. In Giovanni Motta, Francesco Rizzo, and James A. Storer, editors, *Hyperspectral Data Compression*. Springer, 2006.

[24] NASA Jet Propulsion Laboratory. AVIRIS homepage. http://aviris.jpl.nasa.gov/, October 2007.

[25] D. Landgrebe. Hyperspectral image data analysis. *IEEE Signal Processing Magazine*, 19(1): 17–28, January 2002.

[26] X. Li. Distributed coding of multispectral images: A set theoretic approach. In *Proc. Int'l Conf. Image Processing (ICIP)*, 2004.

[27] A. D. Liveris, Z. Xiong, and C. N. Georghiades. Compression of binary sources with side information at the decoder using LDPC codes. *IEEE Communications Letters*, 6(10):440–442, October 2002.

[28] R. Lockwood, T. Cooley, R. Nadile, J. Gardner, P. Armstrong, A. Payton, T. Davis, S. Straight, T. Chrien, E. Gussin, and D. Makowski. Advanced responsive tactically-effective military imaging spectrometer (ARTEMIS) design. In *Proc. Int'l Symp. Geoscience and Remote Sensing (IGARSS)*, 2006.

[29] D.J.C. MacKay. *Information Theory, Inference, and Learning Algorithms*. Cambridge University Press, 2003.

[30] E. Magli, M. Barni, A. Abrardo, and M. Grangetto. Distributed source coding techniques for lossless compression of hyperspectral images. *EURASIP Journal on Applied Signal Processing*, 2007. Article ID 45493. 13 pages.

[31] A. C. Miguel, A. R. Askew, A. Chang, S. Hauck, and R. E. Ladner. Reduced complexity wavelet-based predictive coding of hyperspectral images for FPGA implementation. In *Proc. Data Compression Conference (DCC)*, 2004.

[32] G. Motta, F. Rizzo, and J. A. Storer. Compression of hyperspectral imagery. In *Proc. Data Compression Conference (DCC)*, 2003.

[33] California Institute of Technology. ASTER spectral library. http://speclib.jpl.nasa.gov/, 2000.

[34] S. Pradhan and K. Ramchandran. Distributed source coding using syndromes (DISCUS): Design and construction. In *Proc. Data Compression Conference (DCC)*, 1999.

[35] R. Puri, A. Majumdar, and K. Ramchandran. PRISM: A video coding paradigm with motion estimation at the decoder. *IEEE Trans. Image Processing*, 16(10):2436–2448, October 2007.

[36] R. Puri and K. Ramchandran. PRISM: A new robust video coding architecture based on distributed compression principles. In *Proc. Allerton Conf. Communications, Control, and Computing*, October 2002.

[37] A. K. Rao and S. Bhargava. Multispectral data compression using bidirectional interband prediction. *IEEE Trans. Geoscience and Remote Sensing*, 34(2):385–397, March 1996.

[38] R. E. Roger and M. C. Cavenor. Lossless compression of AVIRIS images. *IEEE Trans. Image Processing*, 5(5):713–719, May 1996.

[39] M. J. Ryan and J. F. Arnold. The lossless compression of AVIRIS images by vector quantization. *IEEE Trans. Geoscience and Remote Sensing*, 35(3):546–550, May 1997.

[40] J. A. Saghri, A. G. Tescher, and J. T. Reagan. Practical transform coding of multispectral imagery. *IEEE Signal Processing Magazine*, 12(1):32–43, January 1995.

[41] A. Said and W. A. Pearlman. A new, fast, and efficient image codec using set partitioning in hierarchical trees. *IEEE Trans. Circuits and Systems for Video Technology*, 6(3):243–250, June 1996.

[42] A. Sehgal, A. Jagmohan, and N. Ahuja. Scalable video coding using Wyner–Ziv codes. In *Proc. Picture Coding Symposium (PCS)*, December 2004.

[43] V. Shettigara and G. Fowler. Target re-acquisition capability of hyperspectral technology. In *Proc. Land Warfare Conference*, pages 257–265, October 2005.

[44] D. Slepian and J. K. Wolf. Noiseless coding of correlated information sources. *IEEE Trans. Information Theory*, 19(4):471–480, July 1973.

[45] R. B. Smith. Introduction to hyperspectral imaging. Technical report, MicroImages, 2001. http://www.microimages.com/getstart/pdf/hyprspec.pdf.

[46] C. Tang, N.-M. Cheung, A. Ortega, and C. S. Raghavendra. Efficient inter-band prediction and wavelet based compression for hyperspectral imagery: A distributed source coding approach. In *Proc. Data Compression Conference (DCC)*, 2005.

[47] X. Tang and W. A. Pearlman. Lossy-to-lossless block-based compression of hyperspectral volumetric data. In *Proc. Int'l Conf. Image Processing (ICIP)*, 2004.

[48] S. R. Tate. Band ordering in lossless compression of multispectral images. In *Proc. Data Compression Conference (DCC)*, pages 311–320, 1994.

[49] D. Taubman and M. Marcellin. *JPEG2000: Image Compression Fundamentals, Standards and Practice*. Kluwer, 2002.

[50] V. Thirumalai, I. Tosic, and P. Frossard. Distributed coding of multiresolution omnidirectional images. In *Proc. Int'l Conf. Image Processing (ICIP)*, 2007.

[51] H. van der Werff. *Knowledge based remote sensing of complex objects: recognition of spectral and spatial patterns resulting from natural hydrocarbon seepages*. Ph.D. thesis, Universiteit Utrecht, Utrecht, Netherlands, 2006.

[52] H. Wang, S. D. Babacan, and K. Sayood. Lossless hyperspectral image compression using context-based conditional averages. In *Proc. Data Compression Conference (DCC)*, 2005.

[53] H. Wang, N.-M. Cheung, and A. Ortega. WZS: Wyner-Ziv scalable predictive video coding. In *Proc. Picture Coding Symposium (PCS)*, December 2004.

[54] H. Wang, N.-M. Cheung, and A. Ortega. A framework for adaptive scalable video coding using Wyner-Ziv techniques. *EURASIP Journal on Applied Signal Processing*, 2006. Article ID 60971. 18 pages.

[55] Y. Wang, J. T. Rucker, and J. E. Fowler. Three-dimensional tarp coding for the compression of hyperspectral images. *IEEE Geoscience and Remote Sensing Letters*, 1(2):136–140, April 2004.

[56] R. P. Westerlaken, S. Borchert, R. K. Gunnewiek, and R. L. Lagendijk. Analyzing symbol and bit plane-based LDPC in distributed video coding. In *Proc. Int'l Conf. Image Processing (ICIP)*, 2007.

[57] X. Wu and N. Memon. Context-based, adaptive, lossless image coding. *IEEE Trans. Communications*, 45(4):437–444, April 1997.

[58] X. Wu and N. Memon. Context-based lossless interband compression—extending CALIC. *IEEE Trans. Image Processing*, 9(6):994–1001, June 2000.

[59] A. D. Wyner and J. Ziv. The rate-distortion function for source coding with side information at the decoder. *IEEE Trans. Information Theory*, 22(1):1–10, January 1976.

[60] Z. Xiong, A. Liveris, and S. Cheng. Distributed source coding for sensor networks. *IEEE Signal Processing Magazine*, 21(5):80–94, September 2004.

CHAPTER

Securing Biometric Data

11

Anthony Vetro
Mitsubishi Electric Research Laboratories, Cambridge, MA

Stark C. Draper
Department of Electrical and Computer Engineering, University of Wisconsin, Madison, WI

Shantanu Rane and Jonathan S. Yedida
Mitsubishi Electric Research Laboratories, Cambridge, MA

CHAPTER CONTENTS

Distributed Source Coding: Theory, Algorithms, and Applications

11.1 INTRODUCTION

11.1.1 Motivation and Objectives

Securing access to physical locations and to data is of primary concern in many personal, commercial, governmental, and military contexts. Classic solutions include carrying an identifying document or remembering a password. Problems with carrying a document include forgeries, while problems with a password include poorly chosen or forgotten passwords. Computer-verifiable biometrics, such as fingerprints and iris scans, provide an attractive alternative to conventional solutions. Unlike passwords, biometrics do not have to be remembered, and, unlike identifying documents, they are difficult to forge. However, they have characteristics that raise new security challenges.

The key characteristic differentiating biometrics from passwords is measurement noise. Each time a biometric is measured, the observation differs, at least slightly. For example, in the case of fingerprints, the reading might change because of elastic deformations in the skin when placed on the sensor surface, dust or oil between finger and sensor, or a cut to the finger. Biometric systems must be robust to such variations. Biometric systems deal with such variability by relying on pattern recognition. To perform recognition in current biometric systems, the biometric measured at enrollment is stored on the device for comparison with the "probe" biometric collected later for authentication. This creates a security hole: an attacker who gains access to the device also gains access to the biometric. This is a serious problem since, in contrast to passwords or credit card numbers, an individual cannot generate new biometrics if his or her biometrics are compromised.

The issue of secure storage of biometric data is the central design challenge addressed in this chapter. Useful insight into desirable solution characteristics can be gained through consideration of password-based authentication. In order to preserve the privacy of passwords in the face of a compromised database or personal computer, passwords are not stored "in-the-clear." Instead, a cryptographic "hash" of one's password is stored. The hash is a scrambling function that is effectively impossible to invert. During authentication, a user types in a password anew. Access is granted only if the hash of the new password string matches the stored hash of the password string entered at enrollment. Because of the noninvertibility of the hash, password privacy is not compromised even if the attacker learns the stored hash. Unfortunately, the variability inherent to biometric measurement means that this hashing solution cannot be directly applied to biometric systems—enrollment and probe hashes would hardly ever match.

The aim of the secure biometric systems detailed in this chapter is to develop a hashing technology robust to biometric measurement noise. In particular, we focus on an approach that uses "syndrome" bits from a Slepian–Wolf code [1] as a "secure" biometric. The syndrome bits on their own do not contain sufficient information to deduce the user's enrollment biometric (or "template"). However, when combined with a second reading of the user's biometric, the syndrome bits enable the recovery and verification of the enrollment biometric. A number of other researchers have attempted to develop secure biometric systems with similar characteristics, and we will review some of these proposals in Section 11.2.

11.1.2 Architectures and System Security

There are two fundamental applications for secure biometric technology: access control and key management. In access control, the system modulates access through inspection of a candidate user's biometric. In key management, the system objective is to extract a stable encryption key from the user's biometric. While access control and key management are different goals, the syndrome-encoding and recovery techniques we discuss apply to both. In an access-control application, the recovered biometric is verified by comparison with a stored hash of the original in a manner identical to password-based systems. In a key-management application, the (now recovered) original serves as a shared secret from which an encryption (decryption) key can be generated.

Although secure biometric technology addresses one security threat facing biometric systems, it should be kept in mind that a variety of threats exist at various points in the biometric subsystem chain. For instance, individual modules can be forged or tampered with by attackers. Examples include a fake feature extraction module that produces preselected features that allow an intruder to gain access, or a fake decision-making entity that bypasses the authentication subsystem altogether. In remote authentication settings, where biometric measurements are collected at a remote site, not co-located with the stored enrollment data, other weak points exist. Dishonest entities such as servers that impersonate a user or perform data mining to gather information could be the source of successful attacks. Furthermore, in remote settings, the communication channel could also be compromised and biometric data could be intercepted and modified. Not all these threats are guarded against with secure biometric templates. Some can be dealt with using standard cryptographic techniques. But, in general, system designers need to be aware of all possible points of attack in a particular system.

In view of these threats, a few desirable properties regarding biometric system security are as follows:

- *Availability:* Legitimate users should not be denied access.
- *Integrity:* Forging fake identity should be infeasible.
- *Confidentiality:* Original biometric data should be kept secret.
- *Privacy:* Database cross-matching should reveal little information.
- *Revocability:* Revocation should be easy.

11.1.3 Chapter Organization

Section 11.2 of this chapter describes related work in this area to give readers a sense for alternative approaches to the secure biometrics problem. Section 11.3 formally quantifies the trade-off between security and robustness for the class of secure biometric systems that we consider, and introduces the syndrome-coding-based approach. In Section 11.4, we describe a prototype system developed for iris biometrics. In Sections 11.5 and 11.6, two different approaches for securing fingerprint data are described. The first is based on a statistical modeling of the fingerprint data. The second approach involves transforming the fingerprint data to a representation with statistical properties that are well suited to off-the-shelf syndrome codes. A summary of this new application of distributed source coding is given in Section 11.7, including a discussion on future research opportunities and potential standardization.

11.2 RELATED WORK

One class of methods for securing biometric systems is "transform-based." Transform-based approaches essentially extract features from an enrollment biometric using a complicated transform. Authentication is performed by pattern matching in the transform domain. Security is assumed to come from the choice of a good transform that masks the original biometric data. In some cases, the transform itself is assumed to be kept secret, and design considerations must be made to ensure this secrecy. Particularly when the transform itself is compromised, it is difficult to prove rigorously the security of such systems. Notable techniques in this category include cancelable biometrics [2,3], score matching-based techniques [4], and threshold-based biohashing [5].

This chapter focuses on an alternative class of methods that are based on using some form of "helper data." In such schemes, user-specific helper data is computed and stored from an enrollment biometric. The helper data itself and the method for generating this data can be known and is not required to be secret. To perform authentication of a probe biometric, the stored helper data is used to reconstruct the enrollment biometric from the probe biometric. However, the helper data by itself should not be sufficient to reconstruct the enrollment biometric. A cryptographic hash of the enrollment data is stored to verify bitwise exact reconstruction.

Architectural principles underlying helper data-based approaches can be found in the information-theoretic problem of "common randomness" [6]. In this setting, different parties observe dependent random quantities (the enrollment and the probe) and then through finite-rate discussion (perhaps intercepted by an eavesdropper) attempt to agree on a shared secret (the enrollment biometric). In this context, error-correction coding (ECC) has been proposed to deal with the joint problem of providing security against attackers, while accounting for the inevitable variability between enrollment and probe biometrics. On the one hand, the error-correction capability of an error-correcting code can accommodate variations between multiple measurements of the same biometric. On the other hand, the check bits of the error-correction code perform much the same function as a cryptographic hash of a password on conventional access-control systems. Just as attackers cannot invert the hash and steal the password, they cannot use the check bits to recover and steal the biometric.

An important advantage of helper data-based approaches relative to transform-based approaches is that the security and robustness of helper data-based schemes are generally easier to quantify and prove. The security of transform-based approaches is difficult to analyze since there is no straightforward way to quantify security when the transform algorithm itself is compromised. In helper data-based schemes, this information is known to an attacker, and the security is based on the performance bounds of error-correcting codes, which have been deeply studied.

To the best of our knowledge, Davida, Frankel, and Matt were the first to consider the use of ECC in designing a secure biometrics system for access control [7]. Their approach seems to have been developed without knowledge of the work on common randomness in the information theory community. They describe a system for securely storing a biometric and focus on three key aspects: security, privacy, and robustness. They achieve security by signing all stored data with a digital signature scheme, and they achieve privacy and robustness by using a systematic algebraic error-correcting code to store the data. A shortcoming of their scheme is that the codes employed are only decoded using bounded distance decoding. In addition, the security is hard to assess rigorously and there is no experimental validation using real biometric data.

The work by Juels and Wattenberg [8] extends the system of Davida et al. [7] by introducing a different way of using error-correcting codes. Their approach is referred to as "fuzzy commitment." In the enrollment stage the initial biometric is measured, and a random codeword of an error-correcting code is chosen. The hash of this codeword along with the difference between an enrollment biometric and the codeword are stored. During authentication, a second measurement of the user's biometric is obtained, then the difference between this probe biometric and the stored difference is determined, and error correction is then carried out to recover the codeword. Finally, if the hash of the resulting codeword matches the hash of the original codeword, then access is granted. Since the hash is difficult to invert, the codeword is not revealed. The value of the initial biometric is hidden by subtracting a random codeword from it, so the secure biometric hides both codeword and biometric data. This scheme relies heavily on the linearity/ordering of the encoded space to perform the difference operations. In reality, however, the feature space may not match such linear operations well.

A practical implementation of a fuzzy commitment scheme for iris data is presented in [9]. The authors utilize a concatenated-coding scheme in which Reed–Solomon codes are used to correct errors at the block level of an iris (e.g., burst errors due to eyelashes), while Hadamard codes are used to correct random errors at the binary level (e.g., background errors). They report a false reject rate of 0.47 percent at a key length of 140 bits on a small proprietary database including 70 eyes and 10 samples for each eye. As the authors note, however, the key length does not directly translate into security, and they estimate a security of about 44 bits. It is also suggested in [9] that passwords could be added to the scheme to substantially increase security.

In [10] Juels and Sudan proposed the fuzzy vault scheme. This is a cryptographic construct that is designed to work with unordered sets of data. The "fuzzy vault" scheme essentially combines the polynomial reconstruction problem with ECC.

Briefly, a set of t values from the enrollment biometric are extracted, and a length k vector of secret data (i.e., the encryption key) is encoded using an (n,k) ECC. For each element of the enrollment biometric, measurement-codeword pairs would be stored as part of the vault. Additional random "chaff" points are also stored, with the objective of obscuring the secret data. In order to unlock the vault, an attacker must be able to separate the chaff points from the legitimate points in the vault, which becomes increasingly difficult with a larger number of chaff points. To perform authentication, a set of values from a probe biometric could be used to initialize a codeword, which would then be subject to erasure and error decoding to attempt recovery of the secret data.

One of the main contributions of the fuzzy vault work was to realize that the set overlap noise model described in [10] can effectively be transformed into a standard errors and erasures noise model. This allowed application of Reed–Solomon codes, which are powerful codes and sufficiently analytically tractable to obtain some privacy guarantees. The main shortcoming is that the set overlap noise model is not realistic for most biometrics since feature points typically vary slightly from one biometric measurement to the next rather than either matching perfectly or not matching at all.

Nonetheless, several fuzzy vault schemes applied to various biometrics have been proposed. Clancy et al. [11] proposed to use the $X - Y$ location of minutiae points of a fingerprint to encode the secret polynomial, and they describe a random point-packing technique to fill in the chaff points. The authors estimate 69 bits of security and demonstrate a false reject rate of 30 percent. Yang and Verbauwhede [12] also used the minutiae point location of fingerprints for their fuzzy vault scheme. However, they convert minutiae points to a polar coordinate system with respect to an origin that is determined based on a similarity metric of multiple fingerprints. This scheme was evaluated on a very small database of 10 fingers, and a false reject rate of 17 percent was reported.

There do exist variants of the fuzzy vault scheme that do not employ ECC. For instance, the work of Uludag et al. [13] employs cyclic redundancy check (CRC) bits to identify the actual secret from several candidates. Nandakumar et al. [14] further extended this scheme in a number of ways to increase the overall robustness of this approach. On the FVC2002-DB2 database [15], this scheme achieves a 9 percent false reject rate (FRR) and a 0.13 percent false accept rate (FAR). The authors also estimate 27 to 40 bits of security depending on the assumed distribution of minutiae points.

As is evident from the literature, error-correcting codes indeed provide a powerful mechanism to cope with variations in biometric data. While the majority of schemes have been proposed in the context of fingerprint and iris data, there also exist schemes that target face, signature, and voice data. Some schemes that make use of multibiometrics are also beginning to emerge. Readers are referred to review articles on biometrics and security for further information on work in this area [16, 17].

In the sections that follow, the secure biometrics problem is formulated in the context of distributed source coding. We first give a more formal description of the problem setup, and we then describe solutions using techniques that draw from information theory, probabilistic inference, signal processing, and pattern recognition. We quantify security and robustness and provide experimental results for a variety of different systems.

11.3 OVERVIEW OF SECURE BIOMETRICS USING SYNDROMES

In this section, we present the architectural framework and information-theoretic security analysis for secure biometrics using syndromes. We present notation in Section 11.3.1, address the prototypical problem in Section 11.3.2, develop measures of security in Section 11.3.3, quantify security in Section 11.3.4, and discuss a secure biometrics implementation using error-correcting codes in Section 11.3.5.

11.3.1 Notation

We denote random variables using sans-serif and random vectors using bold sans-serif, x and $\mathbf{\mathsf{x}}$, respectively. The corresponding sample values and vectors are denoted using serifs x and $\mathbf{x}$, respectively. The length of vectors will be apparent from context or, when needed, indicated explicitly as, for example, x^n for the n-length random vector $\mathbf{\mathsf{x}}$. The ith element of a random or sample vector is denoted as x_i or x_i, respectively. Sets are denoted using caligraphic font; for example, the set of sample values of x is denoted $\mathcal{X}$, its n-fold product $\mathcal{X}^n$, and $|\cdot|$ applied to a set denotes its cardinality. We use $H(\cdot)$ to denote entropy; its argument can be either a random variable or its distribution; we use both interchangeably. For the special case of a Bernoulli-p source we use $H_B(p)$ to denote its entropy. Along the same lines, we use $I(\cdot;\cdot)$ and $I(\cdot;\cdot|\cdot)$ to denote mutual and conditional mutual information, respectively.

11.3.2 Enrollment and Authentication

As depicted in Figure 11.1, the secure biometrics problem is realized in the context of a Slepian–Wolf coding framework. In the following, we describe the system operation in terms of an access-control application. During enrollment, a user is selected, and raw biometric $\mathbf{\mathsf{b}}$ is determined by nature. The biometric is a random vector drawn according to some distribution $p_{\mathbf{\mathsf{b}}}(\mathbf{b})$. A joint sensing, feature extraction, and quantization function $f_{\text{feat}}(\cdot)$ maps the raw biometric into the length-n enrollment biometric $\mathbf{\mathsf{x}} = f_{\text{feat}}(\mathbf{\mathsf{b}})$. Next, a function $f_{\text{sec}}(\cdot)$ maps the enrollment biometric $\mathbf{\mathsf{x}}$ into the secure biometric $\mathbf{\mathsf{s}} = f_{\text{sec}}(\mathbf{\mathsf{x}})$ as well as into a cryptographic hash of the enrollment $\mathbf{\mathsf{h}} = f_{\text{hash}}(\mathbf{\mathsf{x}})$. The structure of the encoding function $f_{\text{sec}}(\cdot)$ reveals information about $\mathbf{\mathsf{x}}$ without leaking too much secrecy. In contrast, the cryptographic hash function $f_{\text{hash}}(\cdot)$ has no usable structure and is assumed to leak no information about $\mathbf{\mathsf{x}}$. The access-control point stores $\mathbf{\mathsf{s}}$ and $\mathbf{\mathsf{h}}$, as well as the functions $f_{\text{sec}}(\cdot)$ and $f_{\text{hash}}(\cdot)$. The access-control point does not store $\mathbf{\mathsf{b}}$ or $\mathbf{\mathsf{x}}$.

In the authentication phase, a user requests access and provides a second reading of the biometric $\mathbf{\mathsf{b}}'$. We model the biometrics of different users as statistically

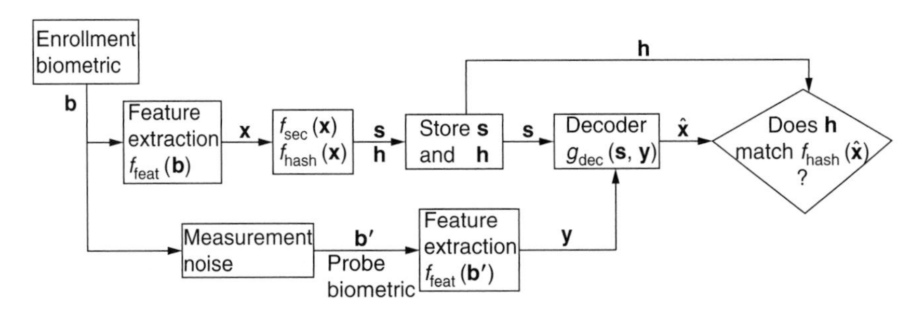

FIGURE 11.1

Block diagram of Slepian–Wolf system for secure biometrics.

independent. Therefore, if the user is not the legitimate user $p_{\mathbf{b}',\mathbf{b}}(\mathbf{b}', \mathbf{b}) = p_{\mathbf{b}}(\mathbf{b}')p_{\mathbf{b}}(\mathbf{b})$. On the other hand, if $\mathbf{b}'$ comes from the legitimate user $p_{\mathbf{b}',\mathbf{b}}(\mathbf{b}', \mathbf{b}) = p_{\mathbf{b}'|\mathbf{b}}(\mathbf{b}'|\mathbf{b})p_{\mathbf{b}}(\mathbf{b})$, where $p_{\mathbf{b}'|\mathbf{b}}(\cdot|\cdot)$ models the measurement noise between biometric readings. The features extracted from this second reading are $\mathbf{y} = f_{\text{feat}}(\mathbf{b}')$. Instead of working with $p_{\mathbf{b}',\mathbf{b}}(\mathbf{b}', \mathbf{b})$, we choose to work with $p_{\mathbf{x},\mathbf{y}}(\mathbf{x}, \mathbf{y})$. The feature extraction function $f_{\text{feat}}(\cdot)$ induces the distribution $p_{\mathbf{x},\mathbf{y}}(\mathbf{x}, \mathbf{y})$ from $p_{\mathbf{b}',\mathbf{b}}(\mathbf{b}', \mathbf{b})$. Per the preceding discussion, if the user is legitimate $p_{\mathbf{x},\mathbf{y}}(\mathbf{x}, \mathbf{y}) = p_{\mathbf{x}}(\mathbf{x})p_{\mathbf{y}|\mathbf{x}}(\mathbf{y}|\mathbf{x})$, and if the user is illegitimate, then $p_{\mathbf{x},\mathbf{y}}(\mathbf{x}, \mathbf{y}) = p_{\mathbf{x}}(\mathbf{x})p_{\mathbf{x}}(\mathbf{y})$.[1]

The decoder $g_{\text{dec}}(\cdot, \cdot)$ combines the secure biometric $\mathbf{s}$ with the probe $\mathbf{y}$ and either produces an estimate of the enrollment $\hat{\mathbf{x}} = g_{\text{dec}}(\mathbf{s}, \mathbf{y})$ or a special symbol $\varnothing$ indicating decoding failure. Finally, the stored $\mathbf{h}$ is compared to $f_{\text{hash}}(\hat{\mathbf{x}})$. If they match, access is granted. If they do not, access is denied.[2]

11.3.3 Performance Measures: Security and Robustness

The probability of authentication error (false rejection) is

$$P_{\text{FR}} = \Pr\left[\mathbf{x} \neq g_{\text{dec}}(\mathbf{y}, f_{\text{sec}}(\mathbf{x}))\right],$$

where $P_{\mathbf{y},\mathbf{x}}(\mathbf{y}, \mathbf{x}) = P_{\mathbf{y}|\mathbf{x}}(\mathbf{y}|\mathbf{x})P_{\mathbf{x}}(\mathbf{x})$. As discussed later, we will find it natural to use a logarithmic performance measure to quantify authentication failure. We use the error exponent

$$E_{\text{FR}} = -\frac{1}{n}\log P_{\text{FR}} \tag{11.1}$$

as this measure.

[1] Figure 11.1 can be thought of as somewhat specific to a single observation. If one had multiple observations of the underlying biometric, one could symmetrize the joint distribution by assuming that each observation of the underlying biometric (including the enrollment) was through a noisy channel. The current setting simplifies the model and is sufficient for our purposes.

[2] In a data encryption application, an encryption key is generated from $\mathbf{x}$ and the matching decryption key from $\hat{\mathbf{x}}$. A cryptographic hash function $f_{\text{hash}}(\cdot)$ is not required; if the reconstruction is not exact, then the generated key will not match the one used to encrypt and decryption will fail.

It must be assumed that an attacker makes many attempts to guess the desired secret. Therefore, measuring the probability that a single attack succeeds is not particularly meaningful. Instead, security should be assessed by measuring how many attempts an attack algorithm must make to have a reasonable probability of success. We formalize this notion by defining an attack as the creation of a list of candidate biometrics. If the true biometric is on the list, the attack is successful. The list size required to produce a successful attack with high probability translates into our measure of security.

Let $\mathcal{L} = \mathcal{A}_{R_{\text{sec}}}(\cdot)$ be a list of $2^{nR_{\text{sec}}}$ guesses for $\mathbf{x}$ produced by the attack algorithm $\mathcal{A}()$ that is parametrized by the rate R_{sec} of the attack and takes as inputs $p_{\mathbf{x}}(\cdot)\, p_{\mathbf{y}|\mathbf{x}}(\cdot|\cdot)$, $f_{\text{sec}}(\cdot)$, $f_{\text{hash}}(\cdot)$, $g_{\text{dec}}(\cdot,\cdot)$, $\mathbf{s}$, and $\mathbf{h}$. The attack algorithm does not have access to a probe generated from the enrollment $\mathbf{x}$ according to $p_{\mathbf{y}|\mathbf{x}}(\cdot|\cdot)$ because it does not have a measurement of the original biometric. From the quantities it does know, a good attack is to generate a list $\mathcal{L}$ of candidate biometrics that match the secure biometric $\mathbf{s}$ (candidate biometrics that do not match $\mathbf{s}$ can be eliminated out of hand). That is, for each candidate $\mathbf{x}_{\text{cand}} \in \mathcal{L}$, $f_{\text{sec}}(\mathbf{x}_{\text{cand}}) = \mathbf{s}$. While the cryptographic hash $f_{\text{hash}}(\cdot)$ is assumed to be noninvertible, we conservatively assume that the secure biometric encoding $f_{\text{sec}}(\cdot)$ is known to the attacker, and we further assume that the attacker can invert the encoding. Hence the list $\mathcal{L}$ can be generated.

Once the list $\mathcal{L}$ is created, a natural attack is to test each $\mathbf{x}_{\text{cand}} \in \mathcal{L}$ in turn to check whether $f_{\text{hash}}(\mathbf{x}_{\text{cand}}) = \mathbf{h}$. If the hashes match, the attack has succeeded. The system is secure against attacks if and only if the list of all possible candidate biometrics matching the secure biometric is so enormous that the attacker will only have computational resources to compute the hashes of a negligible fraction of candidate biometrics. Security thus results from dimensionality reduction: a high-dimensional $\mathbf{x}$ is mapped to a low-dimensional $\mathbf{s}$ by $f_{\text{sec}}(\cdot)$. The size of the total number of candidate biometrics that map onto the secure biometric $\mathbf{s}$ is exponential in the difference in dimensionality.

The probability that a rate-R_{sec} attack is successful equals the probability that the enrollment biometric is on the attacker's list,

$$P_{\text{SA}}(R_{\text{sec}}) = \Pr\left[\mathbf{x} \in \mathcal{A}_{R_{\text{sec}}}\left(p_{\mathbf{x}}(\cdot), p_{\mathbf{y}|\mathbf{x}}(\cdot|\cdot), f_{\text{sec}}(\cdot), f_{\text{hash}}(\cdot), g_{\text{dec}}(\cdot,\cdot), \mathbf{h}, \mathbf{s}\right)\right].$$

The system is said to be "ϵ-secure" to rate-R_{sec} attacks if $P_{\text{SA}}(R_{\text{sec}}) < \epsilon$.

Equivalently, we refer to a scheme with $P_{\text{SA}}(R_{\text{sec}}) = \epsilon$ as having $n \cdot R_{\text{sec}}$ bits of security with confidence $1-\epsilon$. With probability $1-\epsilon$, an attacker must search a key space of $n \cdot R_{\text{sec}}$ bits to crack the system security. In other words, the attacker must make $2^{nR_{\text{sec}}}$ guesses. The parameter R_{sec} is a logarithmic measure of security, quantifying the rate of the increase in security as a function of blocklength n. For instance, 128-bit security requires $nR_{\text{sec}} = 128$. It is because we quantify security with a logarithmic measure that we also use the logarithmic measure of error exponents to quantify robustness in (11.1).

Our objective is to construct an encoder and decoder pair that obtains the best combination of robustness (as measured by P_{FR}) and security (as measured by $P_{\text{SA}}(R_{\text{sec}})$) as a function of R_{sec}. In general, improvement in one necessitates a decrease in the other. For example, if $P_{SA}(0.5)=\epsilon$ and $P_{FR}=2^{-10}$ at one operating point, increasing the security to $0.75n$ might yield another operating point at $P_{SA}(0.75)=\epsilon$ and $P_{FR}=2^{-8}$. With this sense of the fundamental trade-offs involved, we now define the security-robustness region.

Definition 11.1. For any $\epsilon>0$ and any $p_{\mathbf{x},\mathbf{y}}(\mathbf{x},\mathbf{y})$ the security-robustness region $\mathcal{R}_\epsilon$ is defined as the set of pairs (r,γ) for which an encoder-decoder pair $(f_{\text{sec}}(\cdot), g_{\text{dec}}(\cdot,\cdot))$ exists that achieves rate-r security with an authentication failure exponent of γ:

$$\mathcal{R}_\epsilon=\left\{(r,\gamma)\,\middle|\,P_{SA}(r)\leqslant\epsilon,\ \gamma\geqslant-\frac{1}{n}\log P_{FR}\right\}.$$

11.3.4 Quantifying Security

In this section, we quantify an achievable subset of the security-robustness region $\mathcal{R}_\epsilon$. This specifies the trade-off between P_{FR} and $P_{SA}(\cdot)$ in an idealized setting. Our derivation assumes that $\mathbf{x}$ and $\mathbf{y}$ are jointly ergodic and take values in finite sets, $\mathbf{x}\in\mathcal{X}^n$, $\mathbf{y}\in\mathcal{Y}^n$. One can derive an outer bound to the security-robustness region by using upper bounds on the failure exponent (via the sphere-packing bound for Slepian-Wolf coding). Since our prime purpose in this section is to provide a solid framework for our approach, we don't further develop outer bounds here.

We use a rate-R_{SW} random "binning" function (a Slepian-Wolf code [1]) to encode $\mathbf{x}$ into the secured biometric $\mathbf{s}$. Specifically, we independently assign each possible sequence $\mathbf{x}\in\mathcal{X}^n$ an integer selected uniformly from $\{1,2,\ldots,2^{nR_{SW}}\}$. The secure biometric is this index $s=f_{\text{sec}}(\mathbf{x})$. Each possible index $s\in\{1,2,\ldots,2^{nR_{SW}}\}$ indexes a set or "bin" of enrollment biometrics, $\{\tilde{\mathbf{x}}\in\mathcal{X}^n|f_{\text{sec}}(\tilde{\mathbf{x}})=s\}$. The secure biometric can be thought of either as a scalar index s or as its binary expansion, a uniformly distributed bit sequence $\mathbf{s}$ of length nR_{SW}.

During authentication, a user provides a probe biometric $\mathbf{y}$ and claims to be a particular user. The decoder $g_{\text{dec}}(\mathbf{y},\mathbf{s})$ searches for the most likely vector $\hat{\mathbf{x}}\in\mathcal{X}^n$ given $\mathbf{y}$ according to the joint distribution $p_{\mathbf{x},\mathbf{y}}$ such that $\hat{\mathbf{x}}$ is in bin $\mathbf{s}$, that is, $f_{\text{sec}}(\hat{\mathbf{x}})=\mathbf{s}$. If a unique $\hat{\mathbf{x}}$ is found, then the decoder outputs this result. Otherwise, an authentication failure is declared and the decoder returns $\varnothing$.

According to the Slepian-Wolf theorem [1, 18], the decoder will succeed with probability approaching 1 as n increases provided that $R_{SW}>(1/n)H(\mathbf{x}|\mathbf{y})$. Thus, P_{FR} approaches zero for long blocklengths. The theory of error exponents for Slepian-Wolf coding [19] tells us that $-(1/n)\log P_{FR}\geqslant E_{SW}(R_{SW})$, where

$$E_{SW}(R_{SW})=\max_{0\leqslant\rho\leqslant1}\left\{\rho R_{SW}-\frac{1}{n}\log\sum_{\mathbf{y}}p_{\mathbf{y}}(\mathbf{y})\left[\sum_{\mathbf{x}}p_{\mathbf{x}|\mathbf{y}}(\mathbf{x}|\mathbf{y})^{\frac{1}{1+\rho}}\right]^{1+\rho}\right\}. \tag{11.2}$$

If $R_{SW}<(1/n)H(\mathbf{x}|\mathbf{y})$ then $E_{SW}(R_{SW})=0$. For $R_{SW}>(1/n)H(\mathbf{x}|\mathbf{y})$ the error exponent $E_{SW}(R_{SW})$ increases monotonically in R_{SW}. Note that (11.2) holds for any joint distribution, not just independent and identically distributed (i.i.d.) ones. However, if the source and channel are memoryless, the joint distribution is i.i.d., and $p_{\mathbf{x},\mathbf{y}}(\mathbf{x},\mathbf{y}) = \prod_{i=1}^{n} p_{x,y}(x_i, y_i)$. As a result, the second term of (11.2) simplifies considerably to $-\log \sum_y p_y(y)\Big[\sum_x p_{x|y}(x|y)^{\frac{1}{1+\rho}}\Big]^{1+\rho}$.

Next, we consider the probability of successful attack, that is, how well an attacker can estimate $\mathbf{x}$ given the secure biometric $\mathbf{s}$. According to the asymptotic equipartition property [20], under the fairly mild technical condition of ergodicity, it can be shown that conditioned on $\mathbf{s} = f_{\text{sec}}(\mathbf{x})$, $\mathbf{x}$ is approximately uniformly distributed over the typical set of size $2^{H(\mathbf{x}|\mathbf{s})}$. Therefore, with high probability, it will take approximately this many guesses to identify $\mathbf{x}$. We compute $H(\mathbf{x}|\mathbf{s})$ as

$$H(\mathbf{x}|\mathbf{s}) = H(\mathbf{x},\mathbf{s}) - H(\mathbf{s}) \overset{(a)}{=} H(\mathbf{x}) - H(\mathbf{s}) \overset{(b)}{=} H(\mathbf{x}) - nR_{\text{SW}}, \tag{11.3}$$

where (a) follows because $\mathbf{s} = f_{\text{sec}}(\mathbf{x})$, that is, $\mathbf{s}$ is a deterministic function of $\mathbf{x}$, and (b) follows from the method of generating the secure biometric, that is, $\mathbf{s}$ is uniformly distributed over length-nR_{SW} binary sequences (in other words $\mathbf{s}$ is a length-nR_{SW} i.i.d. Bernoulli(0.5) sequence).

Using (11.2) and (11.3), we bound the security-robustness region in the following:

Theorem 11.1. For any $\epsilon > 0$ as $n \to \infty$, an inner bound to the security-robustness region $\mathcal{R}_\epsilon$ defined in Definition 11.1 is found by taking a union over all possible feature extraction functions $f_{\text{feat}}(\cdot)$ and secure biometric encoding rates R_{SW}

$$\mathcal{R}_\epsilon \supset \bigcup_{f_{\text{feat}}(\cdot), R_{\text{SW}}} \left\{ r, \gamma \,\middle|\, r < \frac{1}{n} H(\mathbf{x}) - R_{\text{SW}},\ \gamma < E_{\text{SW}}(R_{\text{SW}}) \right\}$$

where $E_{\text{SW}}(R_{\text{SW}})$ is given by (11.2) for the $p_{\mathbf{x},\mathbf{y}}(\cdot,\cdot)$ induced by the chosen $f_{\text{feat}}(\cdot)$.

PROOF. The theorem is proved by the random-binning encoding and maximum-likelihood decoding construction specified above. The same approach holds for any jointly ergodic sources. The uniform distribution of the true biometric across the conditionally typical set of size $2^{H(\mathbf{x}|\mathbf{s})}$ provides security; cf. Equation (11.3). As long as the rate of the attack $r < \frac{1}{n} H(\mathbf{x}) - R_{\text{SW}}$, then $P_{\text{SA}}(r) < \epsilon$ for any $\epsilon > 0$ as long as n is sufficiently large. Robustness is quantified by the error exponent of Slepian–Wolf decoding given by (11.2). ■

Figure 11.2 plots an example of the security-robustness region for a memoryless insertion and deletion channel that shares some commonalities with the fingerprint channel that we discuss in Section 11.5. The enrollment biometric $\mathbf{x}$ is an i.i.d. Bernoulli sequence with $p_x(1) = 0.05$. The true biometric is observed through the asymmetric binary channel with deletion probability $p_{y|x}(0|1)$ and insertion probability $p_{y|x}(1|0)$. We plot the resulting security-robustness regions for two choices of insertion and deletion probabilities.

We now contrast $P_{\text{SA}}(\cdot)$, the measure of security considered in Theorem 11.1 and defined in Definition 11.1, with the probability of breaking into the system using the classic attack used to calculate the FAR. In the FAR attack, $\mathbf{y}$ is chosen independently

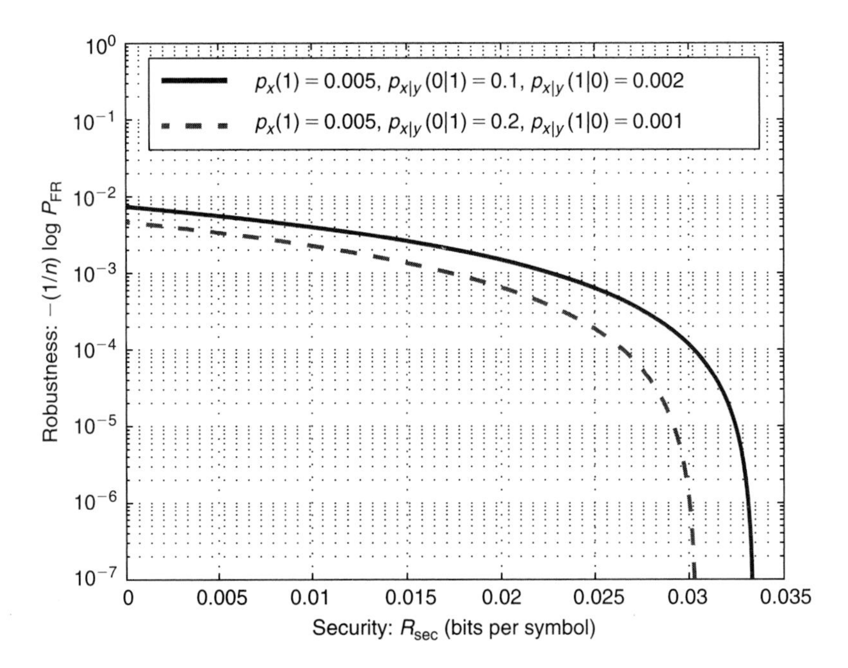

FIGURE 11.2

Example security-robustness regions. The horizontal axis represents the maximum security rate R_{sec} such that $P_{SA}(R_{sec}) < \epsilon$, while the vertical axis represents robustness. The security-robustness region of the system corresponding to the solid curve (all points below the curve) dominates that of the dashed curve.

of $\mathbf{x}$, that is, $p_{\mathbf{y},\mathbf{x}}(\mathbf{y}, \mathbf{x}) = p_{\mathbf{y}}(\mathbf{y})p_{\mathbf{x}}(\mathbf{x})$. This attack fails unless the $\mathbf{y}$ chosen is jointly typical with $\mathbf{x}$, that is, unless the pair $\mathbf{y}$ and (the unobserved) $\mathbf{x}$ look likely according to $p_{\mathbf{y},\mathbf{x}}(\cdot, \cdot)$. Given that a $\mathbf{y}$ is selected that is jointly typical with the enrollment $\mathbf{x}$, the decoder will then successfully decode to $\mathbf{x}$ with high probability, the hash will match, and access will be granted. To find such a $\mathbf{y}$ when picking according to the marginal $p_{\mathbf{y}}(\mathbf{y})$ takes approximately $2^{I(\mathbf{y};\mathbf{x})} = 2^{H(\mathbf{x})-H(\mathbf{x}|\mathbf{y})}$ guesses. We must set $R_{SW} > (1/n)H(\mathbf{x}|\mathbf{y})$, else as discussed above, Equation (11.2) tells us that P_{FR} goes to one. This constraint means that (cf. Equation (11.3)) $H(\mathbf{x}|\mathbf{s}) < H(\mathbf{x}) - H(\mathbf{x}|\mathbf{y})$. Thus, while an FAR-type attack required $2^{H(\mathbf{x})-H(\mathbf{x}|\mathbf{y})}$ guesses, the smarter attack considered in the theorem required $2^{H(\mathbf{x})-nR_{SW}}$, and thus an FAR-type attack will almost always take many more guesses than an attack that makes its guesses conditioned on $\mathbf{s}$.

We again emphasize that an attack that identifies a biometric $\tilde{\mathbf{x}}$ such that $f_{sec}(\tilde{\mathbf{x}}) = \mathbf{s}$ is not necessarily a successful attack. Indeed, our security analysis assumes that an attacker can easily find $\tilde{\mathbf{x}}$ that satisfies $f_{sec}(\tilde{\mathbf{x}}) = \mathbf{s}$. However, if $\tilde{\mathbf{x}} \neq \mathbf{x}$, then $f_{hash}(\tilde{\mathbf{x}}) \neq f_{hash}(\mathbf{x}) = \mathbf{h}$ and access will not be granted. Thus, in the bounds on security provided by Theorem 11.1, it is assumed that the attacker is limited to guesses of $\tilde{\mathbf{x}}$ that satisfy $f_{sec}(\tilde{\mathbf{x}}) = \mathbf{s}$.

11.3.5 Implementation Using Syndrome Coding

In our work, the enrollment biometric $\mathbf{x}$ is binary, and we use a linear code for the encoding function,

$$\mathbf{s} = f_{\text{sec}}(\mathbf{x}) = \mathbf{H}\mathbf{x}, \tag{11.4}$$

where $\mathbf{H}$ is a $k \times n$ binary matrix and addition is mod-2, that is, $a \oplus b = \text{XOR}(a, b)$. Using the language of algebra, the secure biometric $\mathbf{s}$ is the "syndrome" of the set of sequences $\tilde{\mathbf{x}} \in \{0, 1\}^n$ satisfying $\mathbf{H}\tilde{\mathbf{x}} = \mathbf{s}$. This set is also referred to as the "coset" or "equivalence class" of sequences. Note that all cosets are of equal cardinality.[3]

An attacker should limit his set of guesses $\mathcal{A}_{R_{\text{sec}}}$ to a subset of the coset corresponding to the stored $\mathbf{s}$. If all $\mathbf{x}$ sequences were equally likely (which is the case since cosets are of equal size and if $\mathbf{x}$ is an i.i.d. Bernoulli(0.5) sequence), then the attacker would need to check through nearly the entire list to find the true biometric with high probability. For this case and from (11.3), we calculate the logarithm of the list size to be $H(\mathbf{x}) - H(\mathbf{s}) = n - k$, where n and k are the dimensions of the $\mathbf{x}$ and $\mathbf{s}$ vectors, respectively, and are also the dimensions of the $\mathbf{H}$ matrix in (11.4). This follows from the model: $H(\mathbf{x}) = n$ since $\mathbf{x}$ is i.i.d. Bernoulli(0.5) and $H(\mathbf{s}) = k$ since cosets are of equal size and $p_{\mathbf{x}}(\mathbf{x}) = 2^{-n}$ for all $\mathbf{x}$.

If the enrollment biometric $\mathbf{x}$ is not a uniformly distributed i.i.d. sequence—which is going to be the case generally—the attacker need not check through the entire coset corresponding to $\mathbf{s}$. Instead the attacker should intersect the coset with the set of sequences in $\mathcal{X}^n$ that look like biometrics. These are the "typical" sequences [20] determined by the probability measure $p_{\mathbf{x}}(\cdot)$. This intersection is taken into account in Equation (11.3).[4] If the rows of the $\mathbf{H}$ matrix in (11.4) are generated in an independent and identically distributed manner, then step (b) in (11.3) simplifies as follows:

$$H(\mathbf{x}|\mathbf{s}) = H(\mathbf{x}) - H(\mathbf{s}) = H(\mathbf{x}) - \sum_{i=1}^{k} H(s_i) = H(\mathbf{x}) - kH(s). \tag{11.5}$$

In an actual implementation, we generally do not generate the rows of $\mathbf{H}$ in an i.i.d. manner, but rather use a structured code such as a low-density parity-check (LDPC) code. In such situations, (11.3) is a *lower* bound on the security of the system since $H(\mathbf{s}) \leqslant \sum_{i=1}^{k} H(s_i)$ using the chain rule for entropy and the fact that conditioning reduces entropy; the third equality still holds as long as the rows of $\mathbf{H}$ are identically distributed (even if not independent). Furthermore, contrast (11.5) with (11.3). In Equation (11.3), $H(\mathbf{s}) = nR_{\text{SW}}$ because of the random binning procedure. The

[3] It can be shown that any $\tilde{\mathbf{x}}$ in the $\mathbf{s}'$ coset can be written as $\tilde{\mathbf{x}} = \mathbf{x} \oplus \mathbf{z}$ for some $\mathbf{x}$ in the $\mathbf{s}$ coset and where $\mathbf{z}$ is fixed. Thus, $\mathbf{H}\tilde{\mathbf{x}} = \mathbf{H}(\mathbf{x} \oplus \mathbf{z}) = \mathbf{s} + \mathbf{H}\mathbf{z} = \mathbf{s}'$. The $\mathbf{s}'$ coset corresponds to all elements of the $\mathbf{s}$ coset (defined by its syndrome $\mathbf{s}$) shifted by $\mathbf{z}$, and thus the cardinalities of the two cosets are equal.

[4] We note that calculating the intersection may be difficult computationally. However, the security level quantified by Theorem 11.1 is conservative in the sense that it assumes that the attacker can calculate the intersection and produce the resulting list effortlessly.

assumptions of this procedure no longer hold when using linear codes to implement binning.

It is informative to consider estimating (11.5). The second term, $kH(\mathsf{S})$ is easy to estimate since it involves only the entropy of a marginal distribution. An estimation procedure would be to encode many biometrics using different codes, construct a marginal distribution for S, and calculate the entropy of the marginal. Particularly, if the code alphabet is small (say binary), little data is required for a good estimate. The first term $H(\mathbf{X})$ is harder to estimate. Generally, we would need to collect a very large number of biometrics (if n is large) to have sufficient data to make a reliable estimate of the entropy of the n-dimensional joint distribution. Thus, the absolute level of security is difficult to evaluate. However, the analysis provides a firm basis on which to evaluate the comparative security between two systems. The $H(\mathbf{X})$ term is common to both and cancels out in a calculation of relative security—the difference between the individual securities, which is $kH(\mathsf{S}) - k'H(\mathsf{S}')$.

11.4 IRIS SYSTEM

This section describes a prototype implementation of a secure biometrics system for iris recognition based on syndrome coding techniques. Experimental results on the Chinese Academy of Sciences Institute of Automation (CASIA) database [21] are presented.

11.4.1 Enrollment and Authentication

At enrollment, the system performs the following steps. Starting with an image of a user's eye, the location of the iris is first detected, and the torus is then unwrapped into a rectangular region. Next, a bank of Gabor filters are applied to extract a bit sequence. The Matlab implementation from [22] could be used to perform these steps. Finally, the extracted feature vector $\mathbf{x}$ is produced by discarding bits at certain fixed positions that were determined to be unreliable.[5] The resulting $\mathbf{x} = f_{\text{feat}}(\mathbf{b})$ consists of the most reliable bits; in our implementation 1806 bits are extracted. Finally, the bit string $\mathbf{x}$ is mapped into the secure biometric $\mathbf{s}$ by computing the syndrome of $\mathbf{x}$ with respect to a LDPC code. Specifically, a random parity-check matrix $\mathbf{H}$ is selected from a good low-rate degree distribution obtained via density evolution [23] and $\mathbf{s} = \mathbf{H} \cdot \mathbf{x}$ is computed.

To perform authentication, the decoder $g_{\text{dec}}(\cdot, \cdot)$ repeats the detection, unwrapping, filtering, and least-reliable bit dropping processes. The resulting observation $\mathbf{y}$ is used as

[5] Unreliable positions are those positions at which the bit values (0 or 1) are more likely to flip due to the noise contributed by eyelids and eyelashes, and due to a slight misalignment in the radial orientation of the photographed images. The bit positions corresponding to the outer periphery of the iris tend to be less reliable than those in the interior. These bit positions can be determined from the training data.

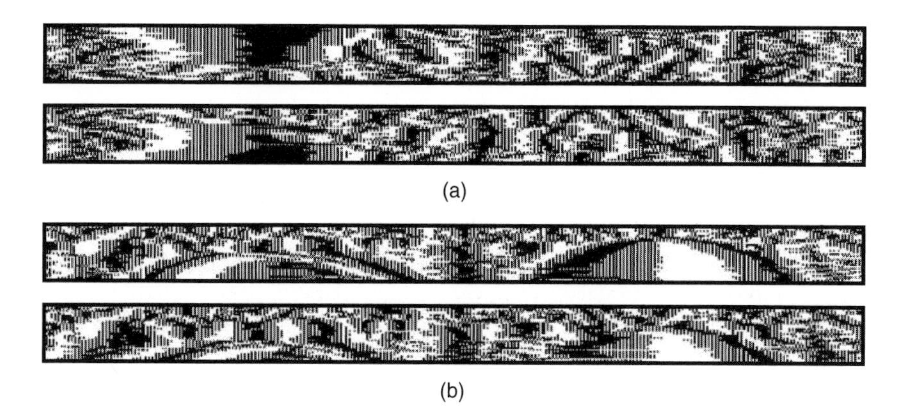

FIGURE 11.3

Sample bit sequences extracted from iris data. (a) Two sample measurements from one user; (b) two sample measurements from a second user.

the input to a belief propagation decoder that attempts to find a sequence $\hat{\mathbf{s}}$ satisfying $\mathbf{H} \cdot \hat{\mathbf{s}} = \mathbf{s}$. If the belief propagation decoder succeeds, then the output $\hat{\mathbf{s}} = g_{\text{dec}}(\mathbf{s}, \mathbf{y})$. Otherwise, an authentication failure (or false rejection) is declared, and the output of $g_{\text{dec}}(\mathbf{s}, \mathbf{y})$ is $\varnothing$.

Sample iris measurements from two different users are shown in Figure 11.3. The bit correlation between different samples of the same user and differences between samples of different users are easily seen. It has also been observed that the bit sequences extracted from the irises contain significant inter-bit correlation. Specifically, let $p_{i,j}$ be the probability of an iris bit taking the value i followed by another bit with the value j. If the bits extracted from an iris were independent and identically distributed, one would expect $p_{i,j} = 1/4$ for all $(i, j) \in \{0, 1\}^2$. Instead, the following probabilities have been measured from the complete dataset:

$$p_{0,0} = 0.319, \quad p_{0,1} = 0.166, \quad p_{1,0} = 0.166, \quad p_{1,1} = 0.349$$

Ignoring the inter-bit memory would result in degraded performance. Therefore, the belief propagation decoder is designed to exploit this source memory. Further details can be found in [24].

11.4.2 Experimental Results

The system is evaluated using the CASIA iris database [21]. The iris segmentation algorithm that was implemented was only able to correctly detect the iris in 624 out of 756 images [22, Chapter 2.4]. Since our emphasis is on the secure biometrics problem and not on iris segmentation, experiments were performed with the 624 iris that were segmented successfully. Furthermore, half of the iris images were used for training.

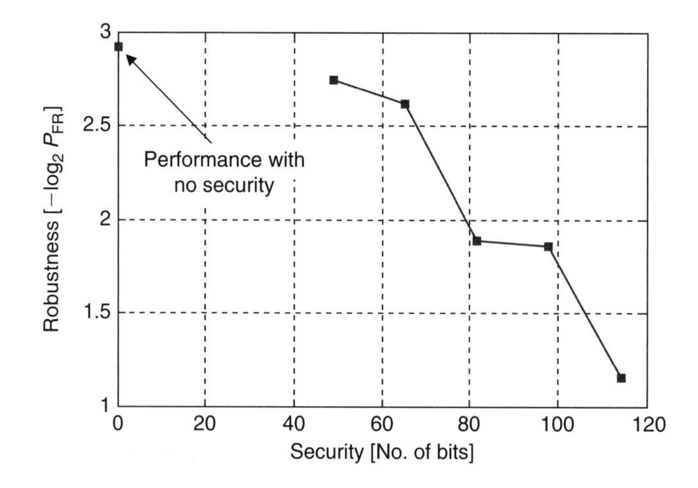

FIGURE 11.4

Performance result of 312 iris images from CASIA database. Horizontal axis represents security, while vertical axis plots robustness in terms of the probability of false rejection. The original length of the bit sequence extracted from an iris is $n = 1806$, while the length of the syndrome is $1806 - t$ bits, where t is plotted along the horizontal axis above. In fact, the actual number of bits of security is smaller than t because of inter-bit correlation in iris biometrics. A detailed explanation appears at the end of this section.

Figure 11.4 reports performance results for the 312 image test set from the CASIA iris database. The horizontal axis represents security, while the vertical axis represents the probability of false rejection for a legitimate user. Better systems correspond to points in the lower right, but as Theorem 11.1 shows theoretically and the figure demonstrates, there is a trade-off between security and robustness. Specifically, if a rate R LDPC code is used, then $\mathbf{s}$ contains $n(1-R)$ bits. Under the idealized model where the iris data consists of i.i.d. Bernoulli (0.5) bits, our approach yields approximately $1806 \cdot R$ bits of security with confidence approaching 1. Increasing R yields higher security, but lower robustness, so the security-robustness region can be estimated by varying this parameter.

Note that if the biometric is stored in the clear, there is a probability of false rejection equal to 0.0012 (i.e., the leftmost point in the graph). Thus, it is shown that, relative to an insecure scheme, with essentially no change in the probability of authentication failure the syndrome-based scheme achieves almost 50 bits of security.

Higher levels of security can be achieved if larger authentication error rates are allowed. As discussed in Section 11.3, the true level of security is more difficult to evaluate. Specifically, the original length of the bit sequence extracted from an iris in the system is 1806, and the length of the syndrome produced by our encoder is $1806 - t$, where t is a point on the horizontal axis of Figure 11.4. If the original biometric is an i.i.d. sequence of Bernoulli(0.5) random bits, then the probability of guessing the true biometric from the syndrome would be about 2^{-t} (i.e., security of t bits). However, as discussed earlier in this section, there is significant inter-bit memory in iris biometrics. In particular, if the source was a first-order Markov source with the measured $p_{i,j}$ statistics, the entropy of an 1806-bit measurement is only $0.9166 \times 1806 = 1655$ bits, or 90 percent of the true blocklength. Since $1806 - t > 90$ percent of 1806 for all reasonable values of P_{FR} in Figure 11.4, this suggests that an attacker with unbounded computational resources might be able to determine the true syndrome more quickly than by randomly searching a key space of size 2^t. That said, we are not aware of any computationally feasible methods of improving upon random guessing, and we believe that the estimated security provided here is still reasonable.

11.5 FINGERPRINT SYSTEM: MODELING APPROACH

In the previous section, we remarked on the difficulties caused by the correlations between bits in an iris biometric. These problems were dealt with by explicitly including the correlations in a belief propagation decoder. For fingerprint data, such problems are more severe. Models for fingerprint biometrics do not obviously map onto blocks of i.i.d. bits as would be ideal for a Slepian–Wolf LDPC code. We present two solutions to this problem. In this section, a "modeling" solution is discussed, in which the relationship between the enrollment biometric and the probe biometric is modeled as a noisy channel. The rest of this section describes a somewhat complex statistical factor graph model for fingerprint data and corresponding graph-based inference decoding techniques.

In Section 11.6, a second "transformation" approach is introduced, in which the fingerprint biometric is transformed, as well as possible, into a block of i.i.d. bits, and then a standard LDPC code and decoder are used. Although these two approaches are described in detail for fingerprint biometrics, other biometrics will have a similar dichotomy of possible approaches. For fingerprints, we have found that the transformation approach gives better results and makes it easier to quantify the security of the system, but both approaches are worth understanding.

11.5.1 Minutiae Representation of Fingerprints

A popular method for working with fingerprint data is to extract a set of "minutiae points" and to perform all subsequent operations on them [25]. Minutiae points have been observed to be stable over many years. Each minutiae is a discontinuity in the ridge map of a fingerprint, characterized by a triplet (x, y, θ) representing its spatial location in two dimensions and the angular orientation. In the minutiae map $\mathbf{M}$ of a fingerprint, $\mathbf{M}(x, y) = \theta$ if there is a minutiae point at (x, y) and $\mathbf{M}(x, y) = \emptyset$ (empty set) otherwise. A minutiae map may be considered as a joint quantization and feature extraction function that operates on the fingerprint image, that is, the output of the $f_{\text{feat}}(\cdot)$ box in Figure 11.1. In Figure 11.5, the minutiae map is visualized using a matrix as depicted in the right-hand plot, where a 1 simply indicates the presence of a minutiae at each quantized coordinate. In this figure, as well as in the model described throughout the rest of this section, the θ coordinate of the minutiae is ignored.

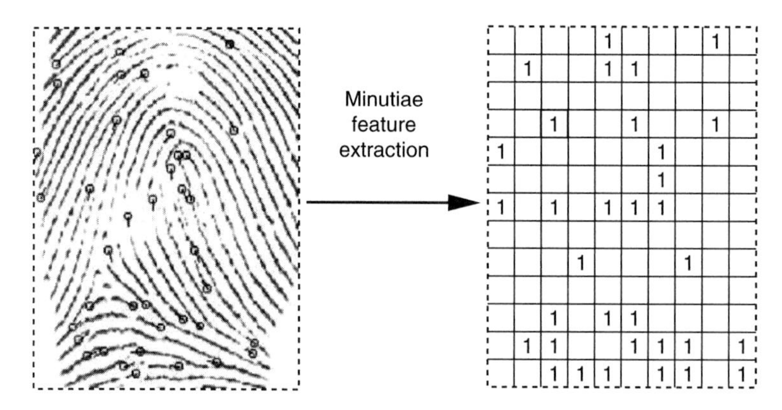

FIGURE 11.5

Fingerprint and extracted feature vector.

It is noted that different fingerprints usually have different numbers of minutiae. Furthermore, the number and location of minutiae could vary depending on the particular extraction algorithm that is used. For some applications, it could be important to account for such factors in addition to typical differences between fingerprint measurements, which will be discussed further in the next subsection. In the work described here, the enrollment feature vector **x** is modeled as a Bernoulli i.i.d.random vector.

11.5.2 Modeling the Movement of Fingerprint Minutiae

In the following, a model for the statistical relationship $p_{\mathbf{y}|\mathbf{x}}(\mathbf{y}|\mathbf{x})$ between the enrollment biometric and the probe biometric is described. This model captures three main effects: (1) movement of enrollment minutiae when observed the second time in the probe, (2) deletions, that is, minutiae observed at enrollment, but not during probe, and (3) insertions, that is, "spurious" minutiae observed in probe but not during enrollment.

Figure 11.6 depicts these three mechanisms in turn. First, minutiae observed at enrollment are allowed to jitter slightly around their locations in the enrollment vector when registered the second time in the probe. This movement is modeled within a local neighborhood, where up to three pixels in either the horizontal or vertical direction (or both) could be accounted for. The size of the local neighborhood depends on the resolution of the minutiae map and how coarsely it is quantized. Second, a minutiae point may be registered in the enrollment reading but not in the probe. Or a minutiae point may be displaced beyond the local neighborhood defined by the movement model. Both count as "deletions." Finally, minutiae points that are not observed at enrollment, but may be in the probe vector, are termed insertions.

The statistical model is formalized using a factor graph [26] as shown in Figure 11.7. The presence of a minutiae point at position t, $t \in \{1, 2, \ldots, n\}$ in the enrollment grid is represented by the binary random variable x_t that takes on the value $x_t = 1$

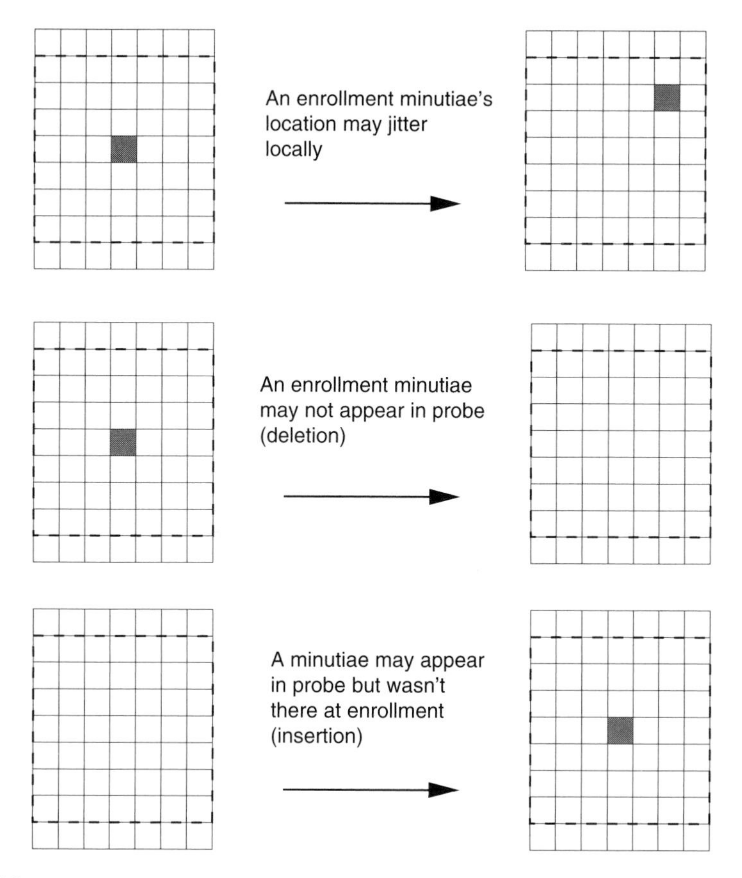

FIGURE 11.6

Statistical model of fingerprints corresponding to local movement, deletion, and insertion.

only if a minutiae is present during enrollment.[6] For simplicity, the figure shows a one-dimensional movement model. All experimental results use a two-dimensional movement model.

The decoder observes two vectors: the probe biometric y_i for $i \in \{1, 2, \ldots, n\}$ and s_j for $j \in \{1, 2, \ldots, k\}$. The decoder's objective is to estimate the hidden x_t enrollment variables.

The factor graph breaks down into three pieces. At the bottom of Figure 11.7 is the code graph representing the $\mathbf{H}$ matrix (cf. (11.4)) that maps $\mathbf{x}$ into $\mathbf{s}$. At the top of Figure 11.7 is the observation $\mathbf{y}$. In between $\mathbf{x}$ and $\mathbf{y}$ is our model of movement,

[6] Note that t indexes a position in the two-dimensional field of possible minutiae locations. The particular indexing used (e.g., raster-scan) is immaterial. The product of the number of rows and the number of columns equals n.

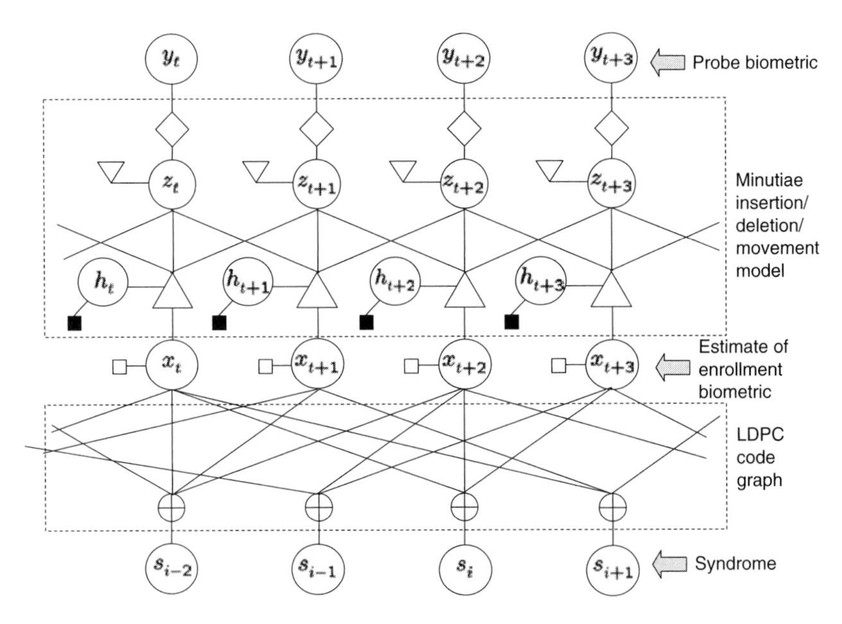

FIGURE 11.7

Factor graph of the minutiae movement model.

deletion, and insertion. Each circle in the figure represents a variable node either observed (**s** and **y**) or unobserved (**x**, **h**, and **z**) that needs to be estimated. The vector **h** is a vector of binary variables, each indicating the current belief (at a given point in the decoding process) whether an enrollment minutiae at position t is deleted. If a probe minutiae is observed at position t (i.e., $y_t = 1$), then z_t indicates the current beliefs of what enrollment locations the minutiae might have come from and $\mathbf{z}_{\mathcal{N}(t)} = \{z_i | i \in \mathcal{N}(t)\}$ are the set of these variables in the neighborhood of enrollment position t.

The constraints between the variables and the priors that define the joint probability function of all system variables are represented by the polygon factor nodes. The constraints enforced by each are as follows. The prior on x_t is $p_{\square}(x_t)$. The prior on deletion is $p_{\blacksquare}(h_t)$. The prior on insertion is $p_{\nabla}(z_t)$. The constraint that each enrollment minutiae is paired with only a single probe minutiae is enforced by the function node $\triangle$. In other words, $\triangle$ says that an enrollment minutiae can move to at most one position in the probe, or it can be deleted. Finally, in the reverse direction, $\Diamond$ constrains probe minutiae either to be paired with only a single enrollment minutiae or to be explained as an insertion. For a more detailed discussion of the statistical model, see [27, 28]. The complete statistical model of the enrollment and probe biometrics is

$$p_{\mathbf{x},\mathbf{y}}(\mathbf{x}, \mathbf{y}) = p_{\mathbf{x}}(\mathbf{x}) p_{\mathbf{y}|\mathbf{x}}(\mathbf{y}|\mathbf{x}) = \sum_{\{h_i\}} \sum_{\{z_i\}} \prod_t p_{\square}(x_t) p_{\blacksquare}(h_t) p_{\nabla}(z_t) \triangle(x_t, h_t, \mathbf{z}_{\mathcal{N}(t)}) \Diamond(z_t, y_t).$$

The above statistical model of the biometrics is combined with the code graph. This yields the complete model used for decoding $p_{\mathsf{x},\mathsf{y},\mathsf{s}}(\mathbf{x}, \mathbf{y}, \mathbf{s}) = p_{\mathsf{x},\mathsf{y}}(\mathbf{x}, \mathbf{y}) \prod_j \oplus(s_j, \mathbf{x})$, where $\oplus(s_j, \mathbf{x})$ indicates that the mod-2 sum of s_j and the x_i connected to syndrome j by the edges of the LDPC code is constrained to equal zero. A number of computational optimizations must be made for inference to be tractable in this graph. See [27, 28] for details.

11.5.3 Experimental Evaluation of Security and Robustness

We use a proprietary Mitsubishi Electric (MELCO) database to evaluate our techniques. The database consists of a set of fingerprint measurements with roughly 15 measurements per finger. One measurement is selected as the enrollment, while decoding is attempted, with the remaining 14 serving as probes. The locations of the minutiae points were quantized to reside in a 70×100 grid, resulting in a blocklength $n = 7000$.

The mean and standard deviation of movement, deletions (p_D), and insertions (p_I) for the MELCO dataset are plotted in Figure 11.8. The label $d = 1$ labels the probability that an enrollment minutiae moved a distance of one pixel in either the vertical or horizontal direction, or both (i.e., the max- or ∞-norm). These parameters are used to set parameter values in the factor graph.

A summary of test results is given in Table 11.1. Results are categorized by the number of minutiae in the enrollment print. To first order, this is a measure of the randomness of the enrollment biometric. As an estimate of $H(\mathsf{X})$, we say that if a fingerprint has, for example, 33 minutiae, its entropy is $7000 \times H_B(33/7000) = 7000 \times 0.0432 = 302$. Each row in the table tabulates results for enrollment biometrics, with the number of minutiae indicated in the first column. The second column indicates how many users had that number of minutiae in their enrollment biometric.

In the security-robustness trade-off developed in Section 11.3.3, it was found that holding all other parameters constant (in particular, the rate of the error-correcting code), security should increase and robustness decrease as the biometric entropy

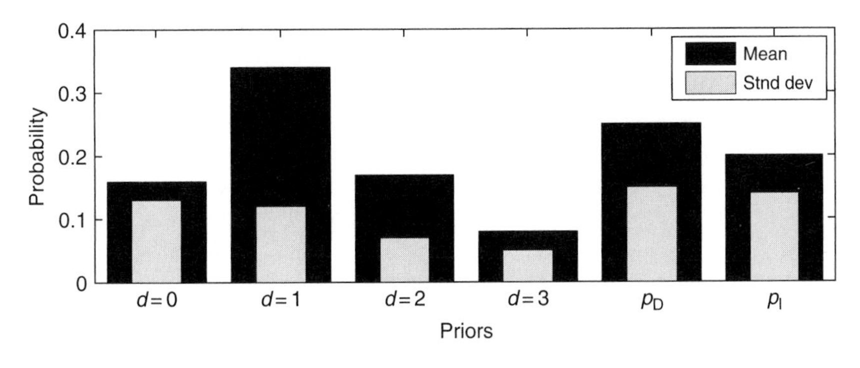

FIGURE 11.8

Empirical movement statistics.

Table 11.1 Test Parameters, FRR and FAR Results for Full-model Decoding Working on MELCO Data at Encoding Rate $R_{LDPC} = 0.94$

Enrollment		False Negatives		False Positives		Security		
# Minutiae	# Users	FRR	# Tested	FAR	# Tested	$H(x)$	$H(s)$	# Bits
31	195	0.11	2736	0.0098	110,000	0.0410	0.682	0.5
32	139	0.13	1944	0.0032	78,000	0.0421	0.693	3.6
33	107	0.15	1506	0.0024	60,000	0.0432	0.701	8.2
34	79	0.20	1101	0.0011	44,000	0.0443	0.711	11.6
35	59	0.32	824	0.0003	33,000	0.0454	0.716	17.2

increases. To test this, we use LDPC codes of rate $R_{LDPC} = 0.94$ and length-7000 for all syndrome calculations. The second and third groups of columns, labeled FRR and FAR, respectively, bear out the theoretic analysis. As the number of enrollment minutiae in a given fingerprint increases, the FRR goes up while the FAR drops. All nonenrollment probes of the given user are used to calcuate FRR. Summing the # Tested column under FRR gives 8111, which is roughly equal to the sum of the number of users (579) times the number of probes per user (roughly 14). To calculate the FRR, we test the enrollment biometric uniformly against other users' biometrics. Note that for all results it is assumed that the fingerprints in the database are prealigned.[7]

The final group of columns in Table 11.1 is labeled Security. Here, we quantify the information-theoretic security for the prototype. From (11.5) and recalling that the length of the biometric is $n = 7000$, the number of bits of security is

$$\begin{aligned} H(\mathbf{x}|\mathbf{s}) &= H(\mathbf{x}) - kH(s) \\ &= 7000H(x) - 7000(1 - R_{LDPC})H(s). \end{aligned} \tag{11.6}$$

Equation (11.6) follows from our model that the underlying source is i.i.d. so $H(\mathbf{x}) = 7000H(x)$ and because we use syndrome codes via (11.4) the number of syndromes $k = 7000(1 - R_{LDPC})$. Using $R_{LDPC} = 0.94$ and substituting the values for $H(x)$ and $H(s)$ from the different rows of Table 11.1 into (11.6) gives the bits of security for this system, which are tabulated in the last column of the table.

[7] We align fingerprints using a simple greedy minutiae-matching approach over a number of vertical and horizontal shifts (there was no rotational offset in the dataset). More generally, alignment would have to be done blindly prior to syndrome decoding. This is not as difficult as it may seem at first. For instance, many fingers have a "core point" and orientation in their pattern that can be used to define an inertial coordinate system in which to define minutiae locations. Doing this independently at enrollment and at verification would yield approximate prealignment. The movement part of the factor graph model is to be able to compensate for small residual alignment errors.

11.5.4 Remarks on the Modeling Approach

This section describes a secure fingerprint biometrics scheme in which an LDPC code graph was augmented with a second graph that described the "fingerprint channel" relating the enrollment to the probe biometric. A number of improvements are possible. For example, we implement an LDPC code designed for a binary symmetric channel (BSC). This design is not tuned to the fingerprint channel model. One possible improvement is to refine the design of the LDPC to match that channel. In general, however, while the "fingerprint channel" is a reasonable model of the variations between the enrollment and probe fingerprints, the techniques developed are specific to the feature set, and the resulting inference problem is complex and nonstandard. In addition, higher levels of security are desired. For these reasons, we take a different approach in the next section that aims to redesign the feature extraction algorithm to yield biometric features that are well matched to a standard problem of syndrome decoding.

11.6 FINGERPRINT SYSTEM: TRANSFORMATION APPROACH

In this section we aim to revamp the feature extraction algorithm to produce biometric features with statistics well matched to codes designed for the BSC. Since the construction of LDPC codes for the BSC is a deeply explored and well-understood topic, we are immediately able to apply that body of knowledge to the secure biometrics problem. We believe this is a more promising approach, in part because the design insights we develop can be applied to building transforms for other biometric modalities. In contrast, the biometric channel model developed in Section 11.5.2 is specific to fingerprints and minutiae points. In addition, the system we describe for fingerprints in this section achieves a higher level of information-theoretic security.

The transform-based secure fingerprint biometrics scheme is depicted in Figure 11.9. In Section 11.5, the function $f_{\text{feat}}(\cdot)$ extracted minutiae maps from the enrollment and probe fingerprints. Here, in addition to minutiae extraction, the $f_{\text{feat}}(\cdot)$ box also encompasses a feature transformation algorithm that converts the 2D minutiae maps to 1D binary feature vectors. The central idea is to generate binary feature vectors that are i.i.d. Bernoulli(0.5), independent across different users but such that different measurements of the same user are related by a binary symmetric channel with crossover probability p (BSC-p), where p is much smaller than 0.5. This is one of the standard channel models for LDPC codes, and therefore standard LDPC codes can be used for Slepian-Wolf coding of the feature vectors. We emphasize that the feature transformation we now present is made public and is *not* assumed to provide any security—in contrast to some of the transform-based techniques discussed in Section 11.2.

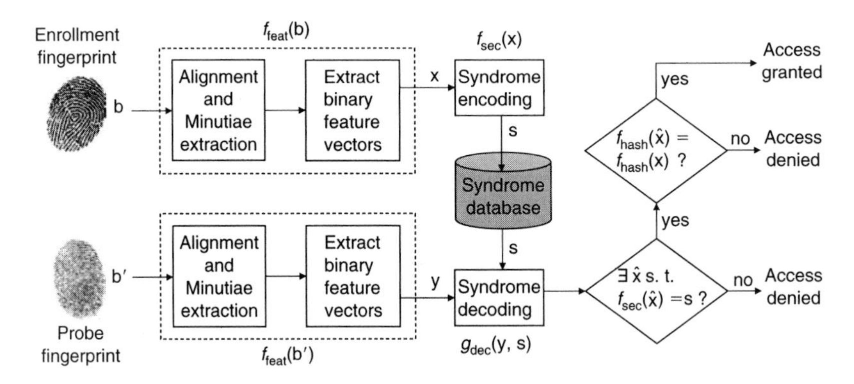

FIGURE 11.9

Robust feature extraction is combined with syndrome coding to build a secure fingerprint biometrics system.

11.6.1 Desired Statistical Properties of Feature Vectors

We aim to have a feature vector that possesses the following properties:

1. A bit in a feature vector representation is equally likely to be a 0 or a 1. Thus,

$$\Pr\{x_i = 0\} = \Pr\{x_i = 1\} = 1/2 \text{ and } H(x_i) = 1 \text{ bit for all } i \in \mathcal{I} = \{1, 2, \ldots, n\}.$$

2. Different bits in a given feature vector are independent of each other, so that a given bit provides no information about any other bit. Thus, the pairwise entropy $H(x_i, x_j) = H(x_i) + H(x_j) = 2$ bits for all $i \neq j$ where $i, j \in \mathcal{I}$. This property, along with the first property, ensures that the feature vector cannot be compressed further; that is, it presents the maximum possible uncertainty for an attacker who has to guess a portion of a feature vector given some other portion.

3. Feature vectors **x** and **y** from different fingers are independent of each other, so that one person's feature vector provides no information about another person's feature vector. Thus, the pairwise entropy $H(x_i, y_j) = H(x_i) + H(y_j) = 2$ bits for all $i, j \in \mathcal{I}$.

4. Feature vectors **x** and **x**′ obtained from different readings of the same finger are statistically related by a BSC-p. If p is small, it means that the feature vectors are robust to repeated noisy measurements with the same finger. Thus, $H(x'_i|x_i) = H(p)$ for all $i \in \mathcal{I}$.

The last property ensures that a Slepian–Wolf code with an appropriately chosen rate makes it possible to estimate the enrollment biometric when provided with feature vectors from the enrollee. At the same time, the chosen coding rate makes it extremely difficult (practically impossible) to estimate the enrollment biometric when provided with feature vectors from an attacker or from a different user. To show that the resulting biometrics system is information-theoretically secure, proceed just as in

(11.3) to obtain

$$\begin{aligned} H(\mathbf{x}|\mathbf{s}) &= H(\mathbf{x}, \mathbf{s}) - H(\mathbf{s}) = H(\mathbf{x}) - H(\mathbf{s}) \\ &= H(\mathbf{x}) - nR_{\mathrm{SW}} = n\,(H(x_i) - R_{\mathrm{SW}}) \\ &= n(1 - R_{\mathrm{SW}}) = nR_{\mathrm{LDPC}} > 0 \end{aligned} \tag{11.7}$$

where the last two equalities follow from properties 1 and 2, and R_{LDPC} is the rate of the LDPC code used. Thus, the higher the LDPC code rate, the smaller is the probability of successful attack conditioned on an observation of **s**. Moreover, $H(\mathbf{x}|\mathbf{s}) > 0$ and hence $nR_{\mathrm{SW}} < H(\mathbf{x})$ implies that, if properties 1–4 are satisfied, the system has positive information-theoretic security for any LDPC code rate.

11.6.2 Feature Transformation Algorithm

To extract n bits from a minutiae map, it suffices to ask n "questions," each with a binary answer. A general framework to accomplish this is shown in Figure 11.10. First, n operations are performed on the biometric to yield a nonbinary feature representation that is then converted to binary by thresholding. As an example, one can project the minutiae map onto n orthogonal basis vectors and quantize the positive projections to 1s and negative projections to 0s.

In the implementation we now describe, the n operations count the number of minutiae points that fall in randomly chosen cuboids in $X - Y - \Theta$ space (x-position, y-position, θ-minutiae-orientation), as shown in Figure 11.10(b). To choose a cuboid, an

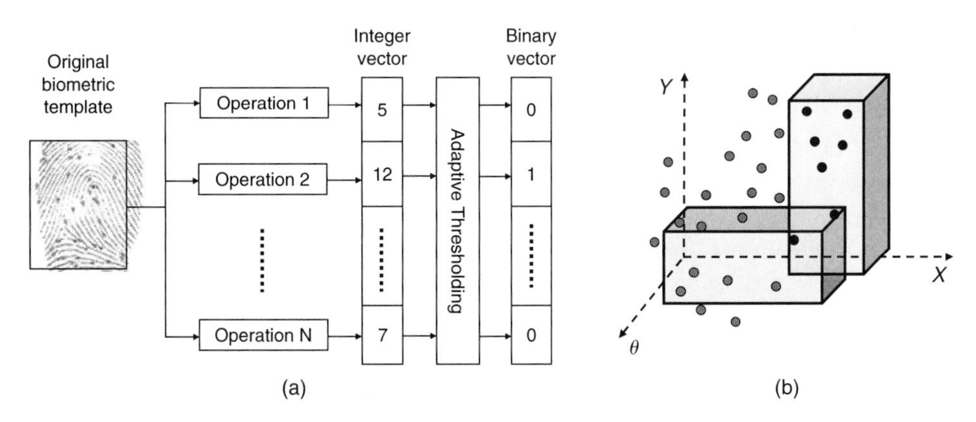

FIGURE 11.10

(a) n questions can be asked by performing n operations on the biometric followed by thresholding. In our scheme, the operation involves counting the minutiae points in a randomly generated cuboid. (b) To obtain a binary feature vector, the number of minutiae points in a cuboid is thresholded with respect to the median number of minutiae points in that cuboid calculated over the entire dataset. Overlapping cuboid pairs will result in correlated bit pairs. For details about eliminating bit pairs with very high correlation, the reader is referred to [29].

origin is selected uniformly at random in $X - Y - \Theta$ space, and the dimensions along the three axes are also chosen at random.

Next, define the threshold as the median of the number of minutiae points in the chosen cuboid, measured across the complete training set. A similar method is used for face recognition in [30]. The threshold value may differ for each cuboid based on its position and volume. If the number of minutiae points in a randomly generated cuboid exceeds the threshold, then a 1-bit is appended to the feature vector; otherwise a 0-bit is appended. We consider the combined operation of (a) generating a cuboid and (b) thresholding as equivalent to posing a question with a binary answer. With n such questions we get an n-bit feature vector.

The simplest way to generate feature vectors is to use the same questions for all users. In the sequel, we consider a more advanced approach in which the questions are user-specific. The rationale behind using user-specific questions is that some questions are more robust (reliable) than others. In particular, a question is robust if the number of minutiae points in a cuboid is much greater or much less than the median calculated over the entire dataset. Thus, even if there is spurious insertion or deletion of minutiae points when a noisy measurement of the same fingerprint is provided at a later time, the answer to the question (0 or 1) is less likely to change. On the other hand, if the number of minutiae points is close to the median, the 0 or 1 answer to that question is less reliable. Thus, more reliable questions result in a BSC-p intra-user channel with low p. Different users have a different set of robust questions, and we propose to use these while constructing the feature vector. We emphasize that for the purposes of security analysis, the set of questions used in the system is assumed public. An attacker who steals a set of syndromes and poses falsely as a user will be given the set of questions appropriate to that user. Our security analysis is not based in any way on the obscurity of the questions, but rather on the information-theoretic difficulty of recovering the biometric given only the stolen syndromes.

For a given user i, the *average* number of minutiae points $\bar{m}_{i,j}$ in a given cuboid C_j is calculated over repeated noisy measurements of the same fingerprint. Let m_j and σ_j be the median and standard deviation of the number of minutiae points in C_j over the dataset of all users. Then, let $\Delta_{i,j} = (\bar{m}_{i,j} - m_j)/\sigma_j$. The magnitude, $|\Delta_{i,j}|$ is directly proportional to the robustness of the question posed by cuboid C_j for user i. The sign of $\Delta_{i,j}$ determines whether the cuboid C_j should be placed into $\mathcal{L}_{0,i}$, a list of questions with a 0 answer for user i, or into $\mathcal{L}_{1,i}$, a list of questions with a 1 answer for user i. Both of these lists are sorted in the decreasing order of $|\Delta_{i,j}|$. Now, a fair coin is flipped to choose between $\mathcal{L}_{0,i}$ and $\mathcal{L}_{1,i}$ and the question at the top of the chosen list is stored on the device. After n coin flips, approximately $n/2$ of the most robust questions from each list will be stored on the device. This process is repeated for each enrolled user i.

11.6.3 Experimental Evaluation of Security and Robustness

In the following experiments, we use the same Mitsubishi Electric fingerprint database as described in the previous section, which contains minutiae maps of 1035 fingers with 15 fingerprint samples taken from each finger. The average number of minutiae points in a single map is approximately 32. As before, all fingerprints are prealigned. To measure the extent to which the desired target statistical properties in Section 11.6.1 are achieved, we examine the feature vectors obtained from the minutiae maps according to the method described in Section 11.6.2. The n most robust questions were selected to generate the feature vectors, with n ranging from 50 to 350. Figure 11.11 shows the statistical properties of the feature vectors for $n = 150$. As shown in Figure 11.11(a), the histogram of the average number of 1-bits in the feature vectors is clustered around $n/2 = 75$. Figure 11.11(b) shows that the pairwise entropy measured between bits of different users is very close to 2 bits. Thus, bits are nearly pairwise independent and nearly uniformly distributed, approximating property 1.

In order to measure the similarity or dissimilarity of two feature vectors, the normalized Hamming distance (NHD) is used. The NHD between two feature vectors $\mathbf{x}$ and $\mathbf{y}$, each having length n, is calculated as follows:

$$\text{NHD}(\mathbf{x}, \mathbf{y}) = \frac{1}{n} \sum_{i=1}^{n} (x_i \oplus y_i)$$

where $\oplus$ is summation modulo 2. The plot of Figure 11.12(a) contains three histograms: (1) the intra-user variation is the distribution of the average NHD measured pairwise over 15 samples of the same finger, (2) the inter-user variation is the distribution of the NHD averaged over all possible pairs of users, each with his own specific set of questions, and (3) the attacker variation is the NHD for the case in which an attacker attempts to identify himself as a given user i, while using a different fingerprint $j \neq i$ but using the 150 robust questions of user i. As seen in the figure, there is a clean separation between the intra-user and inter-user NHD distributions, and a small overlap between the intra-user and attacker distributions. One way to ascertain the effectiveness of the feature vectors is to choose different threshold NHDs in Figure 11.12(a) and plot the intra-user NHD against the inter-user NHD. This trade-off between intra-user NHD

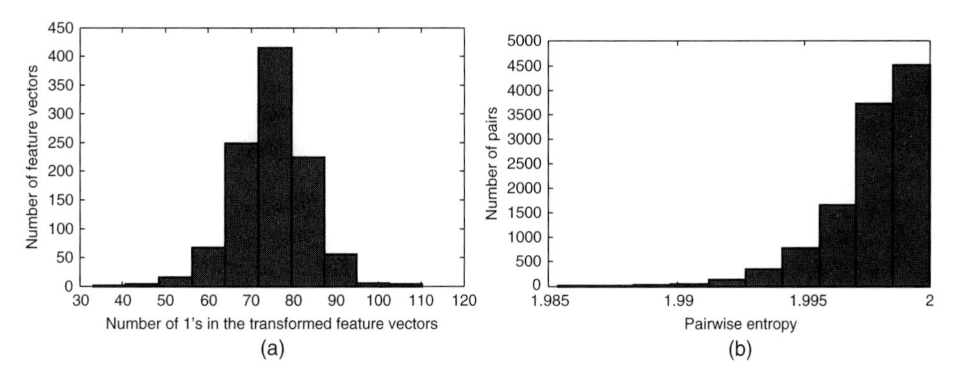

FIGURE 11.11

(a) Histogram of the number of ones in the feature vectors for $n = 150$ is clustered around $n/2 = 75$. (b) The pairwise entropy measured across all pairs and all users is very close to 2 bits.

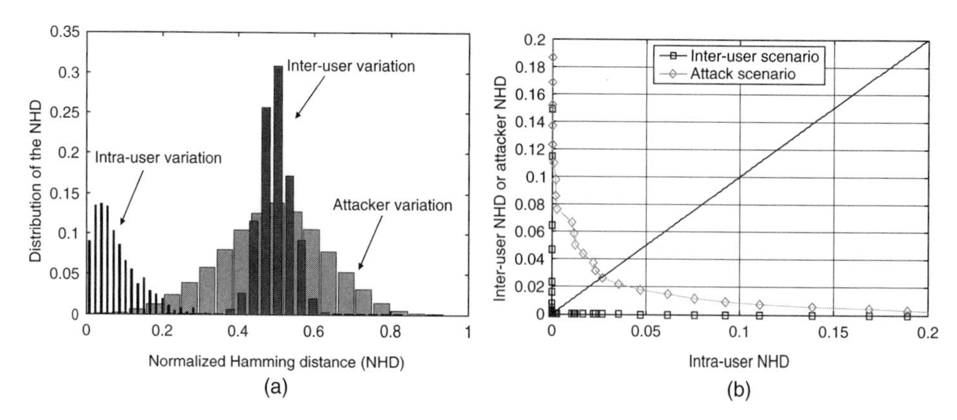

FIGURE 11.12

(a) The normalized Hamming distance (NHD) between feature vectors shows clear separation within and across users. (b) The trade-off between intra-user NHD and inter-user NHD is plotted by sweeping a threshold NHD across the histograms in Figure 11.12(a). For $n = 150$, the equal error rate is 0.027 when the attacker has access to the victim's questions and is nearly zero when the attacker is impersonating a victim without knowing his specific questions.

and inter-user NHD is shown in Figure 11.12(b) both for the case in which every user employs specific questions and for the case in which an attacker uses the questions stolen from the user being impersonated. A metric for evaluating plots such as Figure 11.12(b) is the equal error rate (EER), which is defined as the point at which intra-user NHD equals inter-user NHD. A lower EER indicates a superior trade-off. Figure 11.13 plots the EER for various values of n. Observe that user-specific questions provide a significantly lower EER than using the same questions for all users regardless of the robustness of the questions. Even if the attacker is provided with the user-specific questions, the resulting EER is lower than the case in which everybody has the same questions.

Based on the separation of intra-user and inter-user distributions, we expect that a syndrome code designed for a BSC-p, with appropriate $p < 0.5$ would authenticate almost all genuine users while rejecting almost all impostors. Table 11.2 shows the FRR and FAR[8] for overall syndrome coding with different values of n and p. These FAR and FRR values are measures of the security-robustness trade-off of the distributed biometric coding system. The LDPC code rate is chosen so as to provide about 30 bits of security. This restriction on the LDPC code rate in turn places a restriction on how large p can be, especially for small n. Because of this restriction, the FRR is relatively large for $n = 100$. The lowest FRR is achieved for $n = 150$. As n increases, less robust questions need to be employed, so the statistical properties of the feature vectors

[8] While determining the FAR, if an input feature vector $\widehat{\mathbf{a}}$ satisfies the syndrome, it is counted as a false accept. This is a conservative FAR estimate since any $\widehat{\mathbf{a}}$ for which $f_{\text{hash}}(\widehat{\mathbf{a}}) \neq f_{\text{hash}}(\mathbf{a})$ is denied access.

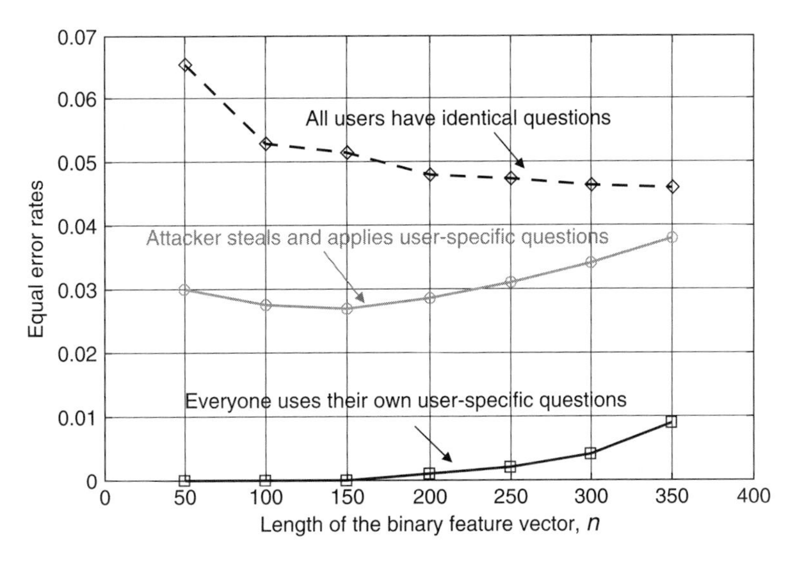

FIGURE 11.13

User-specific questions result in lower EER than common questions, even if the user-specific questions are given to the attacker.

Table 11.2 Syndrome Coding with an Appropriate LDPC Code Gives an Information-theoretically Secure Biometrics System with Low FRR and Extremely Low FAR

n	BSC Crossover Probability, p	R_{LDPC}	FRR after Syndrome Coding	FAR after Syndrome Coding	No. of Bits of Security
100	0.1	0.3	0.23	0.0001	30
150	0.13	0.2	0.11	0.0001	30
200	0.2	0.15	0.14	0.0014	30
250	0.2	0.125	0.15	0.0035	31.25

diverge from those in Section 11.6.1. Thus, the FRR increases again when n becomes too large.

Compare the FRR, FAR, and number of bits of security reported in Table 11.2 with those reported in Section 11.5. We observe that the FRR and FAR are comparable, but the transformation approach described in this section provides a higher number of bits of security compared to the model-based approach of Section 11.5 (see final column of Table 11.1). The security-robustness trade-off has improved because the statistical properties of the transformed feature vectors are intentionally matched to the standard LDPC code for a binary symmetric channel.

11.7 SUMMARY

This chapter demonstrates that the principles of distributed source coding can be successfully applied to the problem of secure storage of biometrics. A Slepian–Wolf framework is used to store a secure version of the biometric template data collected at enrollment and to recover the enrollment template at authentication. The trade-off between security and robustness in this framework is formally defined and discussed, and sample implementations based on iris and fingerprint data validate the theory.

Although iris data tends to be relatively well behaved and exhibits easily modeled sample-to-sample variability (both between samples of the same user and across users), the same cannot be said of fingerprints. It is shown that the fingerprint noise channel is far removed from the standard bit-flipping (e.g., BSC) channel model of communication systems. The design of a secure system for such biometric modalities therefore requires additional attention. Two approaches are discussed. The first design is based on using a sparse binary matrix representation of minutiae locations and developing a model of minutiae movement that can be combined with a graphical representation of a linear code. Although this approach does not yet yield satisfactory performance in terms of security and robustness, it does reveal various factors that affect performance and provides valuable insight that motivates the transform-based approach of Section 11.6.

In the latter approach, a transform is designed to convert the fingerprint feature set into a binary vector with desirable statistical properties, in the sense of being well matched to well-understood channel coding problems. The resultant design yields very low false-acceptance and false-rejection rates. Furthermore, it ensures operation well into the information-theoretically secure region. We believe this to be a powerful concept that will allow extension of this framework to other biometric data. It may also prove useful in resolving performance issues with other Slepian–Wolf-inspired systems.

Besides further improving security and robustness, there are a number of additional open research issues. As one example, the designs presented in this chapter assumed that the biometric data is prealigned. In practice, this is not the case, and biometric data must be aligned blindly, that is, without access to other reference data. One research trajectory is the design of such algorithms. An alternative to blind alignment is the design of a translation- and rotation-invariant feature set. A second aspect of the secure biometrics that has not received much attention concerns multibiometric systems. In these systems multiple biometrics are collected at enrollment and verification—such as both iris and fingerprint. The measurements are fused to improve overall robustness and security. This particular combination and some encouraging results are presented by Nandakumar in [31]. However, the topic has yet to be studied in the context of a Slepian–Wolf coding system.

As the use of biometrics becomes more widespread, the incentive to attack biometric systems will grow. Assuming the technology for securing biometric data is sufficiently mature, it would be natural to standardize the template protection design.

Such work is within the scope of ISO/IEC JTC1/SC37, which is an international standardization committee on biometrics. Open issues to be handled by this committee would range from quantifying the inherent entropy and security limits of biometric data to remote authentication scenarios.

As a final note, the biometric system described in this chapter is one example where a noisy version of an original signal is available at the decoder for the purpose of authentication. This type of setup is extended to the problem of image authentication following similar principles [32]. We believe that there are many such applications in which the principles of distributed source coding can be applied.

REFERENCES

[1] D. Slepian and J. K. Wolf, "Noiseless Coding of Correlated Information Sources," *IEEE Trans. Information Theory*, pp. 471-480, July 1973.

[2] N. Ratha, J. Connell, R. Bolle, and S. Chikkerur, "Cancelable Biometrics: A Case Study in Fingerprints," in *Intl. Conf. on Pattern Recognition*, Hong Kong, Septemper 2006, pp. 370-373.

[3] N. K. Ratha, S. Chikkerur, J. H. Connell, and R. M. Bolle, "Generating Cancelable Fingerprint Templates," *IEEE Transactions on Pattern Analysis and Machine Intelligence*, vol. 29, no. 4, pp. 561-572, April 2007.

[4] K. Sakata, T. Maeda, M. Matsushita, K. Sasakawa, and H. Tamaki, "Fingerprint Authentication Based on Matching Scores with Other Data," in *Lecture Notes in Computer Science*, ser. LNCS, vol. 3832, 2005, pp. 280-286.

[5] A. Teoh, A. Gho, and D. Ngo, "Random Multispace Quantization as an Analytic Mechanism for Biohashing of Biometric and Random Identity Inputs," *IEEE Transactions on Pattern Analysis and Machine Intelligence*, vol. 28, no. 12, pp. 1892-1901, December 2006.

[6] R. Ahlswede and I. Csiszar, "Common Randomness in Information Theory and Cryptography I: Secret Sharing," *IEEE Trans. Information Theory*, vol. 39, no. 4, pp. 1121-1132, July 1993.

[7] G. I. Davida, Y. Frankel, and B. J. Matt, "On Enabling Secure Applications through Off-line Biometric Identification," in *Proc. IEEE Symposium on Security and Privacy*, Oakland, CA, May 1998, pp. 148-157.

[8] A. Juels and M. Wattenberg, "A Fuzzy Commitment Scheme," in *CCS '99: Proceedings of the 6th ACM Conference on Computer and Communications Security*. New York: ACM Press, 1999, pp. 28-36.

[9] F. Hao, R. Anderson, and J. Daugman, "Combining Cryptography with Biometrics Effectively," University of Cambridge, Tech. Rep. UCAM-CL-TR-640, July 2005.

[10] A. Juels and M. Sudan, "A Fuzzy Vault Scheme," in *Proc. International Symposium on Information Theory*, Lausanne, Switzerland, July 2002, p. 408.

[11] T. C. Clancy, N. Kiyavash, and D. J. Lin, "Secure Smartcard-based Fingerprint Authentication," in *ACM SIGMM Workshop on Biometrics Methods and Applications*, Berkeley, CA, November 2003, pp. 45-52.

[12] S. Yang and I. M. Verbauwhede, "Secure Fuzzy Vault-based Fingerprint Verification System," in *Asilomar Conference on Signals, Systems, and Computers*, vol. 1, Asilomar, CA, November 2004, pp. 577-581.

[13] U. Uludag, S. Pankanti, and A. K. Jain, "Fuzzy Vault for Fingerprints," in *Audio- and Video-Based Biometric Person Authentication, 5th International Conference, AVBPA 2005, Hilton*

Rye Town, NY, USA, July 20-22, 2005, Proceedings, ser. Lecture Notes in Computer Science, vol. 3546. Springer, 2005.

[14] K. Nandakumar, A. K. Jain, and S. Pankanti, "Fingerprint-based Fuzzy Vault: Implementation and Performance," *IEEE Transactions on Information Forensics and Security*, vol. 2, no. 4, pp. 744-757, December 2007.

[15] D. Maio, D. Maltoni, J. Wayman, and A. K. Jain, "FVC2002: Second Fingerprint Verification Competition," in *International Conference on Pattern Recognition*, Quebec, Canada, August 2002, pp. 811-814.

[16] U. Uludag, S. Pankanti, S. Prabhakar, and A. K. Jain, "Biometric Cryptosystems: Issues and Challenges," *Proceedings of the IEEE*, vol. 92, no. 6, pp. 948-960, June 2004.

[17] A. K. Jain, S. Pankanti, S. Prabhakar, L. Hong, and A. Ross, "Biometrics: A Grand Challenge," In *Proc. International Conference on Pattern Recognition*, vol. 2, Cambridge, UK, August 2004, pp. 935-942.

[18] T. M. Cover, "A Proof of the Data Compression Theorem of Slepian and Wolf for Ergodic Sources," *IEEE Trans. Inform. Theory*, vol. 21, no. 2, pp. 226-228, March 1975.

[19] R. G. Gallager, "Source Coding with Side Information and Universal Coding," Massachusetts Institute of Tech., Tech. Rep. LIDS P-937, 1976.

[20] T. M. Cover and J. A. Thomas, *Elements of Information Theory*. New York: Wiley, 1991.

[21] "CASIA Iris Image Database collected by Institute of Automation, Chinese Academy of Sciences." [Online]. Available: http://www.sinobiometrics.com.

[22] L. Masek, "Recognition of Human Iris Patterns for Biometric Identification," Bachelor's Thesis, University of Western Australia, 2003.

[23] T. J. Richardson, M. A. Shokrollahi, and R. L. Urbanke, "Design of Capacity-Approaching Irregular Low-density Parity Check Codes," *IEEE Transactions on Information Theory*, vol. 47, no. 2, pp. 619-637, February 2001.

[24] E. Martinian, S. Yekhanin, and J. S. Yedidia, "Secure Biometrics via Syndromes," in *Allerton Conf.*, Monticello, IL, September 2005, pp. 1500-1510.

[25] A. K. Jain, L. Hong, and R. Bolle, "On-line fingerprint verification," *IEEE Transactions on Pattern Analysis and Machine Intelligence*, vol. 19, no. 4, pp. 302-314, April 1997.

[26] F. R. Kschischang, B. J. Frey, and H. Loeliger, "Factor Graphs and the Sum-Product Algorithm," *IEEE Transactions on Information Theory*, vol. 47, no. 2, pp. 498-519, February 2001.

[27] S. C. Draper, A. Khisti, E. Martinian, A. Vetro, and J. S. Yedidia, "Secure Storage of Fingerprint Biometrics Using Slepian-Wolf Codes," in *Inform. Theory and Apps. Work., UCSD*, San Diego, CA, January 2007.

[28] ———, "Using Distributed Source Coding to Secure Fingerprint Biometrics," in *Int. Conf. Acoutics Speech Signal Proc.*, Honolulu, HI, April 2007, pp. II-(129-132).

[29] Y. Sutcu, S. Rane, J. S. Yedidia, S. C. Draper, and A. Vetro, "Feature Transformation for a Slepian-Wolf Biometric System Based on Error Correcting Codes," in *Computer Vision and Pattern Recognition (CVPR) Biometrics Workshop*, Anchorage, AK, June 2008.

[30] T. Kevenaar, G. Schrijen, M. V. der Veen, A. Akkermans, and F. Zuo, "Face Recognition with Renewable and Privacy Preserving Binary Templates," *Fourth IEEE Workshop on Automatic Identification Advanced Technologies*, pp. 21-26, 2005.

[31] K. Nandakumar, "Multibiometric Systems: Fusion Strategies and Template Security," Ph.D. Thesis, Michigan State University, 2008.

[32] Y. C. Lin, D. Varodayan, and B. Girod, "Image Authentication Based on Distributed Source Coding," in *International Conference on Image Processing*, San Antonio, TX, September 2007.

Index

A

B

C

D

E

J

K

L

M

Q

R

T

U

V

W

Z